Media Theory

First edition

David Eppstein · Jean-Claude Falmagne
Sergei Ovchinnikov

Media Theory

Interdisciplinary Applied Mathematics

First edition

David Eppstein
University of California, Irvine
Department of Computer Science
Irvine, 92697-3425
USA
eppstein@ics.uci.edu

Sergei Ovchinnikov
San Francisco State University
Deptartment of Mathematics
Holloway Avenue 1600
San Francisco, 94132
USA
sergei@sfsu.edu

Jean-Claude Falmagne
University of California, Irvine
School of Social Sciences
Department of Cognitive Sciences
Social Science Plaza A 3171
Irvine, 92697-5100
USA
jcf@uci.edu

ISBN 978-3-540-71696-9   e-ISBN 978-3-540-71697-6

DOI 10.1007/978-3-540-71697-6

Library of Congress Control Number: 2007936368

© 2008 Springer-Verlag Berlin Heidelberg

This work is subject to copyright. All rights are reserved, whether the whole or part of the material is concerned, specifically the rights of translation, reprinting, reuse of illustrations, recitation, broadcasting, reproduction on microfilm or in any other way, and storage in data banks. Duplication of this publication or parts thereof is permitted only under the provisions of the German Copyright Law of September 9, 1965, in its current version, and permission for use must always be obtained from Springer. Violations are liable to prosecution under the German Copyright Law.

The use of general descriptive names, registered names, trademarks, etc. in this publication does not imply, even in the absence of a specific statement, that such names are exempt from the relevant protective laws and regulations and therefore free for general use.

Cover Design: KünkelLopka, Heidelberg

Printed on acid-free paper

9 8 7 6 5 4 3 2 1

springer.com

Preface

The focus of this book is a mathematical structure modeling a physical or biological system that can be in any of a number of 'states.' Each state is characterized by a set of binary features, and differs from some other neighbor state or states by just one of those features. In some situations, what distinguishes a state S from a neighbor state T is that S has a particular feature that T does not have. A familiar example is a partial solution of a jigsaw puzzle, with adjoining pieces. Such a state can be transformed into another state, that is, another partial solution or the final solution, just by adding a single adjoining piece. This is the first example discussed in Chapter 1. In other situations, the difference between a state S and a neighbor state T may reside in their location in a space, as in our second example, in which in which S and T are regions located on different sides of some common border.

We formalize the mathematical structure as a semigroup of 'messages' transforming states into other states. Each of these messages is produced by the concatenation of elementary transformations called 'tokens (of information).' The structure is specified by two constraining axioms. One states that any state can be produced from any other state by an appropriate kind of message. The other axiom guarantees that such a production of states from other states satisfies a consistency requirement.

What motivates our interest in this semigroup is, first, that it provides an algebraic formulation for mathematical systems researched elsewhere and earlier by other means. A prominent example is the 'isometric subgraph of a hypercube' (see Djoković, 1973, for an early reference), that is, a subraph in which the distance between vertices is identical to that in the parent hypercube. But there are many other cases. We shall outline some of them in our first chapter, reserving in depth treatment for later parts of this book. Until recently, however, no common algebraic axiomatization of these outwardly different concepts had been proposed. Our purpose is to give here the first comprehensive treatment of such a structure, which we refer to as a 'medium.'

A second, equally importantly reason for studying media, is that they offer a highly convenient representation for a vast class of empirical situations ranging from cognitive structures in education to the study of opinion polls in political sciences and including, conceivably, genetics, to name just a few pointers. They provide an appropriate medium[1] where the temporal evolution of a system can take place. Indeed, it turns out that, for some applications, the set of states of a medium can be profitably cast as the set of states of a random walk. Moreover, under simple hypotheses concerning the stochastic process involved, the asymptotic probabilities of the states are easy to compute and simple to write. Accordingly, some space is devoted to the development

---

[1] There lies the origin of the term.

of a random walk on the set of states of a medium, and to the description of a substantial application to the analysis of an opinion poll.

In this monograph, we study media from various angles: algebraic in Chapters 2, 3, and 4; combinatoric in Chapters 5 and 6; geometric in Chapters 7 to 9; algorithmic in Chapters 10 and 11. Chapters 12 and 13 are devoted to random walks on media and to applications.

Through the book, each chapter is organized into sections containing paragraphs, which often bear titles such as **Definition**, **Example**, or **Theorem**. For simplicity of reference and to facilitate a search through the book, a single numerical system is used. For instance:

2.4.12 Lemma.

2.4.13 Definition.

are the titles of the twelfth and the thirteenth paragraphs of Chapter 2, Section 2.4. We refer to the above lemma as "Lemma 2.4.12."

Defined technical terms are typed in slanted font just once, at the place where they are defined, which is typically within a "Definition" paragraph. Technical terms used before their definition are put between single quotes (at the first mention). The text of theorems and other results are also set in slanted font.

A short history of the results leading to the concept of a medium and ultimately to this monograph can be found in Section 1.9.

In the course of our work, we benefitted from exchanges with many researchers, whose reactions to our ideas influenced our writing, sometimes substantially. We want to thank, in particular, Josiah Carlson, Dan Cavagnaro, Victor Chepoi, Eric Cosyn, Chris Doble, Aleks Dukhovny, Peter Fishburn, Bernie Grofman, Yung-Fong Hsu, Geoff Iverson, Duncan Luce, Louis Narens, Michel Regenwetter, Fred Roberts, Pat Suppes, Nicolas Thiéry, and Hasan Uzun.

A special mention must be made of Jean-Paul Doignon, whose joint work with Falmagne provided much of the foundational ideas behind the concept of a medium. For a long time, we thought that Jean-Paul would be a co-author. However, other commitments prevented him to be one of us. No doubt, had he been a co-author, our book would have been a better one.

Last but not least, Diana, Dina, and Galina deserve much credit for variously letting us be—the relevant one, that is—whenever it seemed that the call of the media was too strong, or for gently drawing us away from them, for our own good sake, when there was an opening. To those three, we are the most grateful.

David Eppstein
Jean-Claude Falmagne
Sergei Ovchinnikov

August 11, 2007

Contents

1 Examples and Preliminaries 1
   1.1 A Jigsaw Puzzle 1
   1.2 A Geometrical Example 4
   1.3 The Set of Linear Orders 6
   1.4 The Set of Partial Orders 7
   1.5 An Isometric Subgraph of $\mathbb{Z}^n$ 8
   1.6 Learning Spaces 10
   1.7 A Genetic Mutations Scheme 11
   1.8 Notation and Conventions 12
   1.9 Historical Note and References 17
       Problems 19

2 Basic Concepts 23
   2.1 Token Systems 23
   2.2 Axioms for a Medium 24
   2.3 Preparatory Results 27
   2.4 Content Families 29
   2.5 The Effective Set and the Producing Set of a State 30
   2.6 Orderly and Regular Returns 31
   2.7 Embeddings, Isomorphisms and Submedia 34
   2.8 Oriented Media 36
   2.9 The Root of an Oriented Medium 38
   2.10 An Infinite Example 39
   2.11 Projections 40
       Problems 45

3 Media and Well-graded Families 49
   3.1 Wellgradedness 49
   3.2 The Grading Collection 52

VIII   Contents

  3.3   Wellgradedness and Media 54
  3.4   Cluster Partitions and Media 57
  3.5   An Application to Clustered Linear Orders 62
  3.6   A General Procedure 68
         Problems 68

4 Closed Media and ∪-Closed Families 73
  4.1   Closed Media 73
  4.2   Learning Spaces and Closed Media 78
  4.3   Complete Media 80
  4.4   Summarizing a Closed Medium 83
  4.5   ∪-Closed Families and their Bases 86
  4.6   Projection of a Closed Medium 94
         Problems 98

5 Well-Graded Families of Relations 101
  5.1   Preparatory Material 102
  5.2   Wellgradedness and the Fringes 103
  5.3   Partial Orders 106
  5.4   Biorders and Interval Orders 107
  5.5   Semiorders 110
  5.6   Almost Connected Orders 114
         Problems 119

6 Mediatic Graphs 123
  6.1   The Graph of a Medium 123
  6.2   Media Inducing Graphs 125
  6.3   Paired Isomorphisms of Media and Graphs 130
  6.4   From Mediatic Graphs to Media 132
         Problems 136

7 Media and Partial Cubes 139
  7.1   Partial Cubes and Mediatic Graphs 139
  7.2   Characterizing Partial Cubes 142
  7.3   Semicubes of Media 149
  7.4   Projections of Partial Cubes 151
  7.5   Uniqueness of Media Representations 154
  7.6   The Isometric Dimension of a Partial Cube 158
         Problems 159

8 Media and Integer Lattices ... 161
- 8.1 Integer Lattices ... 161
- 8.2 Defining Lattice Dimension ... 162
- 8.3 Lattice Dimension of Finite Partial Cubes ... 167
- 8.4 Lattice Dimension of Infinite Partial Cubes ... 171
- 8.5 Oriented Media ... 172
- Problems ... 174

9 Hyperplane arrangements and their media ... 177
- 9.1 Hyperplane Arrangements and Their Media ... 177
- 9.2 The Lattice Dimension of an Arrangement ... 184
- 9.3 Labeled Interval Orders ... 186
- 9.4 Weak Orders and Cubical Complexes ... 188
- Problems ... 196

10 Algorithms ... 199
- 10.1 Comparison of Size Parameters ... 199
- 10.2 Input Representation ... 202
- 10.3 Finding Concise Messages ... 211
- 10.4 Recognizing Media and Partial Cubes ... 217
- 10.5 Recognizing Closed Media ... 218
- 10.6 Black Box Media ... 222
- Problems ... 227

11 Visualization of Media ... 229
- 11.1 Lattice Dimension ... 230
- 11.2 Drawing High-Dimensional Lattice Graphs ... 231
- 11.3 Region Graphs of Line Arrangements ... 234
- 11.4 Pseudoline Arrangements ... 238
- 11.5 Finding Zonotopal Tilings ... 246
- 11.6 Learning Spaces ... 252
- Problems ... 260

12 Random Walks on Media ... 263
- 12.1 On Regular Markov Chains ... 265
- 12.2 Discrete and Continuous Stochastic Processes ... 271
- 12.3 Continuous Random Walks on a Medium ... 273
- 12.4 Asymptotic Probabilities ... 279
- 12.5 Random Walks and Hyperplane Arrangements ... 280
- Problems ... 282

13 Applications 285
13.1 Building a Learning Space 285
13.2 The Entailment Relation 291
13.3 Assessing Knowledge in a Learning Space 293
13.4 The Stochastic Analysis of Opinion Polls 297
13.5 Concluding Remarks 302
Problems 303

Appendix: A Catalog of Small Mediatic Graphs 305

Glossary 309

Bibliography 311

Index 321

1
Examples and Preliminaries

We begin with an example from everyday life, which will serve as a vehicle for an informal introduction to the main concepts of media theory. Several other examples follow, chosen for the sake of diversity, after which we briefly review some standard mathematical concepts and notation. The chapter ends with a short historical notice and the related bibliography. Our purpose here is to motivate the developments and to build up the reader's intuition, in preparation for the more technical material to follow.

1.1 A Jigsaw Puzzle

1.1.1 Gauss in Old Age. Figure 1.1(a) shows a familiar type of jigsaw puzzle, made from a portrait of Carl Friedrich Gauss in his old age. We call a *state* of this puzzle any partial solution, formed by a linked subset of the puzzle pieces in their correct positions. Four such states are displayed in Figure 1.1(a), (b), (c) and (d). Thus, the completed puzzle is a state. We also regard as states the initial situation (the empty board), and any single piece appropriately placed on the board. A careful count gives us 41 states (see Figure 1.1.2). To each of the six pieces of the puzzle correspond exactly two *transformations* which consist in placing or removing a piece. In the first case, a piece is placed either on an empty board, or so that it can be linked to some pieces already on the board. In the second case, the piece is already on the board and removing it either leaves the board empty or does not disconnect the remaining pieces. By convention, these two types of transformations apply artificially to all the states in the sense that placing a piece already on the board or removing a piece that is not on the board leaves the state unchanged.

This is our first example of a 'medium', a concept based on a pair $(\mathcal{S}, \mathcal{T})$ of sets: a set $\mathcal{S}$ states, and a collection $\mathcal{T}$ of transformations capable, in some cases, of converting a state into a different one. The formal definition of such a structure relies on two constraining axioms (see Definition 2.2.1).

Figure 1.1. Four states of a medium represented by the jigsaw puzzle: Carl Friedrich Gauss in old age. The full medium contains 41 states (see Figure 1.2).

By design, none of these transformations is one-to-one. For instance, applying the transformation "adding the upper left piece of the puzzle"—the left part of Gauss's hat and forehead—to either of the states pictured in Figure 1.1(c) or (d) results in the same state, namely (a). In the first case, we have thus a loop. Accordingly, the two transformations associated with each piece are not mutual inverses. However, each of the transformations in a pair can undo the action of the other. We shall say that these transformations are 'reverses' of one another. For a formal definition of 'reverse' in the general case, see 2.1.1.

1.1.2 The Graph of Gauss's Puzzle. When the number of states is finite, it may be convenient to represent a medium by its graph and we shall often do so. The medium of Gauss's puzzle has its graph represented in Figure 1.2 below. As usual, we omit loops.

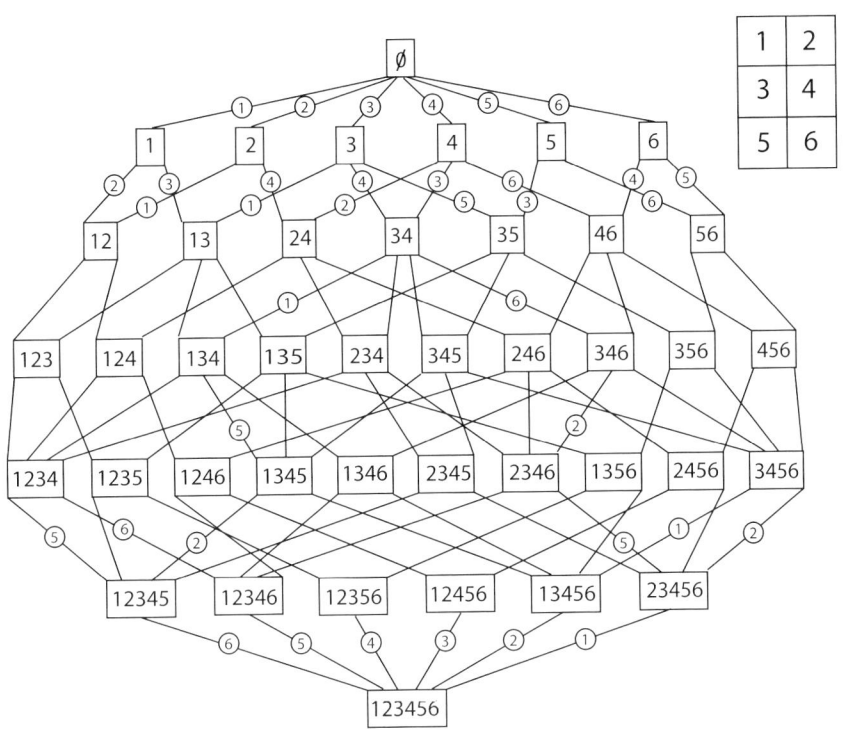

Figure 1.2. Graph of the Gauss puzzle medium. A schematic of the puzzle is a the upper right of the graph, with the six pieces numbered 1,..., 6. Each of the 41 vertices of the graph represent one state of the medium, that is, one partial solution of the puzzle symbolized by a rectangle containing the list of its pieces. Each edge represents a pair of mutually reverse transformations, one adding a piece, and the other removing it. To avoid cluttering the figure, only some of the edges are labeled (by a circle).

An examination of this graph leads to further insight. For any two states S and T, it is possible to find a sequence of transformations whose successive applications from S results in forming T. This 'path' from S to T never strays from the allowed set of states, and can be made minimally short, that is: its length is equal to the number of pieces which are not common to both states. Moreover, any two such paths from S to T will involve exactly the same transformations, but they may be applied in different orders[1]. As an illustration, we have marked in Figure 1.2 two such paths from state $\boxed{34}$ to the completed puzzle by coloring their edges in red and blue, respectively. These two paths are

$$\boxed{34} \xmapsto{1} \boxed{134} \xmapsto{5} \boxed{1345} \xmapsto{2} \boxed{12345} \xmapsto{6} \boxed{123456} \qquad (1.1)$$

$$\boxed{34} \xmapsto{6} \boxed{346} \xmapsto{2} \boxed{2346} \xmapsto{5} \boxed{23456} \xmapsto{1} \boxed{123456}. \qquad (1.2)$$

A medium could be defined from any (standard) jigsaw puzzle according to the rules laid out here[2]. Such media have the remarkable property that their set of transformations is naturally partioned into two classes of equal sizes, namely, one corresponding to the addition of the pieces to the puzzle, and the other one to their removal. In view of this asymmetry which also arises in other situations, we shall talk about 'orientation' to describe such a bipartition (see Definition 2.8.1). Thus, in medium terminology, a transformation is in a given class if and only if its reverse belongs to the other class. There are important cases, however, in which no such natural orientation exists. Accordingly, this concept is not an integral part of the definition of a medium (cf. 2.2.1).

In fact, the next two examples involve media in which no natural orientation of the set of transformations suggests itself.

1.2 A Geometrical Example

1.2.1 An Arrangement of Hyperplanes. Let $\mathcal{A}$ be some finite collection of hyperplanes in $\mathbb{R}^n$. Then $\mathbb{R}^n \setminus (\cup \mathcal{A})$ is the union of the open, convex polyhedral regions bounded by the hyperplanes, some (or all) of which may be unbounded. We regard each polyhedral region as a state, and we denote by $\mathcal{P}$ the finite collection of all the states. From one state P in $\mathcal{P}$, it is always possible to move to another adjacent state by crossing some hyperplane including a facet of P. (We suppose that a single hyperplane is crossed at one time.) We formalize these crossings in terms of transformations of the states. To every hyperplane H in $\mathcal{A}$ corresponds the two ordered pairs (H^-, H^+) and (H^+, H^-) of the two open half spaces H^- and H^+ separated by the hyperplane. These ordered pairs generate two transformations τ_H^+ and τ_H^- of the states, where τ_H^+ transforms a state to an adjacent state in H^+, if possible, and leaves it unchanged otherwise, while τ_H^- transforms a state to an adjacent state in H^-, if possible, and leaves it unchanged otherwise. More formally, applying τ_H^+ to some state P results in some other state Q if $P \subseteq H^-$, $Q \subseteq H^+$,

---

[1] In this particular case, the two paths represented in (1.1) and (1.2) have their transformations in the exact opposite order, but not all pairs of different paths are reversed in this way (Problem 1.1).
[2] In the case of larger puzzles (for instance, 3×3), there might be states with holes: all the pieces are interconnected but there are pieces missing in the middle.

and the polyhedral regions P and Q share a facet which is included in the hyperplane separating H^- and H^+; otherwise, the application of τ_H^+ to P does not change P. The transformation τ_H^- is defined symmetrically. Clearly, the application of τ_H^+ cancels the action of τ_H^- whenever the latter was effective in modifying the state. However, as in the preceding example, τ_H^+ and τ_H^- are not mutual inverses. We say in such a case that τ_H^+ and τ_H^- are mutual reverses. Denoting by $\mathcal{T}$ the set of all such transformations, we obtain a pair $(\mathcal{P}, \mathcal{T})$ which is another example of a medium. A case of five straight lines in $\mathbb{R}^2$ defining fifteen states and ten pairs of mutually reverse transformations is pictured in Figure 1.3. The proof that, in the general case, an arbitrary locally finite hyperplane arrangement defines a medium is due to Ovchinnikov (2006) (see Theorem 9.1.8 in Chapter 9 here).

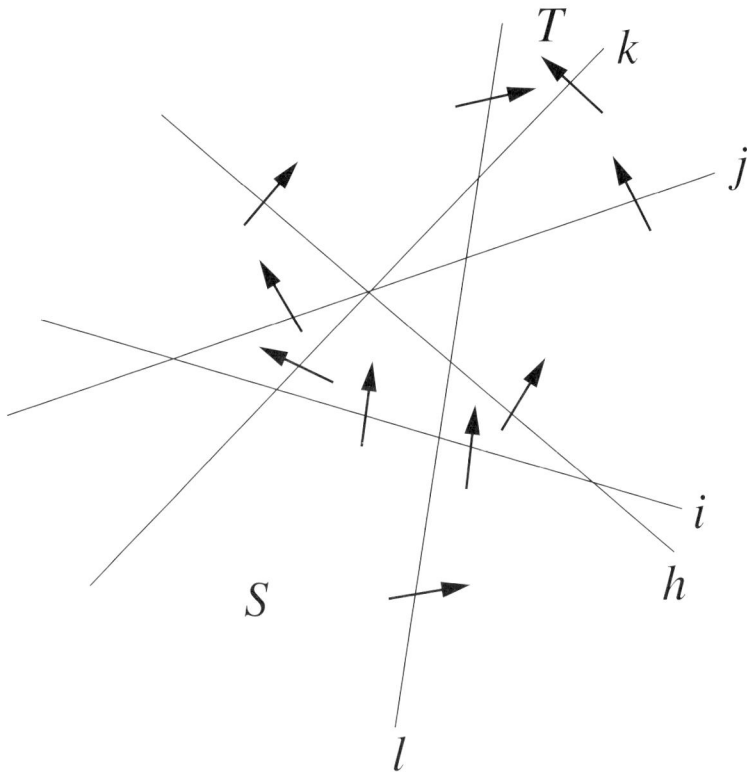

Figure 1.3. A line arrangement in the case of five straight lines in $\mathbb{R}^2$ delimiting fifteen states with ten pairs of transformations. Two direct paths from state S to state T cross the same lines in two different orders: $lihjk$ and $ikjhl$.

1.3 The Set of Linear Orders

In this example, each of the $24 = 4!$ linear orders on the set $\{1,2,3,4\}$ is regarded as a state. A transformation consists of transposing two adjacent numbers: the transformation τ_{ij} replaces an adjacent pair ji by the pair ij, or does nothing if ji does not form an adjacent pair in the initial state. There are thus $6 = \binom{4}{2}$ pairs of transformations τ_{ij}, τ_{ji}. Three of these transformations are 'effective' for the state 3142, namely:

$$3142 \xmapsto{\tau_{13}} 1342$$
$$3142 \xmapsto{\tau_{41}} 3412$$
$$3142 \xmapsto{\tau_{24}} 3124.$$

As in the preceding example, no natural orientation arises here.

1.3.1 The Permutohedron. The graph of the medium of linear orders on $\{1,2,3,4\}$ is displayed in Figure 1.4. Such a graph is sometimes referred to as a *permutohedron* (cf. Bowman, 1972; Gaiha and Gupta, 1977; Le Conte de Poly-Barbut, 1990). Again, we omit loops, as we shall always do in the sequel. The edges of this polyhedron can be gathered into six families of parallel edges; the edges in each family correspond to the same pair of transformations.

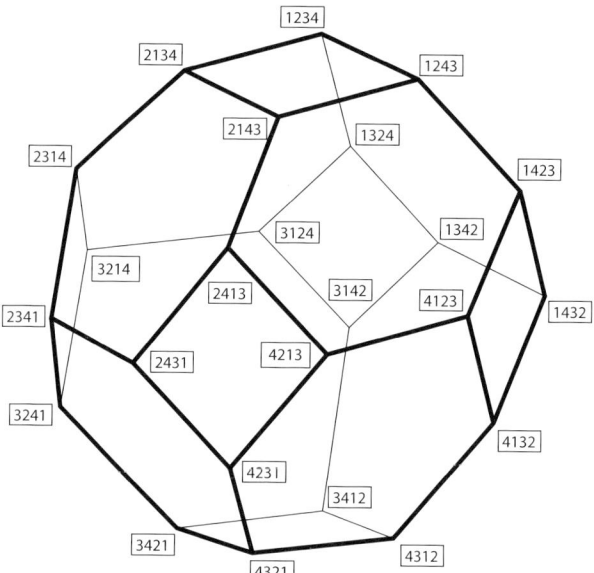

Figure 1.4. Permutohedron of $\{1,2,3,4\}$. Graph of the medium of the set of linear orders on $\{1,2,3,4\}$.

1.3.2 Remark. The family $\mathcal{L}$ of all linear orders on a particular finite set is characteristically associated with the group of permutations on that family. However, as illustrated by the graph of Figure 1.4, in which each set of parallel edges represents a particular the pair of mutually reverse transpositions of two adjacent objects, the concept of a medium is just as compelling as an algebraic structure canonically associated to $\mathcal{L}$.

1.4 The Set of Partial Orders

Consider an arbitrary finite set S. The family $\mathcal{P}$ of all strict partial orders (asymmetric, transitive, cf. 1.8.3, p. 14) on S enjoys a remarkable property: any partial order P can be linked to any other partial order P' by a sequence of steps each of which consists of changing the order either by adding one ordered pair of elements of S (imposing an ordering between two previously-incomparable elements) or by removing one ordered pair (causing two previously related elements to become incomparable), without ever leaving the family $\mathcal{P}$. Moreover, this can always be achieved in the minimal number of steps, which is equal to the 'symmetric difference' between P and P' (cf. Definition 1.8.1; see Bogart and Trotter, 1973; Doignon and Falmagne, 1997, and Chapter 5). To cast this example as a medium, we consider each partial order as a state, with the transformations consisting in the addition or removal of some pair. This medium is thus equipped with a natural orientation, as in the case of the jigsaw puzzle of 1.1.1.

The graph of such a medium is displayed in Figure 1.5 for the family of all partial orders on the set $\{a, b, c\}$. Only the edges corresponding to the transformation $P \mapsto P + \{ba\}$ are indicated. (Note that we sometimes use '+' to denote disjoint union; cf. 1.8.1.) Certain oriented media satisfy an important property: they are 'closed' with respect to their orientation. This property is conspicuous in the graph of Figure 1.5: if P, $P + \{xy\}$ and $P + \{zw\}$ are three partial orders on $\{a, b, c\}$, then $P + \{xy\} + \{zw\}$ is also such a partial order (however, see Problem 1.9).

This medium also satisfies the parallelism property observed in the permutohedron example: each set of parallel edges represents the same pair of mutually reverse transformations of adjacent objects. This property of certain media is explored in Definition 2.6.4 and Theorem 2.6.5.

Chapter 5 contains a discussion of this and related examples of families of relations, such as 'biorders' and 'semiorders' from the standpoint of media theory. (For the partial order example, see in particular Definition 1.8.3 and Theorem 5.3.5.).

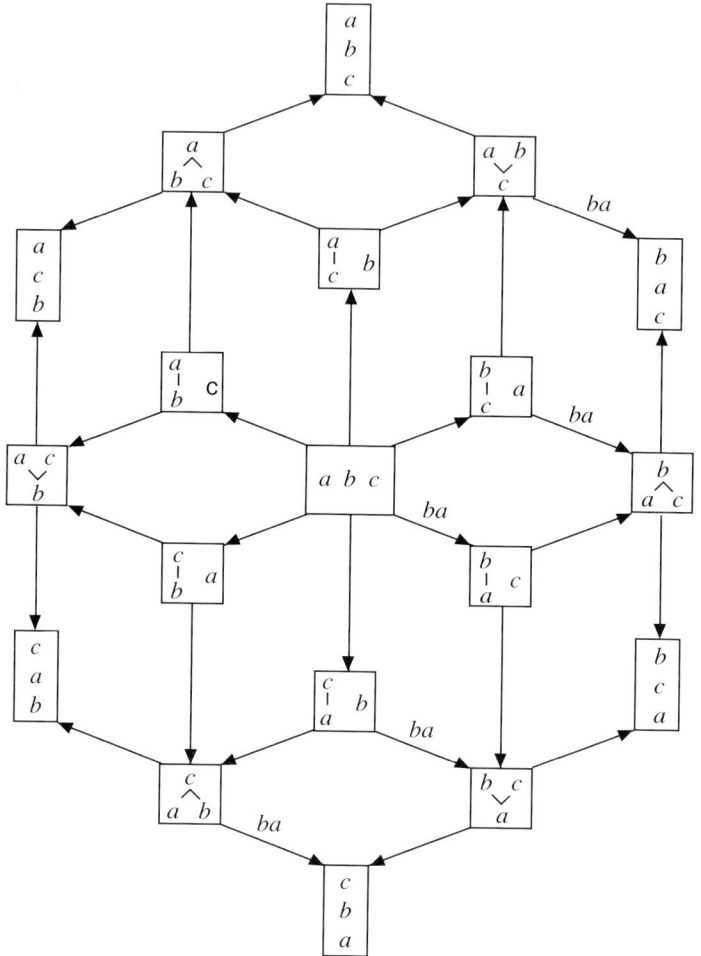

Figure 1.5. Graph of the medium of the set of all partial orders on $\{a, b, c\}$. The orientation of the edges represents the addition of a pair to a partial order. Only one class of edges is labelled, corresponding to the addition of the pair ba (see 1.8.2 for this notation).

1.5 An Isometric Subgraph of $\mathbb{Z}^n$

Perhaps the most revealing geometric representation of a finite oriented medium is as an isometric subgraph of the n-dimensional integer lattice $\mathbb{Z}^n$, for n minimal. By 'isometric' we mean that the distance between vertices in the subgraph is the same as that in $\mathbb{Z}^n$. Such a representation is always possible (cf. Theorems 3.3.4, 7.1.4, and 8.2.2), and algorithms are available for the construction (see Chapter 10).

1.5 An Isometric Subgraph of $\mathbb{Z}^n$

Let $\mathcal{M}$ be a finite oriented medium and let $\mathcal{G} \subset \mathbb{Z}^n$ be its representing subgraph. Each state of the medium $\mathcal{M}$ is represented by a vertex of $\mathcal{G}$, and each pair $(\tau, \tilde{\tau})$ of mutually reverse transformations is associated with a hyperplane $\mathcal{H}$ orthogonal to one of the coordinate axes of $\mathbb{Z}^n$, say q_j. Suppose that $\mathcal{H}$ intersects q_j at the point $(i_1, \ldots, i_j, \ldots, i_n)$. Let us identify $\mathcal{M}$ with $\mathcal{G}$ (thus, we set $\mathcal{M} = \mathcal{G}$). The restriction of the transformation τ to $\mathcal{H} \cap \mathcal{G}$ is a 1-1 function from $\mathcal{H} \cap \mathcal{G}$ onto $\mathcal{H}' \cap \mathcal{G}$, where $\mathcal{H}'$ is a hyperplane parallel to $\mathcal{H}$ and intersecting q_j at the point $(i_1, \ldots, i_j + 1, \ldots, i_n)$. Thus, τ moves $\mathcal{H} \cap \mathcal{G}$ one unit upward. The restriction of τ to $\mathcal{G} \setminus \mathcal{H}$ is the identity function on that set. The reverse transformation $\tilde{\tau}$ moves $\mathcal{H}' \cap \mathcal{G}$ one unit downward, and is the identity on $\mathcal{G} \setminus \mathcal{H}'$.

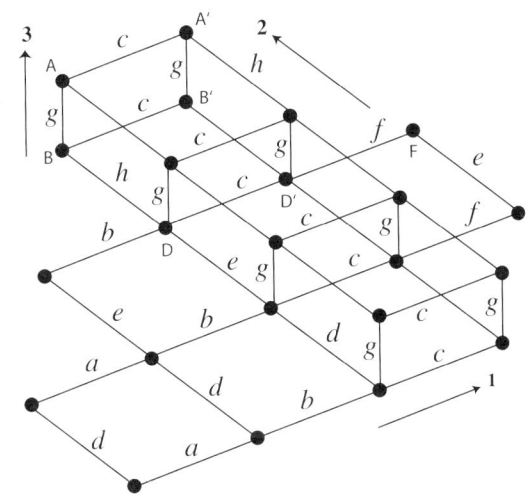

Figure 1.6. An isometric subgraph $\mathcal{D}$ of $\mathbb{Z}^3$. The orientation of the induced medium corresponds to the natural order of the integers and is indicated by the three arrows. To avoid cluttering the graph, the labeling of some edges is omitted.

The oriented graph $\mathcal{D}$ of Figure 1.6, representing a medium with 23 states and 8 pairs of mutually reverse transformations is a special case of this situation. The arrows labeled 1, 2 and 3 indicate the orientations of the axes q_1, q_2 and q_3 of $\mathbb{Z}^3$. The plane $<A, B, D>$ defined by the vertices A, B and D is orthogonal to q_1. The 8 edges marked c correspond to the pair of transformations $(\tau_c, \tilde{\tau}_c)$. The transformation τ_c moves $<A, B, D> \cap \mathcal{D}$ one unit to the right upward, that is, onto $<A', B', D'> \cap \mathcal{D}$, and is represented by loops elsewhere. The transformation $\tilde{\tau}_c$ is the reverse of τ_c.

The medium represented by $\mathcal{D}$ is not closed: the two transformations τ_f and τ_c transform D' into F and B', respectively. But applying τ_c to F gives a loop, and so does the application of τ_f to B'. Note that the subgraph $\mathcal{D}$ is not the unique representation of the medium $\mathcal{M}$ in $\mathbb{Z}^3$. A counter clockwise rotation could render the four edges marked a or b parallel to the third coordinate axis without altering the accuracy of the representation. However, because the graph is oriented we cannot apply a similar treatment to two edges marked f.

The last two examples of this chapter deal with empirical applications[3]. In both of these cases, the medium is equipped with a natural orientation.

1.6 Learning Spaces

1.6.1 Definition. Doignon and Falmagne (1999) formalize the concept of a *knowledge structure* (with respect to a topic) as a family $\mathcal{K}$ of subsets of a basic set Q of items[4] of knowledge. Each of the sets in $\mathcal{K}$ is a *(knowledge) state*, representing the competence of a particular individual in the population of reference. It is assumed that $\varnothing, Q \in \mathcal{K}$. Two compelling learning axioms are:

[K1] If $K \subset L$ are two states, with $|L \setminus K| = n$, then there is a chain of states
$$K_0 = K \subset K_1 \subset \cdots \subset K_n = L$$
such that $K_i = K_{i-1} + \{q_i\}$ with $q_i \in Q$ for $1 \leq i \leq n$. (We use '+' to denote disjoint union.) In words, intuitively: *If the state K of the learner is included in some other state L then the learner can reach state L by learning one item at a time.*

[K2] If $K \subset L$ are two states, with $K \cup \{q\} \in \mathcal{K}$ and $q \notin L$, then $L \cup \{q\} \in \mathcal{K}$. In words: *If item q is learnable from state K, then it is also learnable from any state L that can be reached from K by learning more items.*

A knowledge structure $\mathcal{K}$ satisfying Axioms [K1] and [K2] is called a *learning space* (cf. Cosyn and Uzun, 2005). To cast a learning space as a medium, we take any knowledge state to be a state of the medium. The transformations consist in adding (or removing) an item $q \in Q$ to (from) a state; thus, they take the form of the two functions: $\tau_q : \mathcal{K} \to \mathcal{K} : K \mapsto K + \{q\}$ and $\tilde{\tau}_q : \mathcal{K} \to \mathcal{K} : K \mapsto K \setminus \{q\}$. This results in a 'closed rooted medium' (see Definition 4.1.2, and Theorem 4.2.2). The study of media is thus instrumental in our understanding of learning spaces as defined by [K1] and [K2]. Note that a learning space is known in the combinatorics literature as an 'antimatroid', a structure introduced by Dilworth (1940) (cf. also Edelman and Jamison, 1985; Korte et al., 1991; Welsh, 1995; Björner et al., 1999). An empirical application of these concepts in the schools is reviewed in Section 13.1.

---

[3] In particular, learning spaces provide the theoretical foundation for a widely used internet based system for the assessment of mathematical knowledge.

[4] In a scholarly context, an 'item' might be a type of problem to be solved, such as 'long division' in arithmetic.

1.7 A Genetic Mutations Scheme

The last example of this chapter is artificial. The states of the medium are linear arrangements of genes on a small portion of a chromosome[5]. We consider four pairs of transformations, corresponding to mutations producing chromosomal aberrations observed, for example in the *Drosophila melanogaster* (cf. Villee, 1967). We take the normal state to be the sequence A-B-C, where A, B and C are three genetic segments. The four mutations are listed below.

Table 1.1. Normal state and four types of mutations

Genetic arrangements	Names
A-B-C	normal state
A-B	deletion of segment C
A-B-C-C	duplication of segment C
A-B-C-X	translocation[a] of segment X
B-A-C	inversion of segment AB

[a] From another chromosome.

1.7.1 Mutation Rules. These mutations occur in succession, starting from the normal state A-B-C, according to the five following (fictitious) rules:

[IN] The segment A-B can be inverted whenever C is not duplicated.

[TR] The translocation of the segment X can only occur in the case of a two segment (abnormal) state.

[DE] A single segment C can always be deleted (from any state).

[DU] The segment C can be duplicated (only) in the normal state.

[RE] All the reverses of these four mutations exist, but no other mutations are permitted.

The resulting graph in Figure 1.7 is the graph of a medium if we admit the possibility of reverse mutations in all cases. If such reverse mutations are rare, one can assume, in the framework of the random walk process described in Chapter 12, that some or all the reverse mutations occur with a very low positive probability.

---

[5] This example is inspired by biogenetic theory but cannot be claimed to be fully faithful to it. Our goal here is only to suggest potential applications.

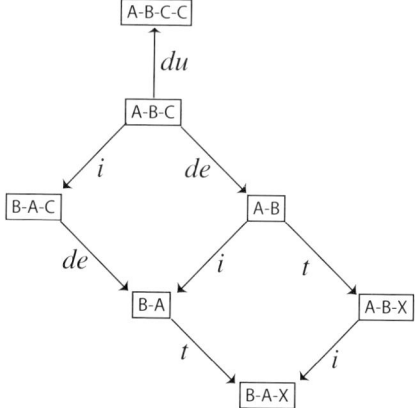

Figure 1.7. Oriented graph of the medium induced by the four mutations listed in Table 1.1, according to the five rules of 1.7.1. In the labelling of the edges, i, du, de and t stand for 'inversion of A-B', 'duplication of C', 'deletion of C' and 'translocation of X', respectively.

1.8 Notation and Conventions

We briefly review the primary mathematical notations and conventions employed throughout this book. A glossary of notation is given on page 309.

1.8.1 Set Theory. Standard logical and set theoretical notation is used throughout. We write $\Leftrightarrow$, as usual, for logical equivalence and $\Rightarrow$ for implication. The notation $\subseteq$ stands for the inclusion of sets, and $\subset$ for the proper (or strict) inclusion. We sometimes denote the union of disjoint sets by $+$ or by the summation sign $\sum$. The union of all the sets in a family $\mathcal{F}$ of subsets is symbolized by

$$\cup \mathcal{F} = \{x \,|\, x \in Y \text{ for some } Y \in \mathcal{F}\}, \tag{1.3}$$

and the intersection of all those sets by

$$\cap \mathcal{F} = \{x \,|\, x \in Y \text{ for all } Y \in \mathcal{F}\}. \tag{1.4}$$

Defined terms and statement of results are set in *slanted* font. The *complement* of a set Y with respect to some fixed ground set $\mathcal{X}$ including Y is the set $\overline{Y} = \mathcal{X} \setminus Y$.

The set of all the subsets, or *power set*, of a set $\mathcal{Z}$ is denoted by $\mathfrak{P}(\mathcal{Z})$. From (1.3) and (1.4), we get $\cup \mathfrak{P}(\mathcal{Z}) = \mathcal{Z}$ and $\cap \mathfrak{P}(\mathcal{Z}) = \varnothing$ (because $\varnothing \in \mathfrak{P}(\mathcal{Z})$). Note that we also write $\mathfrak{P}_F(\mathcal{Z})$ for the set of all **finite** subsets of $\mathcal{Z}$.

The size (or cardinality, or cardinal number) of a set X is written as $|X|$. Two sets having the same cardinal numbers are said to be *equipollent*. The *symmetric difference* of two sets X and Y is the set

$$X \triangle Y = (X \setminus Y) \cup (Y \setminus X).$$

The *symmetric difference distance* of two sets X and Y is defined by

$$d(X,Y) = |X \triangle Y|. \tag{1.5}$$

If $\mathcal{Z}$ is finite and the function d is defined by (1.5) for all $X, Y \in \mathfrak{P}(\mathcal{Z})$, then $(\mathfrak{P}(\mathcal{Z}), d)$ is a metric space. We recall that a *metric space* is a pair $(\mathcal{X}, d)$ where $\mathcal{X}$ is a set, and d is a real valued function on $\mathcal{X} \times \mathcal{X}$ satisfies the three conditions: for all x, y and z in $\mathcal{X}$,

[D1] $d(x, y) \geq 0$, with $d(x, y) = 0$ if and only if $x = y$ (positive definiteness);

[D2] $d(x, y) = d(y, x)$ (symmetry);

[D3] $d(x, z) \leq d(x, y) + d(y, z)$ (triangle inequality).

The *Cartesian product* of two sets X and Y is defined as

$$X \times Y = \{(x, y) \,|\, x \in X \,\&\, y \in Y\}$$

where (x, y) denotes an ordered pair and & means the logical connective 'and.' Writing $\Leftrightarrow$ for 'if and only if,' we thus have

$$(x, y) = (z, w) \quad \Longleftrightarrow \quad (x = z \,\&\, y = w).$$

More generally, $(x_1, \ldots, x_n)$ denotes the ordered n-tuple of the elements $x_1, \ldots, x_n$, and we have

$$X_1 \times \cdots \times X_n = \{(x_1, \ldots, x_n) \,|\, x_1 \in X_1, \ldots, x_n \in X_n\}.$$

The symbols $\mathbb{N}$, $\mathbb{Z}$, $\mathbb{Q}$, and $\mathbb{R}$ stand for the sets of natural numbers, integers, rational numbers, and real numbers, respectively; $\mathbb{N}_0$ and $\mathbb{R}_+$ denote the sets of nonnegative integers and nonnegative real numbers respectively.

1.8.2 Binary Relations, Relative Product. A set R is a *binary relation* if there are two (not necessarily distinct) sets X and Y such that $R \subseteq X \times Y$. Thus, a binary relation is a set of ordered pairs $xy \in X \times Y$, where xy is an abbreviation of (x, y). In such a case, we often write xRy to mean $xy \in R$. The qualifier 'binary' is often omitted. If $R \subseteq X \times X$, then R is said to be a binary relation *on* X. The (*relative*) *product* of two relations R and S is the relation

$$RS = \{xz \,|\, \exists y, xRySz\}$$

(in which $\exists$ denotes the existential quantifier). If $R = S$, we write $R^2 = RR$, and in general $R^{n+1} = R^n R$ for $n \in \mathbb{N}$. By convention, if R is a relation on X, then R^0 denotes the *identity relation* on X:

$$xR^0 y \quad \Longleftrightarrow \quad x = y.$$

Note that when xRy and yRz, we sometimes write $xRyRz$ for short. Elementary properties of relative products are taken for granted. For example: if R, S, T and M are relation, then:

$$S \subseteq M \implies RST \subseteq RMT \tag{1.6}$$

and

$$R(S \cup T) \subseteq RS \cup RT. \tag{1.7}$$

1.8.3 Order Relations. A relation is a *quasi order* on a set X if it is *reflexive* and *transitive* on X, that is, for all x, y, and z in X,

$\qquad xRx$ \hfill (reflexivity)

$\qquad xRy$ & $yRz \implies xRz$ \hfill (transitivity)

(where '$\implies$' means 'implies' or 'only if'). A quasi order R is a *partial order* on X if it is *antisymmetric* on X, that is, for all x and y in X

$$xRy \ \& \ yRx \implies x = y.$$

If R is a partial order on a set X, then the pair (X, R) is referred to as a *partially ordered set*. A relation R is a *strict partial order* on X if it is transitive and asymmetric on X, that is, for all x, y in X,

$$xRy \implies \neg(yRx),$$

where $\neg$ stands for the logical 'not.' The *Hasse diagram* or *covering relation* of a partial order (X, R) is the relation $\check{R} \subseteq R$ such that, for all x, y and z in X, $z\check{R}x$ together with $zRyRx$ implies either $x = y$ or $y = z$. We say then that x covers z. If X is infinite, the Hasse diagram may be empty (see Problem 1.8). Otherwise, $\check{R}$ provide a faithful and economical summary of R. Indeed, we have

$$R = \cup_{n=0}^{\infty} \check{R}^n = \check{R}^0 \cup \check{R} \cup \cdots \cup \check{R}^n \cup \cdots \tag{1.8}$$

The r.h.s. (right hand side) of (1.8) is called the *transitive closure* of $\check{R}$. The Hasse diagram of a relation R is the 'smallest relation' (see Problem 1.4) the transitive closure of which gives back R. In general, we write Q^* for the transitive closure of a relation Q. Thus, we can rewrite the first equality in (1.8) in the compact form $R = \check{R}^*$. The Hasse diagram of a strict partial order can also be defined (see Problem 1.8).

A partial order L on X is a *linear order* if it is is *strongly connected*, that is, for all x, y in X,

$$xLy \quad \text{or} \quad yLx. \tag{1.9}$$

A relation L on X is a *strict linear order* if it is a strict partial order which is *connected*, that is, (1.9) holds for all distinct x, y in X.

Suppose that L is a strict linear order on X. A *L-minimal* element of $Y \subseteq X$ is a point $x \in Y$ such that $\neg(yLx)$ for any $y \in Y$. A strict linear order L on X is a *well-ordering* of X if every nonempty $Y \subseteq X$ has a L-minimal element. In such case, we may say that L *well-orders* X.

We follow Roberts (1979) and call a *strict weak order* on a set X a relation $\prec$ on X satisfying the condition: for all $x, y, z \in X$,

$$x \prec y \quad \Rightarrow \quad \begin{cases} \neg(y \prec x) \text{ and} \\ \text{either } x \prec z \text{ or } z \prec y \text{ (or both)}. \end{cases} \quad (1.10)$$

For the definition of a weak order, see Problem 1.10.

1.8.4 Equivalence Relations, Partitions. A binary relation R is an *equivalence relation* on a set X if it is reflexive, transitive, and symmetric on X, that is, for all x, y in X, we have

$$xRy \quad \Longleftrightarrow \quad yRx.$$

The following construction is standard. Let R be a quasi order on a set X. Define the relation $\sim$ on X by the equivalence

$$x \sim y \quad \Longleftrightarrow \quad (xRy \ \& \ yRx). \quad (1.11)$$

It is easily seen that the relation $\sim$ is reflexive, transitive, and symmetric on X, that is, $\sim$ is an equivalence relation on X. For any x in X, define the set $\langle x \rangle = \{y \in X \mid x \sim y\}$. The family $\mathcal{Z} = X/\sim = \{\langle x \rangle \mid x \in X\}$ of subsets of X is called the *partition* of X induced by $\sim$. Any partition $\mathcal{Z}$ of X satisfies the following three properties

[P1] $Y \in \mathcal{Z}$ implies $Y \neq \emptyset$;
[P2] $Y, Z \in \mathcal{Z}$ and $Y \neq Z$ imply $Y \cap Z = \emptyset$;
[P3] $\cup \mathcal{Z} = X$.

Conversely, any family $\mathcal{Z}$ of subsets of a set X satisfying [P1], [P2], and [P3] is a partition of X. The elements of $\mathcal{Z}$ are called the *classes* of the partition. A partition containing just two classes is sometimes referred to as a *bipartition*.

An example of a bipartition was provided by the family $\{\mathcal{T}^+, \mathcal{T}^-\}$ of our puzzle Example 1.1.1, where $\mathcal{T}^+$ contains all the transformations consisting in adding pieces to the puzzle, and $\mathcal{T}^-$ those containing their reverses, that is, removing those pieces.

1.8.5 Graphs. The language of graph theory is coextensive with that of relations, with the former applying naturally when geometrical representations are used. We will use either of them as appropriate to the situation.

A *directed graph* or *digraph* is a pair (V, A) where V is a set and $A \subseteq V \times V$. The elements of V are referred to as *vertices* and the ordered pairs in A as arcs or *directed edges*. The pair (V, A) may be referred to as a *digraph on V*. Several pictorial representations of graphs have been given already. In Figure 1.6, for example, the vertices are all the partial orders on the set $\{a, b, c\}$, and each of the arcs corresponds to the addition of an ordered pair to a partial order, forming a new partial order. This illustrates the usual convention of representing arcs by arrows and vertices by points or small circles. When the arc vw exists, we say that 'there is an arc from v to w.' Two distinct vertices v,w are called *adjacent* when one or both of the two arcs vw or wv exists. In pictorial representations of graphs, loops—circular arrows joining a vertex to itself—are routinely omitted when they provide redundant information.

As will be illustrated by many examples in this monograph, there are digraphs satisfying the condition: there is an arc from v to w whenever there is an arc from w to v and vice versa. (In the terminology of relations introduced in 1.8.4, we say then that the set of arcs regarded as a relation is symmetric.) In such cases, the digraph is called a *graph* and each pair of arcs (vw, wv) is referred as an *edge* of the graph which we denote by $\{v, w\}$. The permutohedron of Figure 1.4 is a geometric representation of a graph. Note that, as is customary or at least frequent, a single line between two points representing adjacent vertices replaces the pair of arrows picturing the two arcs.

Let s_n be a sequence $v_0, v_1, \ldots, v_n$ of vertices in a digraph such that $v_i v_{i+1}$ is an arc, for $0 \leq i \leq n-1$. Such a sequence is called a *walk* from v_0 to v_n. A *segment* of the walk s_n is a subsequence $s_j, s_{j+1}, \ldots, s_{j+k}$, with $0 \leq j \leq j+k \leq n$. The walk s_n is *closed* if $v_0 = v_n$, and *open* otherwise. A walk whose vertices are all distinct, is a *path*. A closed path is a *circuit* or a *cycle*. The *length* of a walk s_n (whether open or closed), is assigned to be n. Notice that the concepts of walk, path and circuit apply to a graph, with a sequence of edges $\{v_i, v_{i+1}\} = (v_i v_{i+1}, v_{i+1} v_i)$, $0 \leq i \leq n-1$. When the digraph is a graph, there is a path of length n from v to w if and only if there is a path of length n from w to v. In such a case, we define the *(graph theoretical) distance* $\delta(v, w)$ between two distinct vertices v and w to be the length of the shortest path from v to w, if there is such a path, and infinity otherwise. We also define $\delta(v, v) = 0$, for all vertices v. It is easily shown (cf. Problem 1.16) that the pair (V, δ) is a metric space, that is, the function δ satisfies Conditions [D1], [D2] and [D3] in 1.8.1 (with $\delta = d$). To avoid ambiguity when more than one graph is under discussion, we may index the distance function by the graph and write $\delta_G(v, w)$ for the distance between the vertices v and w in the graph $G = (V, A)$. We sometimes need to focus on part of a graph. A graph $H = (W, B)$ is a *subgraph* of a graph $G = (V, A)$ if $W \subseteq V$ and $B \subseteq A$; it is an *induced subgraph* of G if $W \subseteq V$ and $B = (W \times W) \cap A$. A graph $H = (W, B)$ is an *isometric* subgraph of a graph $G = (V, A)$ if it is a subgraph of G and moreover

$$\delta_H(v, w) = \delta_G(v, w) \qquad (v, w \in W).$$

Two graphs (V, A) and (W, B) are *isomorphic* if there is a bijection $\varphi : V \to W$ such that

$$\{\varphi(v), \varphi(w)\} \in B \iff \{v, w\} \in A,$$

for all $v, w \in V$. If (W, B) is a subgraph of a graph (U, C), then we say that φ is an *embedding* of (V, A) into (U, C). This embedding is an *isometric embedding* if (W, B) is an isometric subgraph of (U, C).

In the permutohedron graph of Figure 1.4, the subgraph defined by the subset of the six vertices

$$W = \{2134, 2143, 2413, 2431, 2341, 2314\}$$

(forming an hexagonal face of the permutohedron) is isometric. But removing a single point of W would define a subgraph that would not be isometric (see Problem 1.15 in this connection).

A graph $G = (V, A)$ is *connected* if any two distinct vertices of G are endpoints of a path of G. It is *bipartite* if its vertex set can be partitioned into $V = V_1 + V_2$, such that every edge of G connects a vertex in V_1 with a vertex in V_2. A classical result from König (1916) is that a graph is bipartite if and only if it contains no odd circuit, that is, a circuit with an odd number of edges. A *tree* is a connected graph without cycles. A vertex in a tree is a *leaf* if it is adjacent to a single vertex.

For graph terminology and results, see Busacker and Saaty (1965), Bondy and Murphy (1976), Roberts (1984), or Bondy (1995).

1.9 Historical Note and References

The concept of a medium was introduced by Falmagne (1997), as a generalization of conditions satisfied by certain families of relations, such as the family $\mathcal{P}$ of all (strict) partial orders on a finite set S. A key property of such a family, wellgradedness, already encountered in this chapter is that, for any two partial orders $\prec$ and $\prec'$, there necessarily exists a sequence

$$\prec_1 = \prec, \prec_1, \ldots, \prec_n = \prec' \tag{1.12}$$

of partial orders in $\mathcal{P}$ such that any two consecutive partial orders in (1.12) differ by exactly one pair,

$$|\prec_i \triangle \prec_{i+1}| = 1 \quad (i = 1, \ldots, n-1), \tag{1.13}$$

and, moreover, such a path between $\prec$ and $\prec'$ in $\mathcal{P}$ is minimal in the sense that:

$$|\prec \triangle \prec'| = n. \tag{1.14}$$

This property, which is a focus of Chapter 5, is also satisfied by other families of relations, such as the semiorders and the biorders (cf. Definition 5.1.1, Formulas (5.13) and (5.9)). This was shown by Doignon and Falmagne

(1997), who referred to it as the 'wellgradedness' of such families. On hindsight, it should perhaps have evoked the condition of transitivity of certain semigroups. Curiously, however, the connection between wellgradedness and the axioms specifying a medium was made via the development of random walk models based on such families in view of some applications in the social sciences. In computing the asymptotic probabilities of the states of these random walks (which are formed by the relations in each of these families; see e.g. Falmagne, 1996; Falmagne and Doignon, 1997) it was realized that essentially—but not exactly—the same limit theorem had to be proven. The axiomatization of the concept of a medium was the natural step taken in Falmagne (1997), which also contains a number of basic results. This work was further extended by Falmagne and Ovchinnikov (2002) and Ovchinnikov (2006) (see also Ovchinnikov and Dukhovny, 2000).

Media were later investigated from an algorithmic standpoint by Eppstein and Falmagne (2002). One of their results concerns the existence of a tight bound for the shortest 'reset sequence' (cf. Ginsburg, 1958; Moore, 1956) for a medium. They also describe a near-linear time algorithm for testing whether an orientation is closed, and a polynomial time one for finding a closed orientation if one exists. Those results are contained in Chapter 10.

Concepts essentially equivalents to media were investigated much earlier by graph theorists under the name of 'partial cubes', that is, isometric subgraphs of hypercubes, beginning with Graham and Pollak (1971). Partial cubes were characterized by Djoković (1973) and Winkler (1984) (see Imrich and Klavžar, 2000, which also contains references to other characterizations of partial cubes). The ties between partial cubes and media are described in Chapter 7. Some critical differences between partial cubes and media lie in the language and notation which, in the case of media theory, are akin to automata theory. Obviously, the language and notation one uses are strongly suggestive and may lead to different types of results, as exemplified in this volume.

The random walk models mentioned above were used for the analysis of opinion polls, in particular those polls concerning the presidential election in the US. Such polls are typically performed several times on the same large sample of respondents. The results are referred to as 'panel data.' Among other queries, the respondent are always asked to provide a type of ranking of the candidates. Assuming that the panel data consist in k polls taken at times $t_1, \ldots, t_k$, this means that each respondent has provided a finite sequence $\prec_{t_1}, \ldots, \prec_{t_k}$ of some kind of order relations, such as weak orders. The form of such data suggests an interpretation in terms of visits, by each respondent, of the states of some random walk on the family of all such order relations. In such a framework, these visits take place in real time and $\prec_{t_1}, \ldots, \prec_{t_k}$ is regarded as a sequence of 'snapshots' revealing an aspect of the opinion of a respondent at times $t_1, \ldots, t_k$. Various random walk models have been used to analyze the results of the 1992 US presidential election opposing Clinton, Bush and Perot (Falmagne et al., 1997; Regenwetter et al., 1999; Hsu and

Regenwetter, in press). The general theory of such random walks on media is developed in Falmagne (1997) and Falmagne et al. (2007). Our last two chapters include an exposition of this work.

A major application of media theory is to learning spaces[6], which were defined on p. 10. As is established in Theorem 4.2.2, a learning space is essentially a 'rooted, closed medium.' Learning spaces provide a model for the structural organization of the feasible knowledge states in a topic, such as algebra or chemistry. The particular learning space associated with a topic is at the core of an algorithm for the assessment of knowledge in that topic. It also guide and monitors the students' learning. Such a system is currently used by many schools and colleges in the US and abroad. Further discussion of this topic will be found in Chapter 13.

In view of the wide diversity of the examples covered in this chapter and later in this book, it seems likely that media have other useful applications.

Problems

Many of the questions asked below will be dealt with in depth in the chapters of this book. Some of the problems cannot be solved in formal sense without using the axioms specifying a medium, which can be found in Definition 2.2.1. In such cases the reader should rely on the intuitive conception of a medium developed in this chapter to analyze the problem and attempt a formalization. We propose such exercises as a useful preparation for the rest of this volume.

1.1 In the medium of the Gauss puzzle 1.1.1, identify two states S and T, and two distinct minimally short paths from S to T that do not have their transformations in exact opposite orders. What are conditions (necessary and sufficient) that garantee that the transformations will be in exact opposite orders?

1.2 A facet of the permutohedron on four elements is either a square or a regular hexagon. Describe the facets of a permutohedron on five elements.

1.3 Certain oriented media, such as that of our first example of the puzzle, are closed for their orientation: suppose that S is any state of the medium, and that $\tau : S \mapsto S\tau$ and $\mu : S \mapsto S\mu$ are any two 'positive' transformations, then whenever $S\tau$ and $S\mu$ are defined as states distinct from S, then $(S\tau)\mu = (S\mu)\tau = S\tau\mu = S\mu\tau$ is also a state of the medium. Such a medium is said to be 'closed' (cf. Chapter 4). Under which condition is the medium induced by a finite set of straight line in the plane (in the sense of Section 1.2.1 and Figure 1.3) a closed medium? Can you prove your response?

---

[6] The exact connection between learning spaces and media was recognized only recently.

1.4 Let R be a strict partial order. Suppose that $\mathcal{H}$ is a family of relations such that, for each $Q \in \mathcal{H}$, the transitive closure Q^* of Q is equal to R. Verify that we have then $\cap \mathcal{H} = \check{R}$. (So, the phrase 'the smallest relation the transitive closure of which gives back R' makes sense in 1.8.3.)

1.5 Verify the implication (1.6).

1.6 Prove the inclusion (1.7). Why don't we have the equality?

1.7 The medium represented by its graph in Figure 1.7 is equipped with an orientation. This medium is not closed under that orientation. Also, it does not contain the state $\boxed{\text{A-B-C-X}}$. Modify the rules so as to obtain a closed medium containing the eight states of Figure 1.7, the state $\boxed{\text{A-B-C-X}}$, plus at most two other states.

1.8 Define the Hasse diagram of a strict partial order. Exactly when is the Hasse diagram of a partial order or strict partial order empty?

1.9 We have seen that the medium of all partial orders on the set $\{a, b, c\}$ was closed with respect to the natural orientation of that medium, namely: if P, $P + \{xy\}$ and $P + \{zw\}$ are partial orders in the same medium, then so is $P + \{xy\} + \{zw\}$. Prove or disprove (by a counterexample) that that closedness property holds for all families of partial orders on a finite set.

1.10 In the spirit of the distinction between a partial order and a strict partial order (Definition 1.8.3), define the concept of a weak order $\precsim$ on a set X.

1.11 Let $\mathcal{H}$ be the collection of all the Hasse diagrams of the collection of all the partial order on a given finite set. Is $\mathcal{H}$ well-graded? Hint: Examine Equations (1.12), (1.13) and (1.14) in this connection.

1.12 Let $\mathcal{F}$ and $\mathcal{G}$ be the families of all the partial orders on the sets $\{a, b, c\}$ and $\{a, b, x\}$. Describe $\mathcal{F} \setminus \mathcal{G}$. Does it form a medium?

1.13 Let $\mathcal{F}$ and $\mathcal{G}$ be the families of all the partial orders on two distinct overlapping sets. Describe $\mathcal{F} \cap \mathcal{G}$. Does it form a medium?

1.14 Consider a planar political map of the world (indicating the countries' boundaries). In the style of the hyperplane arrangement of 1.2.1, define the states to be the countries, and let the transformations be the crossings of a single border between two countries. Does this construction define a medium? Why or why not?

1.15 Define a isometric subgraph (W, B) of the permutohedron (L, A) of Figure 1.4, with $W \subset L$, having a maximal number of point. (Thus, adding a single point to W would destroy the isometricity.)

1.16 Let (A, V) be a graph, and let δ be the distance on that graph as defined in 1.8.5. Prove that the pair (A, δ) is a metric space, that is, verify that the three conditions [D1], [D2] and [D3] are satisfied.

2
Basic Concepts

We first introduce the general framework of 'token systems' and a convenient language for our definitions. In this context, we then formulate two independent axioms specifying the concept of a medium and begin the development of the main results. A glossary of notations can be found on page 309.

2.1 Token Systems

2.1.1 Definition. Let $\mathcal{S}$ be a set of *states*. A *token (of information)* is a function $\tau : S \mapsto S\tau$ mapping $\mathcal{S}$ into itself. We write $S\tau = \tau(S)$, and $S\tau_1\tau_2\cdots\tau_n = \tau_n(\cdots\tau_2(\tau_1(S))\cdots)$ for the function composition. By definition, the identity function τ_0 on $\mathcal{S}$ is not a token. Let $\mathcal{T}$ be a set of tokens on $\mathcal{S}$, with $|\mathcal{S}| \geq 2$ and $\mathcal{T} \neq \varnothing$. The pair $(\mathcal{S}, \mathcal{T})$ is then called a *token system*.

Let V and S be two states. Then V is *adjacent* to S if $S \neq V$ and $S\tau = V$ for some token τ in $\mathcal{T}$. A token $\tilde{\tau}$ is a *reverse* of a token τ if for any two adjacent states S and V, we have

$$S\tau = V \iff V\tilde{\tau} = S. \tag{2.1}$$

It is easily verified that a token has at most one reverse. If the reverse $\tilde{\tau}$ of τ exists, then τ and $\tilde{\tau}$ are mutual reverses. We have thus $\tilde{\tilde{\tau}} = \tau$, and adjacency is a symmetric relation on $\mathcal{S}$ (Problem 2.2).

2.1.2 Definition. A *message* is a (possibly empty) string of elements of the set of tokens $\mathcal{T}$. A nonempty message $\boldsymbol{m} = \tau_1\ldots\tau_n$ defines a function $S \mapsto S\tau_1\cdots\tau_n$ on the set of states $\mathcal{S}$. By abuse of notation, we also write then $\boldsymbol{m} = \tau_1\cdots\tau_n$ for the corresponding function. No ambiguity will arise from this double usage. When $S\boldsymbol{m} = V$ for some states S, V and some message $\boldsymbol{m}$, we say that $\boldsymbol{m}$ *produces* V from S. The *content* of a message $\boldsymbol{m} = \tau_1\ldots\tau_n$ is the set $\mathcal{C}(\boldsymbol{m}) = \{\tau_1,\ldots,\tau_n\}$ of its tokens. We write $\ell(\boldsymbol{m}) = n$ to denote the *length* of the message $\boldsymbol{m}$. (We have thus $|\mathcal{C}(\boldsymbol{m})| \leq \ell(\boldsymbol{m})$.) A message $\boldsymbol{m}$

is *effective* (resp. *ineffective*) for a state S if $Sm \neq S$ (resp. $Sm=S$) for the corresponding function $S \mapsto Sm$. A message $m = \tau_1 \ldots \tau_n$ is *stepwise effective* for S if $S\tau_1 \cdots \tau_k \neq S\tau_0 \cdots \tau_{k-1}$, $1 \leq k \leq n$. A message which is both stepwise effective and ineffective for some state is called a *return message* or, more briefly, a *return* (for that state).

A message $m = \tau_1 \ldots \tau_n$ is *inconsistent* if it contains both a token and its reverse, that is, if $\tau_j = \tilde{\tau}_i$ for some pair of distinct indices i and j; otherwise, it is called *consistent*. A message consisting of a single token is thus consistent. Two messages $\mathbf{m}$ and $\mathbf{n}$ are *jointly consistent* if $\mathbf{mn}$ (or, equivalently, $\mathbf{nm}$) is consistent. A consistent message which is stepwise effective for some state S and does not have any of its token occuring more than once is said to be *concise* (for S). A message $m = \tau_1 \ldots \tau_n$ is *vacuous* if the set of indices $\{1, \ldots, n\}$ can be partitioned into pairs $\{i, j\}$, such τ_i and τ_j are mutual reverses.

Note that when a message m is empty, we define $m = \tau_0$, the identity on $\mathcal{S}$ (which, as mentioned above, is not a token). In such a case, m is a place holder symbol that can be deleted, as in: 'let mn be a message in which m is either a concise message or is empty' (that is $mn = n$).

2.2 Axioms for a Medium

2.2.1 Definition. A token system $(\mathcal{S}, \mathcal{T})$ is called a *medium* (on $\mathcal{S}$) if the following two axioms are satisfied.

[Ma] For any two distinct states S, V in $\mathcal{S}$, there is a concise message producing V from S.

[Mb] Any return message is vacuous.

A medium $(\mathcal{S}, \mathcal{T})$ is *finite* if $\mathcal{S}$ is a finite set. For some comments on these axioms, see Remark 2.2.8.

2.2.2 Lemma. *In a medium, each token has a unique reverse.*

Proof. For any token τ, we have $S\tau = V$ for some distinct states S and V. Axiom [Ma] implies that there is a concise message m producing S from V. Hence, τm is a return for S. By Axiom [Mb], τm must be vacuous, and so $\tilde{\tau} \in \mathcal{C}(m)$. As m is concise, we must have $m = \tilde{\tau}$. The uniqueness results from the definition of a reverse of a token (see 2.1.1). □

2.2.3 Definition. The *reverse* of a message $m = \tau_1 \ldots \tau_n$ is defined by $\widetilde{m} = \tilde{\tau}_n \ldots \tilde{\tau}_1$. The following facts are straightforward (see Problem 2.1): (i) if m is stepwise effective for S, then $Sm = V$ implies $V\widetilde{m} = S$; (ii) $\tau \in \mathcal{C}(m)$ if and only if $\tilde{\tau} \in \mathcal{C}(\widetilde{m})$; (iii) if m is consistent, so is $\widetilde{m}$.

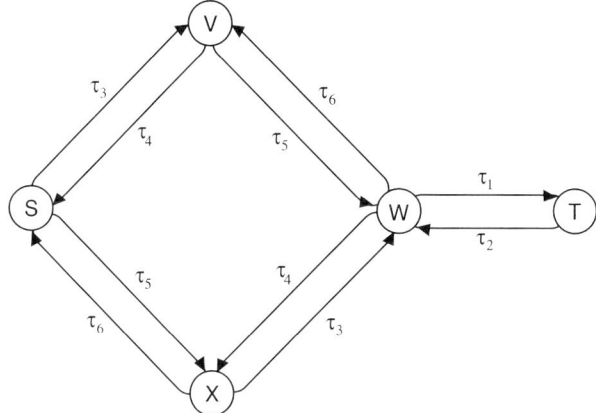

Figure 2.1. Digraph of a medium with set of states $\mathcal{S} = \{S, V, W, X, T\}$ and set of tokens $\mathcal{T} = \{\tau_i \,|\, 1 \leq i \leq 6\}$.

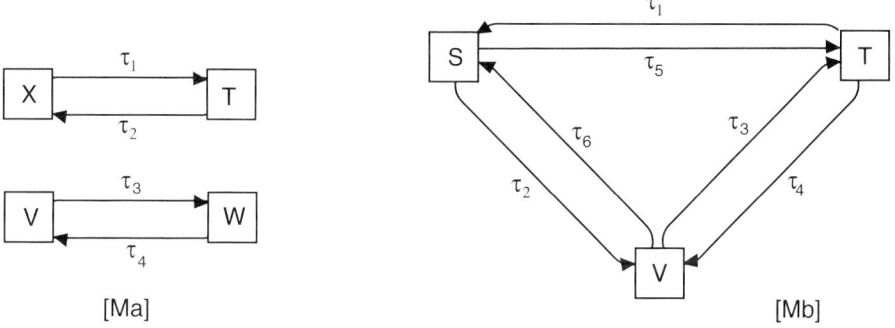

Figure 2.2. Digraphs of two token systems establishing the independence of the two Axioms [Ma] and [Mb]. Each digraph is labelled by the failing axiom.

2.2.4 Example. Figure 2.1 displays the digraph of a medium with set of states $\mathcal{S} = \{S, V, W, X, T\}$ and set of tokens $\mathcal{T} = \{\tau_i \,|\, 1 \leq i \leq 6\}$. It is clear that $\tilde{\tau}_1 = \tau_2$, $\tilde{\tau}_3 = \tau_4$, and $\tilde{\tau}_5 = \tau_6$.

2.2.5 Theorem. *The axioms [Ma] and [Mb] are independent.*

Proof. Each of the two digraphs in Figure 2.2 on page 25 defines a token system satisfying one of the two axioms defining a medium. The labels [Ma] and [Mb] attached to the two digraphs indicates the failing axiom (Problem 2.3). □

2.2.6 Convention. Except when stated otherwise, we assume implicitly for the rest of this chapter that we have fixed a token system in which Axioms [Ma] and [Mb] hold.

2.2.7 Lemma. *Let $\boldsymbol{n}$ and $\boldsymbol{m}$ be two consistent messages producing the same state S, and suppose that these messages are stepwise effective for some not necessarily distinct states T and V, respectively. Then $\boldsymbol{n}$ and $\boldsymbol{m}$ are jointly consistent.*

Proof. If $T = V$, the message $\boldsymbol{n}\widetilde{\boldsymbol{m}}$ is a return for T which, by [Mb], must be vacuous. Suppose that $\boldsymbol{nm}$ is inconsistent. There is then some token τ that occurs in $\boldsymbol{n}$ and in $\widetilde{\boldsymbol{m}}$. Since $\boldsymbol{n}\widetilde{\boldsymbol{m}}$ is vacuous, the token $\widetilde{\tau}$ must occur more than once in it. This is impossible since $\boldsymbol{n}$ and $\widetilde{\boldsymbol{m}}$ are consistent.

Suppose that $T \neq V$. From Axiom [Ma], we know that there exists a concise message $\boldsymbol{w}$ producing T from V. Thus $\boldsymbol{n}\widetilde{\boldsymbol{m}}\boldsymbol{w}$ is a return for T. By [Mb], $\boldsymbol{n}\widetilde{\boldsymbol{m}}\boldsymbol{w}$ is vacuous. If $\boldsymbol{nm}$ is not consistent, there is some token $\tau \in \mathcal{C}(\boldsymbol{n})$ with $\widetilde{\tau} \in \mathcal{C}(\boldsymbol{m})$, which implies that τ appears at least twice in $\boldsymbol{n}\widetilde{\boldsymbol{m}}$. But as $\boldsymbol{n}\widetilde{\boldsymbol{m}}\boldsymbol{w}$ is vacuous and each of $\boldsymbol{n}$ and $\boldsymbol{m}$ is consistent, the token $\widetilde{\tau}$ must appear at least twice in $\boldsymbol{w}$, contradicting the hypothesis that $\boldsymbol{w}$ is concise. □

2.2.8 Remark. Earlier discussions of the concept of a medium (Falmagne, 1997; Falmagne and Ovchinnikov, 2002) were based on a weaker form of [Ma] assuming only the existence, for any two distinct states S and V, of a consistent message producing V from S. The two Lemmas 2.2.2 and 2.2.7 were taken as axioms, and a form of Axiom [Mb] was also used.

All the examples of media encountered so far were finite ones. Our first infinite example is given below.

2.2.9 Example. Let $(\mathbb{Z}^2, \mathcal{T})$ be the medium with the set of transformations $\mathcal{T}$ contains all pairs of tokens τ_{ij}, $\widetilde{\tau}_{ij}$ defined by

$$\tau_{ij} : \mathbb{Z}^2 \to \mathbb{Z}^2 : (k, q) \mapsto (k, q)\tau_{ij} = \begin{cases} (k+1, q) & \text{if } i = k, j = 1, \\ (k, q+1) & \text{if } i = q, j = 2, \\ (k, q) & \text{otherwise,} \end{cases}$$

$$\widetilde{\tau}_{ij} : \mathbb{Z}^2 \to \mathbb{Z}^2 : (k, q) \mapsto (k, q)\widetilde{\tau}_{ij} = \begin{cases} (k-1, q) & \text{if } i = k, j = 1, \\ (k, q-1) & \text{if } i = q, j = 2, \\ (k, q) & \text{otherwise.} \end{cases}$$

It is easily checked that Axioms [Ma] and [Mb] hold. Note that both the set of states and the set of tokens are infinite (see Problem 2.9).

2.3 Preparatory Results

Some simple consequences of the axioms are gathered in the next lemma.

2.3.1 Lemma. (i) *No token can be identical to its own reverse.*

(ii) *Any consistent, stepwise effective message (for some state) is concise.*

(iii) *For any two adjacent states S and V, there is exactly one token producing V from S.*

(iv) *Let $\boldsymbol{m}$ be a message that is concise for some state, then*

$$\ell(\boldsymbol{m}) = |\mathcal{C}(\boldsymbol{m})|, \qquad (2.2)$$

and

$$\mathcal{C}(\boldsymbol{m}) \cap \mathcal{C}(\widetilde{\boldsymbol{m}}) = \varnothing. \qquad (2.3)$$

(v) *No token can be a 1-1 function.*

(vi) *Suppose that $\boldsymbol{m}$ and $\boldsymbol{n}$ are stepwise effective for S and V, respectively, with $S\boldsymbol{m} = V$ and $V\boldsymbol{n} = W$. Then $\boldsymbol{m}\boldsymbol{n}$ is stepwise effective for S, with $S\boldsymbol{m}\boldsymbol{n} = W$.*

(vii) *Any vacuous message which is stepwise effective for some state is a return message for that state.*

Notice that the last statement is a partial converse of Axiom [Mb].

Proof. (i) Suppose that $\tau = \tilde{\tau}$. As $\tau \neq \tau_0$, the identity function on $\mathcal{S}$, we must have $S\tau = V$ for some distinct states S and V. This implies $S\tilde{\tau} = V$, and so both τ and $\tilde{\tau}$ produce V, contradicting Lemma 2.2.7.

(ii) Suppose that $\boldsymbol{m}$ is a consistent, stepwise effective message producing V from S, and that some token τ occurs at least twice in $\boldsymbol{m}$. We have thus $S\boldsymbol{m} = \boldsymbol{n}_1\tau\boldsymbol{n}_2\tau\boldsymbol{n}_3 = V$ for some consistent stepwise effective messages $\boldsymbol{n}_1\tau$ and $\boldsymbol{n}_2\tau\boldsymbol{n}_3$. (Some of $\boldsymbol{n}_1$, $\boldsymbol{n}_2$ and $\boldsymbol{n}_3$ could of course be empty.) This implies that $S\boldsymbol{m} = S\boldsymbol{n}_1\tau\boldsymbol{n}_2\tau = V'$, for some state V'. We conclude that the messages $\boldsymbol{n}_1\tau$ and $\tilde{\tau}\tilde{\boldsymbol{n}}_2$ are consistent, stepwise effective messages producing the state $W = S\boldsymbol{n}_1\tilde{\tau}$ from S and V, respectively. These two messages are not jointly consistent, contradicting Lemma 2.2.7.

Suppose that $S\tau_1 = S\tau_2 = V$. The message $\tau_1\tilde{\tau}_2$ is a return for S. By [Mb], this message is vacuous, so $\{\tau_1, \tilde{\tau}_2\}$ is a pair of mutually reverse tokens. It follows that $\tau_1 = \tilde{\tilde{\tau}}_2 = \tau_2$.

(iv) Equations (2.2) and (2.3) stem readily from the definition of a concise message.

(v) Suppose that $S\tau = V$ for some token τ and two adjacent states S and V. If $V\tau = W \neq V$ for some state W, then $V = S\tau = W\tilde{\tau}$, a contradiction of Lemma 2.2.7, because, by definition, τ is a consistent message. Hence, $S\tau = V\tau = V$, and so τ is not a 1-1 function.

(vi) This is clear.

(vii) Let $\boldsymbol{m}$ be a vacuous message which is stepwise effective for some state S, with $S\boldsymbol{m} = V$. Suppose that $S \neq V$. By Axiom [Ma] there is a concise message n producing S from V. Thus, mn is a return for S, which must be vacuous by [Mb]. Since m is vacuous, n must be vacuous. This cannot be true because n is concise. □

We introduce two graph-theoretical concepts. The first one has been used informally in Chapter 1.

2.3.2 Definition. The *(adjacency) graph* of a medium is constructed as follows: each of its vertices represents a state of the medium, and each of its edges links two vertices representing adjacent states (cf. Definition 2.1.1). Such graphs are undirected.

The graphs of Figures 1.2 and 1.4 are adjacency graphs, whereas the graph of Figure 2.1 is not.

2.3.3 Definition. For a medium $(\mathcal{S}, \mathcal{T})$, we define the function

$$\delta : \mathcal{S} \times \mathcal{S} \to \mathbb{N}_0 : (S, V) \mapsto \delta(S, V) = \begin{cases} 0 & \text{if } S = V \\ |\mathcal{C}(m)| & \text{if } Sm = V, \text{ with } m \\ & \text{a concise message.} \end{cases}$$

It is easily verified that δ is the graph theoretical distance on the graph of $(\mathcal{S}, \mathcal{T})$ (cf. 1.8.5).

While the composition of tokens is not commutative in a medium, a weaker form of it holds, expressed by the next result, which is crucial.

2.3.4 Theorem. *Let m and n be two distinct concise messages transforming some state S. Then*

$$Sm = Sn \iff \mathcal{C}(m) = \mathcal{C}(n).$$

Proof. Suppose that $Sm = Sn = V$. Then, $V\tilde{n} = S$, which yields $Sm\tilde{n} = S$. Thus, $m\tilde{n}$ is ineffective for S and also, by Lemma 2.3.1(vi), stepwise effective for that state; so, $m\tilde{n}$ is a return for S. By Axiom [Mb], $m\tilde{n}$ must be vacuous. Take any $\tau \in \mathcal{C}(m)$. Because $m\tilde{n}$ is vacuous, we must have $\tilde{\tau} \in \mathcal{C}(m\tilde{n})$; in fact, $\tilde{\tau}$ must be in $\mathcal{C}(\tilde{n})$ because m is consistent. This implies $\tau \in \mathcal{C}(n)$, yielding $\mathcal{C}(m) \subseteq \mathcal{C}(n)$, and by symmetry, $\mathcal{C}(n) \subseteq \mathcal{C}(m)$.

Conversely, suppose that $\mathcal{C}(m) = \mathcal{C}(n)$, with $Sm = V \neq Sn = W$. By Axiom [Ma], there is a concise message p producing W from V. Thus, $mp\tilde{n}$ is a return for S. Axiom [Mb] implies that $mp\tilde{n}$ is vacuous. But this cannot be since p is concise, and $\tau \in \mathcal{C}(m)$ if and only if $\tilde{\tau} \in \mathcal{C}(\tilde{n})$. We conclude that we must have $V = W$. □

2.4 Content Families

2.4.1 Definition. Let $(\mathcal{S}, \mathcal{T})$ be a medium. For any state S, define the *(token) content* of S as the set $\widehat{S}$ of all tokens each of which is contained in at least one concise message producing S; formally:

$$\widehat{S} = \{\tau \in \mathcal{T} \mid \exists V \in \mathcal{S} \text{ and } \boldsymbol{m} \text{ concise such that } V\boldsymbol{m} = S \text{ and } \tau \in \mathcal{C}(\boldsymbol{m})\}.$$

We refer to the family $\widehat{\mathcal{S}}$ of all the contents of the states in $\mathcal{S}$ as the *content family* of the medium $(\mathcal{S}, \mathcal{T})$.

2.4.2 Example. As an illustration, we construct the content family of the medium represented in Figure 2.1. We obtain

$$\widehat{S} = \{\tau_2, \tau_4, \tau_6\}, \quad \widehat{V} = \{\tau_2, \tau_3, \tau_6\}, \quad \widehat{W} = \{\tau_2, \tau_3, \tau_5\},$$
$$\widehat{X} = \{\tau_2, \tau_4, \tau_5\}, \quad \widehat{T} = \{\tau_1, \tau_3, \tau_5\}.$$

Note the following fact regarding this family. Take any two adjacent states in the Example of Figure 2.1, say $X\tau_3 = W$. We have

$$\widehat{W} \setminus \widehat{X} = \{\tau_3\} = \{\tilde{\tau}_4\} \text{ and } \widehat{X} \setminus \widehat{W} = \{\tau_4\} = \{\tilde{\tau}_3\}.$$

2.4.3 Theorem. *For any token τ and any state S, we have either $\tau \in \widehat{S}$ or $\tilde{\tau} \in \widehat{S}$ (but not both); so, $|\widehat{S}| = |\widehat{V}|$ for any two states S and V. Moreover, if $\mathcal{S}$ is finite, then $|\widehat{S}| = |\mathcal{T}|/2$ for any $S \in \mathcal{S}$.*

Proof. Since τ is a token, there are states $Q \neq W$ such that $Q\tau = W$. By Axiom [Ma], for any state S, there are concise messages $\boldsymbol{m}$ and $\boldsymbol{n}$ such that $S = Q\boldsymbol{m} = W\boldsymbol{n}$. Thus, $\tau \boldsymbol{n}\widetilde{\boldsymbol{m}}$ is a return for Q, which must be vacuous by [Mb]. Therefore, $\tilde{\tau} \in \mathcal{C}(\boldsymbol{n})$ or $\tilde{\tau} \in \mathcal{C}(\widetilde{\boldsymbol{m}})$, which yields $\tilde{\tau} \in \widehat{S}$ or $\tau \in \widehat{S}$. As a consequence of Lemma 2.2.7, we cannot have both $\tilde{\tau} \in \widehat{S}$ and $\tau \in \widehat{S}$; so, $|\widehat{S}| = |\widehat{V}|$ for any two states S and V. The last statement is obvious. □

2.4.4 Theorem. *If $S\boldsymbol{m} = V$ for some concise message $\boldsymbol{m}$ (thus $S \neq V$), then $\widehat{V} \setminus \widehat{S} = \mathcal{C}(\boldsymbol{m})$, and so $\widehat{V} \triangle \widehat{S} = \mathcal{C}(\boldsymbol{m}) + \mathcal{C}(\widetilde{\boldsymbol{m}}) \neq \varnothing$.*

Proof. By definition, $\widehat{V}$ contains all the tokens from concise messages producing V; so, we must have $\mathcal{C}(\boldsymbol{m}) \subseteq \widehat{V}$. As we also have $V\widetilde{\boldsymbol{m}} = S$, the same argument yields $\mathcal{C}(\widetilde{\boldsymbol{m}}) \subseteq \widehat{S}$. Because, by Theorem 2.4.3, $\widehat{S}$ cannot contain both a token and its reverse, we obtain $\mathcal{C}(\boldsymbol{m}) \subseteq \widehat{V} \setminus \widehat{S}$.

To prove the converse inclusion, suppose that $\tau \in \widehat{V} \setminus \widehat{S}$ for some token τ. Then τ must occur in some concise message producing V. Without loss of generality, we may assume that $W\tau \boldsymbol{n} = V$ for some state W, with $\tau \boldsymbol{n}$ concise.

Suppose that $W \neq S$ and let $\boldsymbol{q}$ be a concise message producing S from W. As the message $\boldsymbol{m}\widetilde{\boldsymbol{n}}\widetilde{\boldsymbol{\tau}}\boldsymbol{q}$ is a return for S, it must be vacuous by [Mb]. Thus, we must have
$$\tau \in \mathcal{C}(\boldsymbol{m}) \cup \mathcal{C}(\widetilde{\boldsymbol{n}}) \cup \mathcal{C}(\boldsymbol{q}).$$
We cannot have either $\tau \in \mathcal{C}(\boldsymbol{q})$ (because this would imply $\tau \in \widehat{S}$), or $\tau \in \mathcal{C}(\widetilde{\boldsymbol{n}})$ (because this would yield $\tau, \widetilde{\tau} \in \mathcal{C}(\tau\boldsymbol{n})$, with $\tau\boldsymbol{n}$ concise, a contradiction). We conclude that $\tau \in \mathcal{C}(\boldsymbol{m})$. The last equation of the theorem results from Lemma 2.3.1(iv), Eq. (2.3).

The case $S = W$ is left to the reader. □

Any state is defined by its content. Indeed, we have:

2.4.5 Theorem. *For any two states S and V, we have*
$$S = V \iff \widehat{S} = \widehat{V}. \tag{2.4}$$

Proof. Suppose that $\widehat{S} = \widehat{V}$ for some $S \neq V$. (The other implication is trivial.) Let $\boldsymbol{q}$ be a concise message producing V from S. From Theorem 2.4.4, we get $\widehat{V} \triangle \widehat{S} = \mathcal{C}(\boldsymbol{q}) + \mathcal{C}(\widetilde{\boldsymbol{q}}) \neq \varnothing$, contradicting our hypothesis that $\widehat{S} = \widehat{V}$. We conclude that (2.4) holds. □

The content family of a medium satisfies a strong structural property which will be investigated in Chapter 3.

2.5 The Effective Set and the Producing Set of a State

Theorem 2.4.5 entails a characterization of each state of a medium by its content. In fact, only a possibly small subset of the content is needed to specify a state, enabling a more economical coding.

2.5.1 Definition. The *effective set* and the *productive set* of a state S in a medium $(\mathcal{S}, \mathcal{T})$ are the two subsets of tokens respectively defined by
$$S^{\mathcal{E}} = \{\tau \in \mathcal{T} \mid S\tau \neq S\}$$
$$S^{\mathcal{P}} = \{\tau \in \mathcal{T} \mid T\tau = S \neq T \text{ for some } T \in \mathcal{S}\}.$$
We have thus $S^{\mathcal{P}} \subseteq \widehat{S}$ and $S^{\mathcal{E}} \subseteq \mathcal{T} \setminus \widehat{S}$.

The importance of these two concepts is that any state in a medium is characterized by either of these two sets. The next theorem is conceptually related to a result pertaining to well–graded families (cf. Doignon and Falmagne, 1999, Theorem 2.8; see also Definition 5.2.1 and Theorem 5.2.2 in our Chapter 5). The connection is spelled out in Theorem 5.2.5.

2.5.2 Theorem. *For any two states S and T in a medium $(\mathcal{S},\mathcal{T})$, the following five conditions are equivalent.*

(i) $S = T$.
(ii) $S^{\mathcal{E}} = T^{\mathcal{E}}$.
(iii) $S^{\mathcal{P}} = T^{\mathcal{P}}$.
(iv) $S^{\mathcal{E}} \subseteq \widehat{T}$.
(v) $S^{\mathcal{P}} \subseteq \mathcal{T} \setminus \widehat{T}$.

Proof. Clearly, Condition (i) implies all the others; also, (ii) and (iii) imply (iv) and (v), respectively. It remains to show that each of (iv) and (v) implies (i). We use contraposition. If $S \neq T$, there exists a concise message $\tau_1 \ldots \tau_n$ producing T from S. By Theorem 2.4.4, this implies both $\tilde{\tau}_1 \in S^{\mathcal{E}} \cap (\mathcal{T} \setminus \widehat{T})$ and $\tau_1 \in S^{\mathcal{P}} \cap \widehat{T}$, contradicting Conditions (iv) and (v), respectively. □

2.6 Orderly and Regular Returns

Our presentation follows Falmagne and Ovchinnikov (2007).

2.6.1 Definition. If $\boldsymbol{m}$ and $\boldsymbol{n}$ are two concise messages producing, from a state S, the same state $V \neq S$, then $\boldsymbol{m}\widetilde{\boldsymbol{n}}$ is a return for S. We call $\boldsymbol{m}\widetilde{\boldsymbol{n}}$ an *orderly return* for S. It is clear from Theorem 2.4.4 that the length of an orderly return must be even (Problem 2.15).

We begin with a result of general interest for orderly returns.

2.6.2 Theorem. *Let S, N, Q and W be four distinct states of a medium and suppose that*

$$N\tau = S, \quad W\mu = Q, \quad S\boldsymbol{q} = N\boldsymbol{q}' = Q, \quad S\boldsymbol{w}' = N\boldsymbol{w} = W \quad (2.5)$$

for some tokens τ and μ and some concise messages $\boldsymbol{q}, \boldsymbol{q}', \boldsymbol{w}$ and $\boldsymbol{w}'$ (see Figure 2.3). Then, the four following conditions are equivalent:

(i) $\ell(\boldsymbol{q}) + \ell(\boldsymbol{w}) \neq \ell(\boldsymbol{q}') + \ell(\boldsymbol{w}')$ *and* $\mu \neq \tilde{\tau}$;
(ii) $\tau = \mu$;
(iii) $\mathcal{C}(\boldsymbol{q}) = \mathcal{C}(\boldsymbol{w})$ *and* $\ell(\boldsymbol{q}) = \ell(\boldsymbol{w})$;
(iv) $\ell(\boldsymbol{q}) + \ell(\boldsymbol{w}) + 2 = \ell(\boldsymbol{q}') + \ell(\boldsymbol{w}')$.

Moreover, any of these conditions implies that $\boldsymbol{q}\tilde{\mu}\widetilde{\boldsymbol{w}}\tau$ is an orderly return for S with $S\boldsymbol{q}\tilde{\mu} = S\tilde{\tau}\boldsymbol{w} = W$. The converse does not hold.

Proof. We prove (i) ⇒ (ii) ⇔ (iii) ⇒ (iv) ⇒ (i).

(i) ⇒ (ii). Suppose that $\tau \neq \mu$. The token $\tilde{\tau}$ must occur exactly once in either $\boldsymbol{q}$ or in $\widetilde{\boldsymbol{w}}$. Indeed, we have $\mu \neq \tilde{\tau}$, both $\boldsymbol{q}$ and $\boldsymbol{w}$ are concise, and the message $\tau\boldsymbol{q}\tilde{\mu}\widetilde{\boldsymbol{w}}$ is a return for S, and so is vacuous by [Ma]. It can be verified that each of the two mutually exclusive, exhaustive cases:

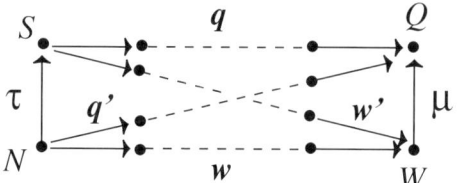

Figure 2.3. Illustration of the conditions listed in (2.5).

[a] $\tilde{\tau} \in \mathcal{C}(q) \cap \mathcal{C}(w')$; and [b] $\tilde{\tau} \in \mathcal{C}(\widetilde{w}) \cap \mathcal{C}(\widetilde{q}')$ lead to

$$\ell(q) + \ell(w) = \ell(q') + \ell(w'), \tag{2.6}$$

contradicting (i). Thus, we must have $\tau = \mu$.

We only prove Case [a]. The other case is treated similarly. Since $\tilde{\tau}$ is in $\mathcal{C}(q)$, neither τ nor $\tilde{\tau}$ can be in $\mathcal{C}(q')$. Indeed, both q and q' are concise and $q\widetilde{q}'\tau$ is a return for S. It follows that both $\tilde{\tau}q'$ and q are concise messages producing Q from S. By Theorem 2.4.4, we must have $\mathcal{C}(\tilde{\tau}q') = \mathcal{C}(q)$, which implies $\ell(\tilde{\tau}q') = \ell(q)$, and so

$$\ell(q) = \ell(q') + 1. \tag{2.7}$$

A argument along the same lines shows that

$$\ell(w) + 1 = \ell(w'). \tag{2.8}$$

Adding (2.7) and (2.8) and simplifying, we obtain (2.6). The proof of Case [b] is similar.

(ii) $\Leftrightarrow$ (iii). If $\mu = \tau$, it readily follows (since both q and w are concise and $Sq\tilde{\tau}\widetilde{w}\tau = S$) that any token in q must have a reverse in $\widetilde{w}$ and vice versa. This implies $\mathcal{C}(q) = \mathcal{C}(w)$, which in turn imply $\ell(q) = \ell(w)$, and so (iii) holds. As $q\tilde{\mu}\widetilde{w}\tau$ is vacuous, it is clear that (iii) implies (ii).

(iii) $\Rightarrow$ (iv). Since (iii) implies (ii), we have $\tau \in \widehat{Q} \setminus \widehat{N}$ by Theorem 2.4.4. But both q and q' are concise, so $\tau \in \mathcal{C}(q') \setminus \mathcal{C}(q)$. As $\tau q\widetilde{q}'$ is vacuous for N, we must have $\mathcal{C}(q) + \{\tau\} = \mathcal{C}(q')$, yielding

$$\ell(q) + 1 = \ell(q'). \tag{2.9}$$

A similar argument gives $\mathcal{C}(w) + \{\tau\} = \mathcal{C}(w')$ and

$$\ell(w) + 1 = \ell(w'). \tag{2.10}$$

Adding (2.9) and (2.10) yields (iv).

(iv) $\Rightarrow$ (i). As (iv) is a special case of the first statement in (i), we only have to prove that $\mu \neq \tilde{\tau}$. Suppose that $\mu = \tilde{\tau}$. We must assign the token $\tilde{\tau}$ consistently so to ensure the vacuousness of the messages $q\widetilde{q}'\tau$ and $\tau w'\widetilde{w}$. By Theorem 2.4.4, $\mathcal{C}(q) = \widehat{Q} \setminus \widehat{S}$. Since $\tilde{\tau} \in \widehat{Q}$ and, by Theorem 2.4.3, $\tilde{\tau} \notin \widehat{S}$, the only possibility is $\tilde{\tau} \in \mathcal{C}(\boldsymbol{q}) \setminus \mathcal{C}(\boldsymbol{q}')$. For similar reasons $\tau \in \mathcal{C}(\boldsymbol{w}) \setminus \mathcal{C}(\boldsymbol{w}')$. We obtain the two concise messages $\tilde{\tau}\boldsymbol{q}'$ and $\boldsymbol{q}$ producing Q from S, and the two concise messages $\boldsymbol{w}$ and $\tau\boldsymbol{w}'$ producing W from N. This gives $\ell(\boldsymbol{q}) = \ell(\tilde{\tau}\boldsymbol{q}')$ and $\ell(\boldsymbol{w}) = \ell(\tau\boldsymbol{w}')$. We obtain so $\ell(\boldsymbol{q}) = \ell(\boldsymbol{q}') + 1$ and $\ell(\boldsymbol{w}) = \ell(\boldsymbol{w}') + 1$, which leads to $\ell(\boldsymbol{q}) + \ell(\boldsymbol{w}) = \ell(\boldsymbol{q}') + \ell(\boldsymbol{w}') + 2$ and contradicts (iv). Thus, (iv) implies (i). We conclude that the four conditions (i)-(iv) are equivalent.

We now show that, under the hypotheses of the theorem, (ii) implies that $\boldsymbol{q}\tilde{\mu}\tilde{\boldsymbol{w}}\tau$ is an orderly return for S with $S\boldsymbol{q}\tilde{\mu} = S\tilde{\tau}\boldsymbol{w} = W$. Both $\boldsymbol{q}$ and $\boldsymbol{w}$ are concise by hypothesis. We cannot have μ in $\mathcal{C}(\boldsymbol{q})$ because then $\tilde{\mu}$ is in $\mathcal{C}(\tilde{\boldsymbol{q}})$ and the two concise messages $\tilde{\boldsymbol{q}}$ and $\tau = \mu$ producing S are not jointly consistent, yielding a contradiction of Lemma 2.2.7. Similarly, we cannot have $\tilde{\mu}$ in $\mathcal{C}(\boldsymbol{q})$ since the two concise messages $\boldsymbol{q}$ and μ producing Q would not be jointly consistent. Thus, $\boldsymbol{q}\tilde{\mu}$ is a concise message producing W from S. For like reasons, with $\tau = \mu$, $\tilde{\tau}\boldsymbol{w}$ is a concise message producing W from S. We conclude that, with $\tau = \mu$, the message $\boldsymbol{q}\tilde{\mu}\tilde{\boldsymbol{w}}\tau$ is an orderly return for S. The example of Figure 2.4, in which we have

$$\mu \neq \tau, \quad \boldsymbol{q} = \alpha\tilde{\tau}, \quad \boldsymbol{w} = \tilde{\mu}\alpha, \quad \boldsymbol{w}' = \alpha\tilde{\tau}\tilde{\mu}, \quad \text{and} \quad \boldsymbol{q}' = \alpha,$$

displays the orderly return $\alpha\tilde{\tau}\tilde{\mu}\tilde{\alpha}\mu\tau$ for S. It serves as a counterexample to the implication: if $\boldsymbol{q}\tilde{\mu}\tilde{\boldsymbol{w}}\tau$ is an orderly return for S, then $\tau = \mu$. □

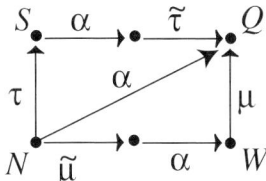

Figure 2.4. The medium represented above (cf. Problem 2.19) shows that, under the conditions of Theorem 2.6.2, the hypothesis that $\boldsymbol{q}\tilde{\mu}\tilde{\boldsymbol{w}}\tau$ is an orderly return for S does not imply $\tau = \mu$, with $\boldsymbol{q} = \alpha\tilde{\tau}$, $\boldsymbol{w} = \tilde{\mu}\alpha$, $\boldsymbol{q}' = \alpha$, and $\boldsymbol{w}' = \alpha\tilde{\tau}\tilde{\mu}$.

2.6.3 Remark. In fact, each of the conditions (i)-(iv) in Theorem 2.6.2 implies that, not only $\boldsymbol{q}\tilde{\mu}\tilde{\boldsymbol{w}}\tau$, but also various other messages are orderly return (see Problem 2.8).

In 2.6.1, the concept of an orderly return was defined with respect to a particular state. The next definition and theorem concern a situation in which a vacuous message is an orderly return with respect to everyone of the produced states. In such a case, any token occurring in a return must have its reverse at the exact 'opposite' place in the return. Several examples of such situations are provided by the 'hexagons' in the permutohedron of Figure 1.4 on p. 6.

2.6.4 Definition. Let $\tau_1 \ldots \tau_{2n}$ be an orderly return for some state S. For any $1 \leq i \leq n$, the two tokens τ_i and τ_{i+n} are called *opposite*. A return $\tau_1 \ldots \tau_{2n}$ from S is *regular* if it is orderly and, for $1 \leq i \leq n$, the message $\tau_i \tau_{i+1} \ldots \tau_{i+n-1}$ is concise for $S\tau_1 \ldots \tau_{i-1}$.

2.6.5 Theorem. *Let $m = \tau_1 \ldots \tau_{2n}$ be an orderly return for some state S. Then the following three conditions are equivalent.*
  (i) *The opposite tokens of m are mutual reverses.*
  (ii) *The return m is regular.*
  (iii) *For $1 \leq i \leq 2n - 1$, the message $\tau_i \ldots \tau_{2n} \ldots \tau_{i-1}$ is an orderly return for the state $S\tau_1 \ldots \tau_{i-1}$.*

Proof. We prove (i) $\Rightarrow$ (ii) $\Rightarrow$ (iii) $\Rightarrow$ (i). In what follows $S_i = S\tau_0\tau_1 \ldots \tau_i$ for $0 \leq i \leq 2n$, so $S_0 = S_{2n} = S$.

(i) $\Rightarrow$ (ii). Since m is an orderly return, for $1 \leq j \leq n$, there is only one occurrence of the pair $\{\tau_j, \tilde{\tau}_j\}$ in m. Since $\tilde{\tau}_j = \tau_{j+n}$, there are no occurrences of $\{\tau_j, \tilde{\tau}_j\}$ in $p = \tau_i \cdots \tau_{i+n-1}$, so it is a concise message for S_{i-1}.

(ii) $\Rightarrow$ (iii). Since m is a regular return, any message $p = \tau_i \cdots \tau_{i+n-1}$ is concise, so any token of this message has a reverse in the message $q = \tau_{i+n} \ldots \tau_{2n} \ldots \tau_{i-1}$. Since p is concise and $\ell(q) = n$, the message q is concise. It follows that pq is an orderly return for the state S_{i-1}.

(iii) $\Rightarrow$ (i). Since the message $\tau_i \ldots \tau_{2n} \ldots \tau_{i-1}$ is an orderly return for S_{i-1}, the messages $q = \tau_{i+1} \ldots \tau_{i+n-1}$ and $q' = \tau_i \ldots \tau_{i+n-1}$ are concise for the states $S' = S_i$ and $N = S_{i-1}$, respectively, and produce the state $Q = S_{i+n-1}$. Likewise, the messages $w = \tilde{\tau}_{i-1} \ldots \tilde{\tau}_{2n} \ldots \tilde{\tau}_{i+n}$ and $w' = \tilde{\tau}_i \ldots \tilde{\tau}_{2n} \ldots \tilde{\tau}_{i+n}$ are concise for the states $N = S_{i-1}$ and $S' = S_i$, respectively, and produce the state $W = S_{i+n}$. It is clear that $\ell(q) + \ell(w) + 2 = \ell(q') + \ell(w')$. By Theorem 2.6.2, $\tau_{i+n} = \tilde{\tau}_i$. $\square$

2.6.6 Remark. Examples of media with returns that are not regular and yet satisfy Conditions (i)-(iv) of Theorem 2.6.2 are easily constructed (see Problem 2.13).

Following Ovchinnikov (2006), we now introduce a couple of general tools for the study of media.

2.7 Embeddings, Isomorphisms and Submedia

2.7.1 Definition. Let $(\mathcal{S}, \mathcal{T})$ and $(\mathcal{S}', \mathcal{T}')$ be token systems. A pair (α, β) of 1-1 functions $\alpha : \mathcal{S} \to \mathcal{S}'$, $\beta : \mathcal{T} \to \mathcal{T}'$ such that

$$S\tau = T \iff \alpha(S)\beta(\tau) = \alpha(T) \qquad (S, T \in \mathcal{S}, \ \tau \in \mathcal{T})$$

is called an *embedding* of the token system $(\mathcal{S}, \mathcal{T})$ into the token system $(\mathcal{S}', \mathcal{T}')$. If the functions α and β are bijections, then these two token systems are said

2.7 Embeddings, Isomorphisms and Submedia

to be *isomorphic* and the embedding (α, β) is referred to as an *isomorphism* of (S, T) *onto* (S', T').

It is easily verified that, in such a case, if one of the token systems is a medium, so is the other (cf. Problem 2.10). Recall that if (S, T) is a medium and $\tau_1, \tau_2 \in T$ with $S\tau_1 = S\tau_2 \neq S$ for some $S \in S$, then $\tau_1 = \tau_2$ by Lemma 2.3.1(iii).

Thus, if (α, β) is an isomorphism of a medium onto another medium, we have $\beta(\tilde{\tau}) = \widetilde{\beta(\tau)}$. Indeed, for a given τ there are two distinct states S and T such that $S\tilde{\tau} = T$. Then

$$\alpha(S)\beta(\tilde{\tau}) = \alpha(T) \iff$$
$$S\tilde{\tau} = T \iff T\tau = S \iff \alpha(T)\beta(\tau) = \alpha(S) \iff \alpha(S)\widetilde{\beta(\tau)} = \alpha(T),$$

and so $\beta(\tilde{\tau}) = \widetilde{\beta(\tau)}$.

We extend the function β of an isomorphism (α, β) to the semigroup of messages by defining

$$\beta(\tau_1 \ldots \tau_n) = \beta(\tau_1) \ldots \beta(\tau_n)$$

for any message $\boldsymbol{m} = \tau_1 \ldots \tau_n$. It is clear that the image of a concise message by the function β is a concise message.

Take a medium (S, T) together with a subset Q of S consisting of at least two states. Let T' be the collection of all restrictions to Q of tokens in T. The pair (Q, T') is not necessarily a medium. The medium represented by its graph in Figure 2.1 provides an example: taking the restrictions of the tokens to the subset $\{S, T\}$ does not yield a medium (cf. Problem 2.11).

We introduce the appropriate concepts for the discussion of such matters.

2.7.2 Definition. Let (S, T) be a token system and let Q be a nonempty subset of S. For any $\tau \in T$, the function

$$\tau_Q : Q \to Q : S \mapsto S\tau_Q = \begin{cases} S\tau & \text{if } S\tau \in Q, \\ S & \text{otherwise} \end{cases}$$

is called the *reduction*[1] of τ to Q. Note that the reduction of a token to a subset Q of states may be the identity on Q. A token system (Q, T_Q) where $T_Q = \{\tau_Q\}_{\tau \in T} \setminus \{\tau_0\}$ is the set of all reductions of tokens in T to Q different from the identity τ_0 on Q is referred to as the *reduction* of (S, T) to Q. We call (Q, T_Q) a *token subsystem* of (S, T). If both (S, T) and (Q, T_Q) are media, then (Q, T_Q) is a *submedium* of (S, T). (This definition implies that (S, T) is a submedium of itself.)

---

[1] The concept of 'reduction' is closely related to, but distinct from, the concept of 'restriction' of a function to a subset of its domain. The standard interpretation of 'restriction' would not remove pairs $(S, S\tau)$ in which $S \in Q$ but $S\tau \in (S \setminus Q)$.

2.7.3 Remarks. (a) The above definition readily implies that if $(S, \mathcal{T})$ and $(\mathcal{Q}, \mathcal{T}_\mathcal{Q})$ are two media, with $(\mathcal{Q}, \mathcal{T}_\mathcal{Q})$ a reduction of $(S, \mathcal{T})$ to $\mathcal{Q}$, then $\tau_\mathcal{Q} = \mu_\mathcal{Q}$ only if $\tau = \mu$, for any $\tau_\mathcal{Q}$ and $\mu_\mathcal{Q}$ in $\mathcal{T}_\mathcal{Q}$.

(b) The image $(\alpha(S), \beta(\mathcal{T}))$ of a token system $(S, \mathcal{T})$ under some embedding $(\alpha, \beta) : (S, \mathcal{T}) \to (S', \mathcal{T}')$ is not, in general, the reduction of $(S', \mathcal{T}')$ to $\alpha(S)$ (cf. Problem 2.12). As shown by Theorem 2.7.4, however, this is true in the case of media.

2.7.4 Theorem. *Suppose that (α, β) is an embedding of a medium $(S, \mathcal{T})$ into a medium $(S', \mathcal{T}')$. Then the reduction $(\alpha(S), \mathcal{T}'_{\alpha(S)})$ of $(S', \mathcal{T}')$ to $\alpha(S)$ is isomorphic to $(S, \mathcal{T})$.*

Proof. For $\tau \in \mathcal{T}$, we use the abbreviation $\beta'(\tau) = \beta(\tau)_{\alpha(S)}$ (thus, the reduction of $\beta(\tau)$ to $\alpha(S)$). Let $S\tau = T \neq S$. Then $\alpha(S)\beta(\tau) = \alpha(T)$ for $\alpha(S) \neq \alpha(T)$ in $\alpha(S)$. Hence β' maps $\mathcal{T}$ into $\mathcal{T}'_{\alpha(S)}$. Let us show that (α, β') is an isomorphism from $(S, \mathcal{T})$ onto $(\alpha(\mathcal{T}), \mathcal{T}'_{\alpha(S)})$.

(i) β' is onto $\mathcal{T}'_{\alpha(S)}$. Suppose that $\tau'_{\alpha(S)} \neq \tau_0$ for some $\tau' \in \mathcal{T}'$. Then, there are $P \neq Q \in S$ such that

$$\alpha(P)\tau'_{\alpha(S)} = \alpha(P)\tau' = \alpha(Q).$$

Let $Q = mP$, where m is a concise message from P. We have

$$\alpha(Q) = \alpha(Pm) = \alpha(P)\beta(m) = \alpha(P)\tau'$$

implying, by Theorem 2.3.4, that $\beta(m) = \tau'$ since $\beta(m)$ is a concise message. Hence, $m = \tau$ for some $\tau \in \mathcal{T}$. Thus, $\beta(\tau) = \tau'$, which implies

$$\beta'(\tau) = \beta(\tau)_{\alpha(S)} = \tau'_{\alpha(S)}.$$

(ii) β' is 1-1. Suppose that $\beta'(\tau) = \beta'(\mu)$. Since $\beta'(\tau)$ and $\beta'(\mu)$ are tokens on $\alpha(S)$ and $(S', \mathcal{T}')$ is a medium, we have $\beta(\tau) = \beta(\mu)$ by Remark 2.7.3 (a). Hence $\tau = \mu$.

(iii) Finally, for any $S, T \in S$ in and $\tau \in \mathcal{T}$

$$S\tau = T \iff \alpha(S)\beta'(\tau) = \alpha(T)$$

because

$$S\tau = T \iff \alpha(S)\beta(\tau) = \alpha(T).$$

□

2.8 Oriented Media

In three of the five examples encountered in our first chapter, the medium was endowed with an 'orientation' dictated by the nature of the tokens. The definition below specifies the idea of an orientation.

2.8 Oriented Media

2.8.1 Definition. An *orientation* of a medium $(\mathcal{S}, \mathcal{T})$ is a partition of its set of tokens into two classes $\mathcal{T}^+$ and $\mathcal{T}^-$ respectively called *positive* and *negative* such that for any $\tau \in \mathcal{T}$, we have

$$\tau \in \mathcal{T}^+ \iff \tilde{\tau} \in \mathcal{T}^-.$$

A medium $(\mathcal{S}, \mathcal{T})$ equipped with an orientation $\{\mathcal{T}^+, \mathcal{T}^-\}$ is said to be *oriented* by $\{\mathcal{T}^+, \mathcal{T}^-\}$ and tokens from $\mathcal{T}^+$ (resp. $\mathcal{T}^-$) are called *positive* (resp. *negative*). Except when specified otherwise, an oriented medium $(\mathcal{S}, \mathcal{T})$ will be implicitly taken to have its orientation denoted by $\{\mathcal{T}^+, \mathcal{T}^-\}$. The *positive* (resp. *negative*) *content* of a state S is the set $\widehat{S}^+ = \widehat{S} \cap \mathcal{T}^+$ (resp. $\widehat{S}^- = \widehat{S} \cap \mathcal{T}^-$) of its positive (resp. negative) tokens. We define the two families

$$\widehat{\mathcal{S}}^+ = \{\widehat{S}^+ \mid S \in \mathcal{S}\} \quad \text{and} \quad \widehat{\mathcal{S}}^- = \{\widehat{S}^- \mid S \in \mathcal{S}\}.$$

Thus, $\widehat{\mathcal{S}}^+$ and $\widehat{\mathcal{S}}^-$ denote the families of all the positive and negative contents, respectively. Any message containing only positive (resp. negative) tokens is called *positive* (resp. *negative*). We say that two messages have the *same sign* when they are both positive, or both negative. A corresponding terminology applies to tokens. Note that a finite medium $(\mathcal{S}, \mathcal{T})$ can be given $2^{|\mathcal{T}|/2}$ different orientations (Problem 2.20).

2.8.2 Remark. Definition 2.7.1 is consistent with a situation in which two oriented media $(\mathcal{S}, \mathcal{T})$ and $(\mathcal{Q}, \mathcal{L})$ are isomorphic but their respective orientations $\{\mathcal{T}^+, \mathcal{T}^-\}$ and $\{\mathcal{L}^+, \mathcal{L}^-\}$ do not match. For example, we can have an isomorphism (α, β) of $(\mathcal{S}, \mathcal{T})$ onto $(\mathcal{Q}, \mathcal{L})$, such that $\tau \in \mathcal{T}^+$ while $\beta(\tau) \in \mathcal{L}^-$. In fact, if $(\mathcal{S}, \mathcal{T})$ and $(\mathcal{Q}, \mathcal{L})$ are isomorphic, they would remain so under any changes of orientations. There are cases in which a more demanding concept is required.

2.8.3 Definition. Two oriented media $(\mathcal{S}, \mathcal{T})$ and $(\mathcal{Q}, \mathcal{L})$ are said to be *sign-isomorphic* if there is an isomorphism (α, β) of $(\mathcal{S}, \mathcal{T})$ onto $(\mathcal{Q}, \mathcal{L})$ such that $\beta(\mathcal{T}^+) = \mathcal{L}^+$ (and so $\beta(\mathcal{T}^-) = \mathcal{L}^-$).

Let $(\mathcal{S}, \mathcal{T})$ be a medium with an orientation $\{\mathcal{T}^+, \mathcal{T}^-\}$. Then any submedium $(\mathcal{Q}, \mathcal{T}_Q)$ of $(\mathcal{S}, \mathcal{T})$ has an *induced orientation* $\{\mathcal{T}_Q^+, \mathcal{T}_Q^-\}$ defined from $\{\mathcal{T}^+, \mathcal{T}^-\}$ by the equivalence

$$\tau_Q \in \mathcal{T}_Q^+ \iff \tau \in \mathcal{T}^+.$$

Except when indicated otherwise, a submedium of an oriented medium is implicitly assumed to be equipped with its induced orientation.

As a consequence of Theorems 2.4.3 and 2.4.5, we have:

2.8.4 Theorem. *For any two states S and V of an oriented medium $(\mathcal{S}, \mathcal{T})$, we have*

$$S = V \iff \widehat{S}^+ = \widehat{V}^+.$$

A similar result holds obviously for negative contents.

Proof. The necessity is trivial. Regarding the sufficiency, note that if both $\widehat{S}^+ = \widehat{V}^+$ and $\widehat{S}^- = \widehat{V}^-$ hold, then

$$\widehat{S} = \widehat{S}^+ + \widehat{S}^- = \widehat{V}^+ + \widehat{V}^- = \widehat{V},$$

and so $S = V$ by Theorem 2.4.5. It suffices thus to prove that $\widehat{S}^+ = \widehat{V}^+$ implies $\widehat{S}^- = \widehat{V}^-$. Suppose that $\widehat{S}^+ = \widehat{V}^+$ and take any $\tau \in \mathcal{T}$. We have successively

$$\begin{aligned}
\tau \in \widehat{S}^- &\iff \widetilde{\tau} \in \mathcal{T}^+ \setminus \widehat{S}^+ && \text{(by Theorem 2.4.3)} \\
&\iff \widetilde{\tau} \in \mathcal{T}^+ \setminus \widehat{V}^+ && \text{(because } \widehat{S}^+ = \widehat{V}^+\text{)} \\
&\iff \tau \in \widehat{V}^- && \text{(by Theorem 2.4.3).}
\end{aligned}$$

$\square$

2.8.5 Theorem. *If S and V are two distinct states in an oriented medium, with $Sm = V$ for some positive concise message m, then $\widehat{S}^+ \subset \widehat{V}^+$.*

Proof. If m is a concise positive message producing V from S, then $\widetilde{m}$ is concise and negative, and $S = V\widetilde{m}$, with $\widehat{S} \setminus \widehat{V} = \mathcal{C}(\widetilde{m})$ by Theorem 2.4.4. Thus, any token in $\widehat{S} \setminus \widehat{V}$ is negative. This implies $\widehat{S}^+ \subset \widehat{V}^+$. $\square$

2.9 The Root of an Oriented Medium

It is natural to ask how an orientation could be systematically constructed so as to reflect properties of the medium. One principle for such a construction is described below.

2.9.1 Definition. The *root* of an oriented medium is a state R such that, for any other state S, any concise message producing S from R is positive. A oriented medium having a root is said to be *rooted*. By abuse of language, we sometimes say that the orientation of a medium is *rooted* (at the state S) if the medium is rooted for that orientation and S is the root.

2.9.2 Theorem. *In an oriented medium $\mathcal{M}$, a state R is a root if and only if $\widehat{R}^+ = \varnothing$. Thus, $\mathcal{M}$ has at most one root and may in fact have no root.*

Proof. (Sufficiency.) Suppose that R is a state, with $\widehat{R}^+ = \varnothing$. Let m be a concise message such that $Rm = S$; thus, $S\widetilde{m} = R$. Recall that $\widehat{R} \setminus \widehat{S} = \mathcal{C}(\widetilde{m})$ by Theorem 2.4.4. Thus, if $\widetilde{m}$ contains a positive token, then $\widehat{R}^+ \neq \varnothing$. Thus, $\widetilde{m}$ is negative, and so m is positive. Since this holds for any message m effective for R, the state R must be a root.

(Necessity.) Suppose that R is a root. By Theorem 2.8.5 and the definition of a root, we have $\widehat{R}^+ \subset \widehat{S}^+$ for any state $S \neq R$. If $\widehat{R}^+$ contains some positive token τ, then τ belongs to the contents of all the states. This implies that $\tilde{\tau}$ is not effective for any state, and so is the identity function, which is not a token by definition. We conclude that $\widehat{R}^+$ must be empty.

For an example of a medium without a root, take the medium of Figure 2.1 with an orientation having the set of positive tokens $\mathcal{T}^+ = \{\tau_2, \tau_4, \tau_5\}$. □

2.9.3 Theorem. *For any medium $(S, \mathcal{T})$ and any state R in S, there exists an orientation making R the root of $(S, \mathcal{T})$.*

Proof. Define such an orientation by setting $\mathcal{T}^- = \widehat{R}$. Then, for any state $S \neq R$ and any concise message $\boldsymbol{m}$ producing S from R, we have $\mathcal{C}(\boldsymbol{m}) = (\widehat{S} \setminus \widehat{R}) \subseteq \mathcal{T}^+$ by Theorem 2.4.4. Thus, $\boldsymbol{m}$ is positive. □

The following type of oriented medium will play a role, in later developments, as a component of larger media.

2.9.4 Definition. A rooted medium $\mathcal{N} = (S, \mathcal{T})$, with $|S| = n \geq 3$ for some $n \in \mathbb{N}$, is a *n-star* if its root is adjacent to all the other $n - 1$ states. (Thus, such a medium has $n - 1$ pairs of tokens.) We say that $\mathcal{N}$ is a *star* if there exist some $n \in \mathbb{N}$ such that $\mathcal{N}$ is a n-star.

The next section contains an example in which all the sets of positive contents are uncountable. We have, however, the following result.

2.9.5 Theorem. *For any medium, there exists an orientation ensuring that the positive contents of all the states are finite sets. In particular, the states of any rooted medium have finite positive contents.*

Proof. Indeed, by Theorem 2.9.3 we can arbitrarily assign any state to be the root R of the medium. We have $\widehat{R}^+ = \varnothing$, and for any other state S, with $R\boldsymbol{m} = S$ for some concise message $\boldsymbol{m}$, we have $\widehat{S}^+ = \mathcal{C}(\boldsymbol{m}) = \widehat{S} \setminus \widehat{R}$, with $\mathcal{C}(\boldsymbol{m})$ finite. □

2.10 An Infinite Example

In the medium of the example below, not only are the set of states and the set of tokens infinite, but so are the contents of all the states; in fact, all theses states are uncountable.

2.10.1 Example. Let $\mathfrak{P}_F(\mathbb{R})$ be the set of all finite subsets of $\mathbb{R}$. We define a medium $(S, \mathcal{T})$ with $S = \{S_X \mid X \in \mathfrak{P}_F(\mathbb{R}), S_X = \mathbb{R} \setminus X\}$ and $\mathcal{T}$ containing, for any real number x, the pair $\tau_x, \tilde{\tau}_x$ of mutually reverse tokens defined by

$$S\tau_x = \begin{cases} S \cup \{x\} & \text{if } S = S_X \text{ for some } X \in \mathfrak{P}_\text{F}, \text{ with } x \in X, \\ S & \text{otherwise;} \end{cases} \quad (2.11)$$

$$S\tilde{\tau}_x = \begin{cases} S \setminus \{x\} & \text{if } S = S_X \text{ for some } X \in \mathfrak{P}_\text{F}, \text{ with } x \notin X, \\ S & \text{otherwise.} \end{cases} \quad (2.12)$$

This example anticipates the main theme of Chapter 3 in which it is shown that such token systems are indeed media (see, in particular, Theorem 3.3.4). Clearly, $\mathcal{S}$ and $\mathcal{T}$ are uncountable, and so are the contents of all the states. Moreover, suppose that we define now an orientation by having the positive tokens and the negative tokens specified by (2.11) and (2.12), respectively. Then, all the positive contents are uncountable, but all the negative ones are finite. Essentially the same counterexample is obtained if one replaces $\mathbb{R}$ by any set $\mathfrak{X}$ of arbitrary cardinality, with the set of states defined by $\{S_X \mid X \in \mathfrak{P}_\text{F}(\mathfrak{X}), S_X = \mathfrak{X} \setminus X\}$. The exact role played by the finite sets is discussed in details in Chapter 3.

2.10.2 Remark. As shown by Example 2.10.1, the positive contents may be uncountable in some cases. Incidentally, in this example, we could take the root to be the set $\mathbb{R}$ of all real numbers, and redefine the orientation accordingly, with the positive tokens becoming the negative ones and vice-versa. We obtain the situation described in the second statement of Theorem 2.9.5, with the positive contents being all the finite sets of real numbers.

2.11 Projections

The concept discussed in this section is due to Cavagnaro (2006). A projection of a medium resemble a submedium in that it summarizes the structure. The similarity stops there. A submedium focusses on a part of a medium by choosing a subset of states and redefining (reducing, cf. Definition 2.7.2) the tokens to that subset. By contrast, a projection gives a macroscopic view of a medium in the form of a new medium whose states are fundamentally different from the original ones.

The interest of such a construction is obvious. A particularly useful implementation arises in the context of learning spaces, a topic already mentioned in Chapter 1 (Definition 1.6.1). We recall that, in this application of media theory, a state of the relevant medium corresponds to a 'knowledge state' of a student, that is, the set of problems that the student is capable of solving. In the case of a placement test, determining the exact knowledge state of a student may be costly, and locating the student in a suitable class of similar states may be sufficient to decide which courses the student is capable of taking. The concept of a projection is then the appropriate tool.

To obtain a projection of a medium $(\mathcal{S}, \mathcal{T})$, one selects a 'symmetric' subset $\mathcal{U}$ of its tokens (thus, $\tau \in \mathcal{U}$ if and only if $\tilde{\tau} \in \mathcal{U}$), which is used to define an equivalence relation on $\mathcal{S}$. The corresponding equivalence classes form the states of the new medium under construction. The tokens in $\mathcal{U}$ are then redefined so that they act on these equivalence classes on a manner that is consistent with their action in $(\mathcal{S}, \mathcal{T})$.

2.11.1 Example. A good intuition of the concept of projection can be gathered from studying the medium pictured by its graph in Figure 2.5 together with one of its possible projections. This particular medium, which is represented at the bottom of the figure, has been encountered earlier: it is the medium of all the linear orders on the set $\{1, 2, 3, 4\}$ (cf. Figure 1.4, page 6). So, each of the 24 linear orders is a state, and a token consists of the transposition of two adjacent element in a linear order. There are thus $12 = \binom{4}{2} \times 2$ tokens. Denoting this medium by $(\mathcal{S}, \mathcal{T})$, we define the symmetric subset of tokens $\mathcal{U} = \{\tau_{13}, \tau_{31}, \tau_{23}, \tau_{32}\} \subset \mathcal{T}$, with $\tau_{ij} : \boldsymbol{x}ij\boldsymbol{y} \mapsto \boldsymbol{x}ji\boldsymbol{y}$ and $\boldsymbol{x}$ or $\boldsymbol{y}$ possibly empty (so, $1234\tau_{32} = 1324$). These two pairs of tokens are represented by the red and blue edges in the graph at the bottom of the figure.

We then gather in the same equivalence class two linear orders l and l' whenever there is a concise message $\boldsymbol{m}$ producing l from l' and such that $\mathcal{U} \cap \mathcal{C}(\boldsymbol{m}) = \varnothing$. For example, the two linear orders 4123 and 2134 are in the same equivalence class because $4123\tau_{21}\tau_{24}\tau_{14}\tau_{34} = 2134$. (Note that, as a consequence of Axiom [Mb], if $\boldsymbol{n}$ is another concise message producing l from l', then necessarily also $\mathcal{U} \cap \mathcal{C}(\boldsymbol{n}) = \varnothing$.) We obtain the four equivalence classes represented by the graphs in each of the four boxes at the top of the figure. Each of these boxes represents one of the new states and is a vertex of the graph representing the projection. The red and blue edges are consistent with the edges of the same color in the bottom graph, and represents the pairs of tokens of the projection. It is clear that this graph represents a medium.

We now generalize this example and justify this construction.

2.11.2 Definition. Suppose that $(\mathcal{S}, \mathcal{T})$ is a medium. Let $\mathcal{U}$ be a nonempty, symmetric subset of $\mathcal{T}$, that is: $\tau \in \mathcal{U}$ if and only if $\tilde{\tau} \in \mathcal{U}$ for any token $\tau \in \mathcal{T}$. Define a relation $\sim$ on $\mathcal{S}$ by

$$S \sim T \iff \widehat{S} \triangle \widehat{T} \subseteq \mathcal{T} \setminus \mathcal{U}. \tag{2.13}$$

2.11.3 Lemma. *As defined by (2.13), $\sim$ is an equivalence relation on $\mathcal{S}$.*

Proof. The relation $\sim$ is clearly reflexive and symmetric. The transitivity of $\sim$ follows from the fact that, for any four sets X, Y, Z and W, we have

$$(X \triangle Y \subseteq W) \wedge (Y \triangle Z \subseteq W) \implies (X \cup Y \cup Z) \setminus (X \cap Y \cap Z) \subseteq W$$
$$\implies X \triangle Z \subseteq W.$$

Indeed, suppose that $S \sim T$ and $T \sim R$. Setting $X = \widehat{S}$, $Y = \widehat{T}$, $Z = \widehat{R}$ and $W = \mathcal{T} \setminus \mathcal{U}$, the above implication yield $\widehat{S} \triangle \widehat{R} \subseteq \mathcal{T} \setminus \mathcal{U}$, and so $S \sim R$. □

42   2 Basic Concepts

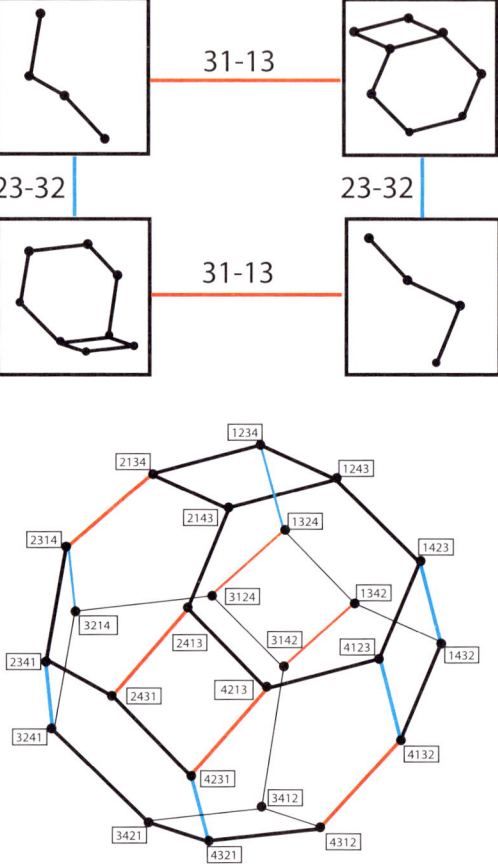

Figure 2.5. Projection of the medium of the linear orders on the set $\{1, 2, 3, 4\}$ by the two pairs of tokens (τ_{13}, τ_{31}) and (τ_{23}, τ_{32}), with $\tau_{ij} : \boldsymbol{n}yx\boldsymbol{m} \mapsto \boldsymbol{n}ji\boldsymbol{m}$ and $\boldsymbol{n}$ or $\boldsymbol{m}$ possibly empty. The graph of the original medium is the permutohedron pictured at the bottom of the figure, with the colors red and blue marking the two pairs of tokens. the projection is pictured at the top (cf. Figure 1.4, page 6).

In the next definition, we start from a medium $(\mathcal{S}, \mathcal{T})$ and use the equivalence relation $\sim$ induced by some $\mathcal{U} \subseteq \mathcal{T}$ via (2.13) to gather the states into equivalence classes that will form the states of a new medium. Notice that any token in $\mathcal{U}$, when it is effective, always maps a state from one equivalence class to one in some other equivalence class. With this in mind, we redefine the tokens in the set $\mathcal{U}$ so that they apply to the classes themselves, serving thus as the tokens of the new medium and completing the construction.

2.11.4 Definition. Suppose that $(\mathcal{S}, \mathcal{T})$ is a medium. Let $\mathcal{U}$ be a nonempty symmetric subset of $\mathcal{T}$ and let $\sim$ be the equivalence relation on $\mathcal{S}$ induced by $\mathcal{U}$ via (2.13). Define

$$[S] = \{T \in \mathcal{S} \mid S \sim T\}, \qquad (S \in \mathcal{S});$$
$$\mathcal{S}_{|\mathcal{U}} = \{[S] \mid S \in \mathcal{S}\},$$
$$\tau_{|\mathcal{U}} : \mathcal{S}_{|\mathcal{U}} \to \mathcal{S}_{|\mathcal{U}} : [S] \mapsto [S]\tau_{|\mathcal{U}}, \qquad (\tau \in \mathcal{U});$$

with

$$[S]\tau_{|\mathcal{U}} = \begin{cases} [T] & \text{if } S \not\sim T \text{ and } \exists (Q, W) \in [S] \times [T], \, Q\tau = W, \\ [S] & \text{otherwise}, \end{cases} \qquad (2.14)$$

$$\mathcal{T}_{|\mathcal{U}} = \{\tau_{|\mathcal{U}} \mid \tau \in \mathcal{U}\}.$$

We refer to the pair $(\mathcal{S}_{|\mathcal{U}}, \mathcal{T}_{|\mathcal{U}})$ as the *projection* of $(\mathcal{S}, \mathcal{T})$ under $\mathcal{U}$.

2.11.5 Lemma. (i) For each $\tau \in \mathcal{U}$, $\tau_{|\mathcal{U}}$ is a well-defined function.
(ii) $(\mathcal{S}_{|\mathcal{U}}, \mathcal{T}_{|\mathcal{U}})$ is a token system.
(iii) $\widetilde{\tau_{|\mathcal{U}}} = \tilde{\tau}_{|\mathcal{U}}$ for any $\tau \in \mathcal{U}$.

Proof. (i) Suppose that $Q, M \in [S]$, with $Q\tau = R$, $M\tau = W$, $Q \not\sim R$ and $M \not\sim W$. We have to show that $[R] = [W]$. Using Theorem 2.4.4 (with $\tau = m$), $Q\tau = R$ and $M\tau = W$ yield

$$\{\tau\} = \widehat{R} \setminus \widehat{Q} = \widehat{W} \setminus \widehat{M} \subseteq \widehat{R} \cap \widehat{W}. \qquad (2.15)$$

Since $Q \sim M$, we deduce from (2.15) that $\widehat{R} \triangle \widehat{W} = \widehat{Q} \triangle \widehat{M} \subseteq \mathcal{T} \setminus \mathcal{U}$; so $R \sim W$, and $[R] = [W]$ follows.
(ii) This follows immediately from (i) and the fact that $\mathcal{U} \neq \varnothing$.
(iii) For $\tau \in (\mathcal{T} \setminus \mathcal{U})$ and $S \not\sim T$, we have

$$S\widetilde{\tau_{|\mathcal{U}}} = T \Leftrightarrow T\tau_{|\mathcal{U}} = S \Leftrightarrow T\tau = S \Leftrightarrow S\tilde{\tau} = Q \Leftrightarrow S\tilde{\tau}_{|\mathcal{U}} = T.$$

The other case is also straightforward. □

Except for details of formulation, the next result is due to Cavagnaro (2006).

2.11.6 Theorem. Let $(\mathcal{S}, \mathcal{T})$ be a medium and let $\mathcal{U}$ be a symmetric subset of $\mathcal{T}$. Then, the three following statements are true.
(i) The projection $(\mathcal{S}_{|\mathcal{U}}, \mathcal{T}_{|\mathcal{U}})$ of $(\mathcal{S}, \mathcal{T})$ under $\mathcal{U}$ is a medium.
(ii) For each state $S \in \mathcal{S}$ such that $|[S]| > 1$, the token subsystem $([S], \mathcal{T}_{[S]})$ of $(\mathcal{S}, \mathcal{T})$ is a submedium of $(\mathcal{S}, \mathcal{T})$.
(iii) Suppose that the set $\{\mathcal{T}^+, \mathcal{T}^-\}$ is an orientation of $(\mathcal{S}, \mathcal{T})$. Define $\mathcal{T}^+_{|\mathcal{U}} = \{\tau_{|\mathcal{U}} \in \mathcal{T}_{|\mathcal{U}} \mid \tau \in \mathcal{T}^+\}$ and $\mathcal{T}^-_{|\mathcal{U}} = \{\tau_{|\mathcal{U}} \in \mathcal{T}_{|\mathcal{U}} \mid \tau \in \mathcal{T}^-\}$; then, the set $\{\mathcal{T}^+_{|\mathcal{U}}, \mathcal{T}^-_{|\mathcal{U}}\}$ is an orientation of the projection $(\mathcal{S}_{|\mathcal{U}}, \mathcal{T}_{|\mathcal{U}})$ of $(\mathcal{S}, \mathcal{T})$. We have

$$\tau_{|\mathcal{U}} \in \mathcal{T}^+_{|\mathcal{U}} \iff \widetilde{\tau_{|\mathcal{U}}} \in \mathcal{T}^-_{|\mathcal{U}} \iff \tau \in \mathcal{T}^+ \qquad (\tau_{|\mathcal{U}} \in \mathcal{T}_{|\mathcal{U}}).$$

The orientations of $(\mathcal{S}, \mathcal{T})$ and $(\mathcal{S}_{|\mathcal{U}}, \mathcal{T}_{|\mathcal{U}})$ are thus consistent: the set $\mathcal{U}$ 'imports' the orientation of $(\mathcal{S}, \mathcal{T})$ into the projection $(\mathcal{S}_{|\mathcal{U}}, \mathcal{T}_{|\mathcal{U}})$.

Proof. (i) By Lemma 2.11.5 (ii), the pair $(\mathcal{S}_{|\mathcal{U}}, \mathcal{T}_{|\mathcal{U}})$ is a token system. We prove that $(\mathcal{S}_{|\mathcal{U}}, \mathcal{T}_{|\mathcal{U}})$ satisfies [Ma] and [Mb].

[Ma]. Take any two distinct states $[S]$ and $[T]$ in the projection $(\mathcal{S}_{|\mathcal{U}}, \mathcal{T}_{|\mathcal{U}})$. Let $\boldsymbol{m}$ be a concise message producing T from S in the medium $(\mathcal{S}, \mathcal{T})$. Such a message exists by [Ma] applied to $(\mathcal{S}, \mathcal{T})$. We decompose the message $\boldsymbol{m}$ in terms of the equivalence classes (forming states in $\mathcal{S}_{|\mathcal{U}}$) visited along the way. For some positive integer n, we obtain the decomposition form

$$S\boldsymbol{m} = S_0\boldsymbol{m} = S_0\boldsymbol{m}_0\tau_1\boldsymbol{m}_1\ldots\boldsymbol{m}_{n-1}\tau_n\boldsymbol{m}_n = T, \quad \text{with}$$
$$\cup_{i=0}^{n} \mathcal{C}(\boldsymbol{m}_i) \subseteq (\mathcal{T} \setminus \mathcal{U}), \quad \{\tau_1, \ldots, \tau_n\} \subseteq \mathcal{U} \tag{2.16}$$

and

$$S_0\boldsymbol{m}_0 = S_1 \in [S_1] = [S], \quad \text{(initial class)}$$
$$S_0\boldsymbol{m}_0\tau_1 = S_2 \in [S_2] \neq [S_1] \quad \text{(second class visited)}$$
$$\ldots$$
$$S_0\boldsymbol{m}_0\tau_1\boldsymbol{m}_1\ldots\tau_{i-1}\boldsymbol{m}_{i-1}\tau_i = S_{2i} \in [S_{2i}] \neq [S_{2i-1}], \quad (i\text{th class visited})$$
$$\ldots$$
$$S_0\boldsymbol{m}_0\tau_1\boldsymbol{m}_1\ldots\tau_n\boldsymbol{m}_n = S_{2n+1} \in [S_{2n+1}] = [T], \quad \text{(final class)}$$

Note that all the component messages $\boldsymbol{m}_i$, $0 \leq i \leq n$, are either concise or empty. On the other hand, at least τ_1 must exist because $[S] \neq [T]$. This decomposition makes it clear that the message $\boldsymbol{m}$ producing T from S in $(\mathcal{S}, \mathcal{T})$ may be compacted, with its tokens $\tau_1, \ldots, \tau_n$ in $\mathcal{U}$ redefined by (2.14) as a concise message $\boldsymbol{p} = \tau_{1|\mathcal{U}}\ldots\tau_{n|\mathcal{U}}$ producing $[T]$ from $[S]$ in the projected token system $(\mathcal{S}_{|\mathcal{U}}, \mathcal{T}_{|\mathcal{U}})$. The conciseness of $\boldsymbol{p}$ follows from that of $\boldsymbol{m}$: the consistency of $\boldsymbol{m}$ entails that of $\boldsymbol{p}$ by Lemma 2.11.5(iii); none of the tokens in $\boldsymbol{p}$ repeats itself because none of the tokens in $\boldsymbol{m}$ does; and the effectiveness of some $\tau_i \in \mathcal{C}(\boldsymbol{m})$ for some state S, implies that of $\tau_{i|\mathcal{U}}$ for $[S]$.

[Mb]. Let $\boldsymbol{p} = \tau_{1|\mathcal{U}}\ldots\tau_{n|\mathcal{U}}$ be a message that is stepwise effective for some state $[S]$ in $\mathcal{S}_{|\mathcal{U}}$. Inverting the construction performed above in the proof of [Ma], there clearly exists a corresponding message $\boldsymbol{m} = \boldsymbol{m}_0\tau_1\boldsymbol{m}_1\ldots\tau_n\boldsymbol{m}_n$ in $(\mathcal{S}, \mathcal{T})$ such that each of the component messages $\boldsymbol{m}_i$, $0 \leq i \leq n$ is concise or empty and the two inclusions of (2.16) are satisfied. So, $\boldsymbol{m}$ is stepwise effective for S. Suppose that $\boldsymbol{p}$ is ineffective for $[S]$, that is, $\boldsymbol{p}$ is a return for $[S]$ in $(\mathcal{S}_{|\mathcal{U}}, \mathcal{T}_{|\mathcal{U}})$. Then $\boldsymbol{m}$ is a return for S in $(\mathcal{S}, \mathcal{T})$; we have

$$[S]\tau_{1|\mathcal{U}}\ldots\tau_{n|\mathcal{U}} = [S],$$
$$S\boldsymbol{m}_0\tau_1\boldsymbol{m}_1\ldots\tau_n\boldsymbol{m}_n = S.$$

This implies that $\boldsymbol{m}$ is vacuous because $\boldsymbol{m}$ satisfies [Mb] in $(\mathcal{S}, \mathcal{T})$. Since both $\mathcal{U}$ and $\mathcal{T} \setminus \mathcal{U}$ are symmetric, any token τ_i, $1 \leq i \leq n$, must have its reverse in $\{\tau_1, \ldots, \tau_n\}$. Thus, $\boldsymbol{p}$ must be vacuous by Lemma 2.11.5(iii). We have shown that both [Ma] and [Mb] hold and so $(\mathcal{S}_{|\mathcal{U}}, \mathcal{T}_{|\mathcal{U}})$ is a medium.

(ii) We verify that [Ma] and [Mb] also hold for any token subsystem $([S], \mathcal{T}_{[S]})$ of $(\mathcal{S}, \mathcal{T})$. (We use the notation of 2.7.2.) Starting with [Ma], let T and Q be two states of $[S]$. Thus, $T \sim Q$ and by definition of $\sim$ in (2.13), there exists a concise message $\boldsymbol{m} = \tau_1 \ldots \tau_n$ in $(\mathcal{S}, \mathcal{T})$ producing T from Q, with $\mathcal{C}(\boldsymbol{m}) \subseteq \mathcal{T} \setminus \mathcal{U}$. We have thus $Q\tau_1 \ldots \tau_i = Q_i \in [S]$ for $1 \leq i \leq n$. This implies that $\tau_{1[S]} \ldots \tau_{n[S]}$ is a concise message of $([S], \mathcal{T}_{[S]})$ producing T from Q. So, Axiom [Ma] holds for $(\mathcal{S}, \mathcal{T})$. Finally, let $\boldsymbol{m} = \tau_{1[S]} \ldots \tau_{n[S]}$ be a message of $([S], \mathcal{T}_{[S]})$ which is a return for some $Q \in [S]$. Then, the corresponding message $\tau_1 \ldots \tau_n$ of $(\mathcal{S}, \mathcal{T})$ is also a return for Q. As such, it is vacuous by [Mb] applied to $(\mathcal{S}, \mathcal{T})$, and it follows that $\boldsymbol{m}$ must be vacuous for $([S], \mathcal{T}_{[S]})$. As the token subsystem $([S], \mathcal{T}_{[S]})$ satisfies [Ma] and [Mb], it is a medium.

(iii) This is straightforward. □

2.11.7 Remark. Our presentation of these results differs from that in Cavagnaro (2006). He defines the concept of the 'projection' of a token system $(\mathcal{S}, \mathcal{T})$ directly via an arbitrary equivalence relation $\sim$ on the set of states. Writing $\mathcal{S}_\sim$ for the partition of $\mathcal{S}$ induced by $\sim$, and $[S]$ for the class of the partition containing state S (as we do), he defines for each token $\tau \in \mathcal{S}$ a relation τ^* on $\mathcal{S}_\sim$ by the equivalence

$$[S]\tau^*[Q] \iff \exists S' \in [S] \text{ such that } S'\tau \in [Q]. \tag{2.17}$$

Note that the relations τ^* defined by (2.17) are not necessarily functions (Problem 2.25). Writing $\mathcal{T}_\sim$ for the set of all the relations τ^* different from the identity on $\mathcal{S}_\sim$, he then investigates conditions on $\sim$ entailing that the pair $(\mathcal{S}_\sim, \mathcal{T}_\sim)$ is a medium if the token system $(\mathcal{S}, \mathcal{T})$ is a medium.

Problems

2.1 Verify the facts (i), (ii) and (iii) mentioned in Definition 2.2.3.

2.2 Verify formally that, in a token system, any token has at most one reverse, and that if $\tilde{\tau}$ is the reverse of τ, then $\tilde{\tilde{\tau}} = \tau$, that is, τ and $\tilde{\tau}$ are mutual reverses.

2.3 Verify the independence of the two Axioms [Ma] and [Mb] of a medium (Theorem 2.2.5 and Figure 2.2).

2.4 In the genetic mutation example of Chapter 1, construct the content of all the states of the medium defined by the graph of Figure 1.7.

2.5 Verify formally that the token system formed by the family of all strict partial orders on a finite set Y satisfies Axioms [Ma] and [Mb] of a medium (with the tokens of the two types: $P \mapsto P \cup \{xy\}$ and $P \mapsto P \setminus \{xy\}$).

2.6 (Continuation.) Describe the contents of the states in this example, and compare such a description with the standard definition of a binary relation.

2.7 (Continuation.) With the same definition of the tokens as in Problem 2.5, check whether the token system formed by the family of all strict weak orders on a finite set is a medium. Which of the axioms is not satisfied? How would you redefine the tokens so that a medium obtains?

2.8 Verify that each the four conditions (i)-(iv) of Theorem 2.6.2 imply that, not only $q\widetilde{\mu}\widetilde{w}\tau$, but various other messages are orderly returns (for S or N). Find those messages and verify that they are indeed orderly returns.

2.9 Argue that, in a medium, the set of states and the set of tokens are either both finite, or both infinite.

2.10 Prove that if two token systems are isomorphic and one of them is a medium, so is the other (cf. Definition 2.7.1).

2.11 For the medium defined by its graph on Figure 2.1, describe exactly what is obtained by limiting the set of states to $\{S,T\}$, and taking the restriction of the six tokens to that set. Why don't we obtain a medium? Do we still, at least, have a token system?

2.12 Show by a counterexample that the image $(\alpha(\mathcal{S}), \beta(\mathcal{T}))$ of a token system $(\mathcal{S}, \mathcal{T})$ under some embedding $(\alpha, \beta) : (\mathcal{S}, \mathcal{T}) \to (\mathcal{S}', \mathcal{T}')$ is not necessarily the reduction of $(\mathcal{S}', \mathcal{T}')$ to $\alpha(\mathcal{S})$.

2.13 Construct some examples of media with orderly returns that are not regular. At least one of these example should satisfy Conditions (i)-(iv) of Theorem 2.6.2.

2.14 Use Theorem 2.4.4 to prove Theorem 2.3.4.

2.15 Prove that the length of any orderly return is even.

2.16 This is a continuation of Problem 1.14 of Chapter 1. Consider the token system induced by a political map of the world, the countries representing the states and each possible crossing of a single border representing a transformation. Which axioms of a medium are satisfied?

2.17 A concise message m for S is called *maximally concise for S* if there is no token τ such that $m\tau$ is concise for S. A state M is called a *center*[2] of the medium if all maximally concise messages for M have the same length. Does a center necessary exist? Could there be more than one center in a medium?

2.18 Describe the contents of the states in the medium of Example 2.2.9.

2.19 Verify that the token system represented in Figure 2.4 is a medium.

2.20 Prove that a finite medium $(S, \mathcal{T})$ can be given $2^{|\mathcal{T}|/2}$ different orientations.

2.21 Given an example of a non rooted medium with a countable number of tokens and such that each of its states has a finite positive content. Can we have a medium having some, but not all, of its states having finite positive contents?

2.22 To Axioms [Ma] and [Mb] of a medium, add the Axiom [Mc] stating that the tokens are commutative transformations. Is [Ma] consistent with [Ma] and [Mb]? What are the consequences of a token system satisfying [Ma], [Mb] and [Mc]?

2.23 Take any natural number n which is not a prime, and let $n = n_1^{i_1} \ldots n_k^{i_k}$ be its prime factorization. Let $\mathcal{N}_n$ be the set of all numbers $m \leq n$ which are either prime or have a factorization involving the same prime numbers as n. (Thus, if $n = 18$, then $\mathcal{N}_{18} = \{2, 3, 6, 9, 12, 18\}$.) Define a set $\mathcal{T}_n = \{\tau_{n_{j,h}}\}$ of transformations on $\mathcal{N}_n$ by

$$\tau_{n_{j,h}} : S \mapsto S\tau_{n_{j,h}} = \begin{cases} S \times n_j & \text{if } (S/n_j^{h-1}) \in \mathbb{N} \text{ and } (S/n_j^{h+1}) \notin \mathbb{N} \\ S & \text{otherwise.} \end{cases}$$

Is the pair $(\mathcal{N}_n, \mathcal{T}_n)$ a medium? If so, what can you say about such a medium? Construct the digraph for the case $\mathcal{N}_{24}$. Suppose that we add 1 to $\mathcal{N}_n$ and redefine $\mathcal{T}_n$ appropriately. What would be the answers to the above questions then?

2.24 The Rubik's Cube has been analyzed from the standpoint of group theory (see for example, Joyner, 2002). Is it possible to formalize the transformations of the cube in terms of the tokens of a medium? What would be the states and the tokens?

2.25 Show by a counterexample that the relations defined by (2.17) need not be functions.

2.26 Prove part (iii) of the Projection Theorem 2.11.6.

---

[2] See Chepoi et al. (2002) for a closely related concept.

3
Media and Well-graded Families

While the axioms of a medium do not require the sets of states or tokens to be finite or otherwise limit their cardinality, they nevertheless impose an essential finiteness constraint on the structure. One goal of this chapter is to clarify this issue. A second goal is to study the tight link existing between the concept of a medium and certain families of sets called 'well-graded.' The two goals are related.

We begin by observing that the content of the states in a medium exhibit a strong structural property which can be surmised from examining the contents of the five states in Example 2.2.4. These contents are recalled below:

$$\widehat{S} = \{\tau_2, \tau_4, \tau_6\}, \quad \widehat{V} = \{\tau_2, \tau_3, \tau_6\}, \quad \widehat{W} = \{\tau_2, \tau_3, \tau_5\},$$
$$\widehat{X} = \{\tau_2, \tau_4, \tau_5\}, \quad \widehat{T} = \{\tau_1, \tau_3, \tau_5\}.$$

Notice that the contents of the two adjacent states S and V differ by exactly two tokens, namely τ_3 and τ_4, which are mutual reverses. A similar observation can be made for any two adjacent states in the medium of Figure 2.1: for any two adjacent states Y and Z, we have[1]

$$|\widehat{Y} \triangle \widehat{Z}| = d(\widehat{Y}, \widehat{Z}) = 2.$$

As asserted in Theorem 3.1.7 and as readily follows from Theorem 2.4.4, this equation holds in fact for any medium. Definition 3.1.5 generalizes that property.

3.1 Wellgradedness

We begin by recalling some basic set-theoretical facts. We omit the proofs (cf. 1.8.1 and Problem 3.5 at the end of this chapter).

---
[1] We recall that $d(P, Q)$ denotes the symmetric difference distance between two sets P and Q; cf. 1.8.1, Equation (1.5).

3.1.1 Lemma. *For any three sets P, Q and S, we have*

$$P \triangle Q \subseteq (P \triangle S) \cup (S \triangle Q). \tag{3.1}$$

Consequently, if both $P \triangle S$ and $S \triangle Q$ are finite, so is $P \triangle Q$; thus

$$d(P,Q) \leq d(P,S) + d(S,Q),$$

with moreover

$$d(P,Q) = d(P,S) + d(S,Q) \iff P \cap Q \subseteq S \subseteq P \cup Q. \tag{3.2}$$

3.1.2 Definition. Let $\mathcal{F}$ be a family of subsets of a set X, and let k be a natural number. A *k-graded path* between two distinct sets P and Q (or from P to Q) in $\mathcal{F}$ is a sequence $P_0 = P, P_1, \ldots, P_n = Q$ in $\mathcal{F}$ such that $d(P_i, P_{i+1}) = k$ for $0 \leq i \leq n-1$, and $d(P,Q) = kn$. A 1-graded path is called a *tight path*.

For any two distinct subsets P and Q of X, we refer to the set

$$[P,Q] = \{S \in \mathcal{F} \mid P \subseteq S \subseteq Q\} \tag{3.3}$$

as the *interval (of $\mathcal{F}$) between P and Q*. (Note that such intervals may be empty.) Of particular interest in the sequel are the intervals of the form $[P \cap Q, P \cup Q]$ with $P, Q \in \mathcal{F}$.

For comments and references about these concepts, see Remark 3.1.6. Various facts about tight paths are gathered in the next two lemmas.

3.1.3 Lemma. *Suppose that $P_0 = P, P_1, \ldots, P_n = Q$ is a tight path between two sets P and Q in some family $\mathcal{F}$ of subsets of a set X. Then, for any two indices $i, j \in \{0, 1, \ldots, n\}$, we have $d(P_i, P_j) = |j - i|$.*

Proof. As the result trivially holds for $i = j$, we assume that $i < j$. By the triangle inequality, we have

$$n = d(P_0, P_n) \leq d(P_0, P_i) + d(P_i, P_j) + d(P_j, P_n)$$
$$\leq i + (j - i) + (n - j) = n.$$

The lemma follows. □

3.1.4 Lemma. *Suppose that $P_0 = P, P_1, \ldots, P_n = Q$ is a tight path between two sets P and Q in some family $\mathcal{F}$ with $\cup \mathcal{F} = X$; then, for $0 \leq j \leq k \leq m \leq n$ and $k \leq n-1$, the following four properties hold:*

(i) $\quad P_{k+1} = \begin{cases} \text{either} & P_k + \{x\} \text{ and } x \in Q \setminus P, \\ \text{or} & P_k \setminus \{x\} \text{ and } x \in P \setminus Q, \end{cases}$

for some $x \in X$;

(ii) $\quad P_k \in [P_j \cap P_m, P_j \cup P_m]$;
(iii) $\quad d(P_j, P_m) = d(P_j, P_k) + d(P_k, P_m)$;
(iv) $\quad P_i, P_{i+1}, \ldots, P_j$ is a tight path between P_i and P_j.

Proof. By Lemma 3.1.3, we have $|P_k \triangle P_{k+1}| = d(P_k, P_{k+1}) = 1$. Hence, $P_k \triangle P_{k+1} = \{x\}$ for some $x \in \mathcal{X}$, and (i) follows. Both (iii) and (iv) result easily from Lemma 3.1.3, and (ii) is an easy consequence of (iii) and (3.2). □

3.1.5 Definition. Let $\mathcal{F}$ be a family $\mathcal{F}$ of subsets of a set $\mathcal{X}$. The family $\mathcal{F}$ is *k-graded* if there is a k-graded path between any two of its sets. We say that a 1-graded family is a *well-graded* or a *wg-family* (cf. Doignon and Falmagne, 1997; Falmagne and Doignon, 1997)[2]. Thus, if $\mathcal{F}$ is a wg-family, there is a tight path between any two of its (distinct) sets. Mostly of interest here are well-graded and 2-graded families. We discuss a few important examples of well-graded families of relations in Chapter 5.

For any $x \in \mathcal{X}$, define

$$x^* = \{y \in \mathcal{X} \,|\, \forall P \in \mathcal{F}, y \in P \Leftrightarrow x \in P\}. \tag{3.4}$$

Thus, $y \in x^*$ if and only if $x \in y^*$. Clearly, the family $\{x^* \,|\, x \in \mathcal{X}\}$ is a partition of $\mathcal{X}$. Notice that if $\cup \mathcal{F} \subset \mathcal{X}$, then $\mathcal{X} \setminus \cup \mathcal{F}$ is a class of that partition. The family $\mathcal{F}$ is *discriminative* if, for all $x \in \mathcal{X}$, the subset x^* is a singleton. It is easily shown that, for a family $\mathcal{F}$, being well-graded and being discriminative are independent conditions (cf. Problem 3.2). However, any wg-family $\mathcal{F}$ satisfying $\cup \mathcal{F} = \mathcal{X}$ and $|\cap \mathcal{F}| \leq 1$ is discriminative (see Lemma 3.2.5). This does not hold in general for k-graded families, with $k \geq 2$.

3.1.6 Remark. By and large, our terminology follows Doignon and Falmagne (1997). A concept closely related to that of tight path was introduced earlier and independently by Restle (1959) (revisited in Luce and Galanter, 1963) and Bogart and Trotter (1973), labelled linear arrays and linear sets, respectively. We prefer to reserve the terms 'line' or 'linear' for cases where their use may be dictated by a compelling geometric representation (cf. Problems 3.14, 3.15 and 3.16).

As an application of Definition 3.1.5 to media, we have the following.

3.1.7 Theorem. *The content family $\widehat{\mathcal{S}}$ of the set of states $\mathcal{S}$ of a medium is 2-graded. Specifically, let S and V be any two distinct states, and let $\boldsymbol{m} = \tau_1 \ldots \tau_n$ be a concise message producing V from S, with*

$$S = S_0, S_0 \tau_1 = S_1, S_1 \tau_2 = S_2, \ldots, S_{n-1} \tau_n = S_n = V. \tag{3.5}$$

Then, we have

---
[2] This concept was introduced earlier under a different name; see Kuzmin and Ovchinnikov (1975), Ovchinnikov (1980).

$$\widehat{S}_i \setminus \widehat{S}_{i-1} = \{\tau_i\} \quad \text{and} \quad \widehat{S}_{i-1} \setminus \widehat{S}_i = \{\tilde{\tau}_i\} \qquad (1 \leq i \leq n), \qquad (3.6)$$

and $d(\widehat{S}, \widehat{V}) = 2n$. Moreover, $\widehat{\mathcal{S}}$ is discriminative and satisfies $\cap \widehat{\mathcal{S}} = \varnothing$.

Proof. Let S, V and $\boldsymbol{m} = \tau_1 \ldots \tau_n$ be as in the theorem. Invoking Theorem 2.4.4, we get

$$d(\widehat{S}, \widehat{V}) = |(\widehat{S} \setminus \widehat{V}) \cup (\widehat{V} \setminus S)| = |\mathcal{C}(\boldsymbol{m})| + |\mathcal{C}(\widetilde{\boldsymbol{m}})| = 2n, \qquad (3.7)$$

and so $d(\widehat{S}, \widehat{V}) = 2n$. By the definition of $\boldsymbol{m}$, we have the sequence of states listed in (3.5). Using Theorem 2.4.4 again, we see that their content must satisfy (3.6) and so $d(\widehat{S}_{i-1}, \widehat{S}_i) = 2$ for $1 \leq i \leq n-1$.

To prove that the $\widehat{\mathcal{S}}$ is discriminative, we must show that $\tau^* \neq \mu^*$ for any two distinct tokens τ and μ. We have $S\tau = V$ and $R\mu = W$, for some states $S \neq V$ and $R \neq W$. The case $S = R$ and $V = W$ is prevented by Lemma 2.3.1(iii). Suppose that $S \neq R$ and $V = W$. By [Ma], there is a concise message $\boldsymbol{n}$ producing R from S. So, $\tilde{\tau}\boldsymbol{n}\mu$ is a return message for V, which is vacuous by [Mb]. Since $\tau \neq \mu$, we have $\tau \in \mathcal{C}(\boldsymbol{n})$, implying $\tau \in \widehat{R}$. As we also have $\mu \in \widehat{V}$ (which is distinct from $\widehat{R}$ by Theorem 2.4.5), we get $\tau^* \neq \mu^*$. In the case $S = R$, $V \neq W$, the same argument applied to $\tilde{\tau}$ and $\tilde{\mu}$ gives $\tilde{\tau}^* \neq \tilde{\mu}^*$, which yields again $\tau^* \neq \mu^*$. The argument in the last remaining case ($S \neq R$, $V \neq W$) is similar. The fact that $\cap \widehat{\mathcal{S}} = \varnothing$ follows easily from the definition of the content of state. □

A crucial relationship exists between oriented media and wg-families, which is the focus of this chapter. The following construction will be useful to clarify the relationship.

3.2 The Grading Collection

3.2.1 Definition. Let $\mathcal{F}$ be a family of sets, with $|\mathcal{F}| \geq 2$ and $\mathcal{X} = \cup \mathcal{F}$, and suppose that, for any $x \in \mathcal{X} \setminus \cap \mathcal{F}$, there is some Q and S in $\mathcal{F}$ satisfying $Q \triangle S = \{x\}$. The family $\mathcal{F}$ is then said to be *graded*. For such a family, let $\mathcal{G}_\mathcal{F}^+$ and $\mathcal{G}_\mathcal{F}^-$ be two families of functions

$$\mathcal{G}_\mathcal{F}^+ = \{\gamma_x \,|\, x \in \mathcal{X} \setminus \cap \mathcal{F},\ \gamma_x : \mathcal{F} \to \mathcal{F}\},$$
$$\mathcal{G}_\mathcal{F}^- = \{\tilde{\gamma}_x \,|\, x \in \mathcal{X} \setminus \cap \mathcal{F},\ \tilde{\gamma}_x : \mathcal{F} \to \mathcal{F}\}$$

with the functions γ_x and $\tilde{\gamma}_x$ defined, for every $x \in \mathcal{X}$ by

$$\gamma_x : S \mapsto S\gamma_x = \begin{cases} S + \{x\} & \text{if } x \notin S,\ S \cup \{x\} \in \mathcal{F} \\ S & \text{otherwise,} \end{cases} \qquad (3.8)$$

$$\tilde{\gamma}_x : S \mapsto S\tilde{\gamma}_x = \begin{cases} S \setminus \{x\} & \text{if } x \in S,\ S \setminus \{x\} \in \mathcal{F} \\ S & \text{otherwise.} \end{cases} \qquad (3.9)$$

The family $\mathcal{G}_\mathcal{F} = \mathcal{G}_\mathcal{F}^+ \cup \mathcal{G}_\mathcal{F}^-$ is called the *grading collection* of $\mathcal{F}$.

The next three lemmas explore the concept of gradedness and its relationship to wellgradedness.

3.2.2 Lemma. *Any graded family $\mathcal{F}$ satisfying $|\cap \mathcal{F}| \leq 1$ is discriminative.*

Proof. Take $x \in \mathcal{X} = \cup \mathcal{F}$ and $y \in x^*$ (cf. 3.1.5). If $x \in \mathcal{X} \setminus \cap \mathcal{F}$, we have $S \triangle P = \{x\}$ for some $S, P \in \mathcal{F}$ with, say, $S \setminus P = \{x\}$. This implies $x = y$, and so $x^* = \{x\}$. The case $x \in \cap \mathcal{F}$ follows from (3.4) and $|\cap \mathcal{F}| = 1$. $\square$

3.2.3 Lemma. *Let $\mathcal{F}$ be a graded family of sets with $\mathcal{X} = \cup \mathcal{F}$, and let $\mathcal{G}_\mathcal{F}$ be its grading collection. Then, the following hold.*

(i) *The pair $(\mathcal{F}, \mathcal{G}_\mathcal{F})$ is a token system.*

(ii) *For any $x \in \mathcal{X} \setminus \cap \mathcal{F}$, the tokens γ_x and $\tilde{\gamma}_x$ defined by (3.8) and (3.9) are mutual reverses.*

(iii) *The token system $(\mathcal{F}, \mathcal{G}_\mathcal{F})$ satisfies Axiom [Mb] of a medium, that is, any return message is vacuous.*

Proof. (i) We have $|\mathcal{F}| \geq 2$ by definition. Take any $x \in \mathcal{X} \setminus \cap \mathcal{F}$. It is clear that both γ_x and $\tilde{\gamma}_x$ are mapping $\mathcal{F}$ into itself. Neither of these two functions can be the identity on $\mathcal{F}$. Indeed, since $\mathcal{F}$ is graded, we have $\{x\} = S \triangle P$ for some $S, P \in \mathcal{F}$. We can assume that $\{x\} = S \setminus P$; so

$$S = P\gamma_x = S\gamma_x \quad \text{and} \quad P = S\tilde{\gamma}_x = P\tilde{\gamma}_x. \qquad (3.10)$$

Thus $(\mathcal{F}, \mathcal{G}_\mathcal{F})$ is a token system.

(ii) It is clear that a result similar to (3.10) holds for any $V, W \in \mathcal{F}$ with $\{x\} = V \triangle W$. Accordingly, the tokens γ_x and $\tilde{\gamma}_x$ are mutual reverses.

(iii) Suppose that $\boldsymbol{m}$ is a return message for some $P \in \mathcal{F}$; thus, we have $P\boldsymbol{m} = P$. Let γ_x be any token of $\boldsymbol{m}$, where γ_x and $\tilde{\gamma}_x$ are defined as in (3.8) and (3.9). Because $\boldsymbol{m}$ is ineffective for P, the reverse $\tilde{\gamma}_x$ of γ_x must also occur in $\boldsymbol{m}$. Each of γ_x and $\tilde{\gamma}_x$ may occur more than once. However, since $\boldsymbol{m}$ is stepwise effective for P, the occurrences of γ_x and $\tilde{\gamma}_x$ must alternate. Assume that γ_x occurs in $\boldsymbol{m}$ before any $\tilde{\gamma}_x$ (the argument is similar in the other case). This means that $x \notin P$. But then the last occurrence of $\tilde{\gamma}_x$ in $\boldsymbol{m}$ cannot be followed by any γ_x (since otherwise, we would get $x \notin P$ because of the first γ_x, and $x \in P$ because of the last one). Thus, the number of occurrences of γ_x and $\tilde{\gamma}_x$ is the same. Since this holds for any γ_x occurring in $\boldsymbol{m}$, the message $\boldsymbol{m}$ is vacuous. $\square$

3.2.4 Convention. In the sequel, when a wg-family $\mathcal{F}$ is under discussion without explicit mention of a ground set $\mathcal{X}$ such that $S \in \mathcal{F}$ implies $S \subseteq \mathcal{X}$, we implicitly assume this ground set to be $\mathcal{X} = \cup \mathcal{F}$.

3.2.5 Lemma. *Any wg-family $\mathcal{F}$ is graded. Accordingly, a wg-family $\mathcal{F}$ is discriminative if $|\cap \mathcal{F}| \leq 1$.*

Proof. Let $\mathcal{F}$ be the wg-family, with $\mathcal{X} = \cup\mathcal{F}$. Take any $x \in \mathcal{X} \setminus \cap\mathcal{F}$. Thus, there are two sets Q and S in $\mathcal{F}$ such that $x \in S \setminus Q$. By definition of a wg-family, there is a tight path $Q_0 = Q, Q_1, \ldots, Q_n = S$. Let Q_j be the first set containing x in that tight path. We must have $j \geq 1$ and $Q_j \setminus \{x\} = Q_{j-1}$, yielding $Q_j \triangle Q_{j-1} = \{x\}$. So, $\mathcal{F}$ is graded. The second statement results from Lemma 3.2.2. □

The following result is a foreseeable consequence of Theorems 3.1.7 and Lemma 3.2.5, and a crucial tool.

3.2.6 Theorem. *In an oriented medium* $(\mathcal{S}, \mathcal{T})$, *the family* $\widehat{\mathcal{S}}^+$ *of all positive contents is well-graded and satisfies both* $\cup\widehat{\mathcal{S}}^+ = \mathcal{T}^+$ *and* $\cap\widehat{\mathcal{S}}^+ = \varnothing$. *It is thus also discriminative.*

Proof. Take any two distinct $\widehat{S}^+, \widehat{V}^+ \in \widehat{\mathcal{S}}^+$. From Theorem 3.1.7, we know that for any concise message $\boldsymbol{m} = \tau_1 \ldots \tau_n$ producing V from S, with

$$S = S_0, S_0\tau_1 = S_1, S_1\tau_2 = S_2, \ldots, S_{n-1}\tau_n = S_n = V, \qquad (3.11)$$

we have

$$\widehat{S}_i \setminus \widehat{S}_{i-1} = \{\tau_i\} \text{ and } \widehat{S}_{i-1} \setminus \widehat{S}_i = \{\tilde{\tau}_i\} \qquad (1 \leq i \leq n), \qquad (3.12)$$

and moreover, $d(\widehat{S}, \widehat{V}) = 2n$. The sequence (3.11) induces the sequence $\widehat{S}^+ = \widehat{S}_0^+, \widehat{S}_1^+, \ldots, \widehat{S}_n^+ = \widehat{V}^+$. By Theorem 2.4.3, exactly one of the τ_i and $\tilde{\tau}_i$ in (3.12) is positive. Thus, we have either $\widehat{S}_i^+ \setminus \widehat{S}_{i-1}^+ = \{\tau_i\} \in \mathcal{T}^+$ or $\widehat{S}_{i-1}^+ \setminus \widehat{S}_i^+ = \{\tilde{\tau}_i\} \in \mathcal{T}^+$, and so $d(\widehat{S}_i^+, \widehat{S}_{i+1}^+) = 1$, for $1 \leq i \leq n$, with $d(\widehat{S}^+, \widehat{V}^+) = d(\widehat{S}, \widehat{V})/2 = n$.

For any $\tau^+ \in \mathcal{T}^+$, there are distinct S and T in $\mathcal{S}$ such that $S\tau^+ = T$; so $\tau^+ \in \widehat{T}$, yielding $\cup\widehat{\mathcal{S}}^+ = \mathcal{T}^+$. The fact that the family $\widehat{\mathcal{S}}^+$ satisfies $\cap\widehat{\mathcal{S}}^+ = \varnothing$ results easily from Theorem 3.1.7. Using Lemma 3.2.5, we conclude that $\widehat{\mathcal{S}}^+$ is also discriminative. □

By symmetry, a similar result clearly holds for the negative contents.

3.3 Wellgradedness and Media

The next theorem characterizes the graded families that represent media.

3.3.1 Theorem. *Let* $\mathcal{F}$ *be a graded family of subsets of a set* $\mathcal{X} = \cup\mathcal{F}$ *and let* $\mathcal{G}_\mathcal{F}$ *be its grading collection. Then, the pair* $(\mathcal{F}, \mathcal{G}_\mathcal{F})$ *is an oriented medium with orientation* $\{\mathcal{G}_\mathcal{F}^+, \mathcal{G}_\mathcal{F}^-\}$ *if and only if* $\mathcal{F}$ *is well-graded.*

Proof. (Sufficiency.) We know by Lemma 3.2.3(i) that $(\mathcal{F}, \mathcal{G}_\mathcal{F})$ is a token system. Axiom [Mb] holds by Lemma 3.2.3(iii). We show that the wellgradedness of the family $\mathcal{F}$ implies Axiom [Ma]. For any distinct $S, V \in \mathcal{F}$ with $d(S, V) = n$, there is a sequence of sets $S_0 = S, S_1, \ldots, S_n = V$ such that

$d(S_i, S_{i+1}) = 1$ for $0 \leq i < n$. This means that either $S_{i+1} = S_i + \{x_i\}$ or $S_{i+1} = S_i \setminus \{x_i\}$ for some $x_i \in \mathfrak{X}$. In other terms, we have either $S_{i+1}\gamma_{x_i} = S_i$ or $S_{i+1}\tilde{\gamma}_{x_i} = S_i$. Thus, $V = S\gamma_1 \ldots \gamma_n$, with either $\gamma_i = \gamma_{x_i}$ or $\gamma_i = \tilde{\gamma}_{x_i}$ for $1 \leq i \leq n$. The message $\gamma_1 \ldots \gamma_n$ is concise for S because $d(S, V) = n$. Since for any $x \in \mathfrak{X}$, we have $\gamma_x \in \mathcal{G}_{\mathcal{F}}^+$ and $\tilde{\gamma}_x \in \mathcal{G}_{\mathcal{F}}^-$, the pair $(\mathcal{G}_{\mathcal{F}}^+, \mathcal{G}_{\mathcal{F}}^-)$ is an orientation of the medium $(\mathcal{F}, \mathcal{G}_{\mathcal{F}})$.

(Necessity.) Let $\mathcal{M} = (\mathcal{F}, \mathcal{G}_{\mathcal{F}})$ be the medium of the theorem, with orientation $(\mathcal{G}_{\mathcal{F}}^+, \mathcal{G}_{\mathcal{F}}^-)$. Take any two distinct states $S, V \in \mathcal{F}$. From Axiom [Ma], we know that there exists for some $n \in \mathbb{N}$ a concise message $\boldsymbol{m} = \gamma_1 \ldots \gamma_n$ producing V from S with

$$S\gamma_1 = S_0\gamma_1 = S_1, \ldots, S_{n-1}\gamma_n = V.$$

By the definition of the medium $\mathcal{M}$, we have for each $1 \leq i \leq n$ some $x_i \in \mathfrak{X}$ such that either $\gamma_i = \gamma_{x_i}$ or $\gamma_i = \tilde{\gamma}_{x_i}$, which yields

$$S_i\gamma_{x_i} = S_i + \{x_i\} = S_{i+1} \quad \text{or} \quad S_i\tilde{\gamma}_{x_i} = S_i \setminus \{x_i\} = S_{i+1}. \tag{3.13}$$

Thus, $d(S_i, S_{i+1}) = 1$ for $1 \leq i \leq n$. It remains to show that $d(S, V) = n$. We use induction. We already have shown that $d(S, S_1) = d(S_0, S_1) = 1$. Suppose that $d(S, S_i) = i$ for some $1 \leq i < n$. As indicated in (3.13), we have two possible ways of constructing S_{i+1} from S_i.

Case 1. $S_{i+1} = S_i + \{x_i\}$. Then $x_i \notin S$ because otherwise there must be some token $\gamma_k = \tilde{\gamma}_i$, with $k < i$ and $S_k\gamma_k = S_k \setminus \{x_i\}$. This cannot be because $\boldsymbol{m}$ is concise. So, we have $d(S, S_{i+1}) = i + 1$.

Case 2. $S_{i+1} = S_i \setminus \{x_i\}$. A similar argument based on the conciseness of $\boldsymbol{m}$ leads us to assert that $x_i \in S \cap S_i$, yielding also $d(S, S_{i+1}) = i + 1$. Applying induction, the theorem follows. □

3.3.2 Definition. For any wg-family of sets $\mathcal{F}$ with $\cup \mathcal{F} = \mathfrak{X}$, we denote by $(\mathcal{F}, \mathcal{W}_{\mathcal{F}})$ the oriented medium defined by Theorem 3.3.1, writing thus $\mathcal{W}_{\mathcal{F}} = \mathcal{G}_{\mathcal{W}}$. We call $(\mathcal{F}, \mathcal{W}_{\mathcal{F}})$ the *representing medium* of the family $\mathcal{F}$, oriented by $\{\mathcal{W}_{\mathcal{F}}^+, \mathcal{W}_{\mathcal{F}}^-\}$, with $\mathcal{W}_{\mathcal{F}}^+ = \mathcal{G}_{\mathcal{F}}^+$ and $\mathcal{W}_{\mathcal{F}}^- = \mathcal{G}_{\mathcal{F}}^-$ defined by (3.8) and (3.9).

3.3.3 Theorem. *Any oriented medium* $(\mathcal{S}, \mathcal{T})$ *is sign-isomorphic to the representing medium* $(\widehat{\mathcal{S}}^+, \mathcal{W}_{\widehat{\mathcal{S}}^+})$ *of its family* $\widehat{\mathcal{S}}^+$ *of positive contents, with the isomorphism* (α, β) *defined by*

$$\alpha(S) = \widehat{S}^+ \qquad (S \in \mathcal{S}), \tag{3.14}$$

and for each pair of tokens $(\tau, \tilde{\tau})$ *in* $\mathcal{T}^+ \times \mathcal{T}^-$

$$\beta(\tau) = \gamma_\tau \qquad (\gamma_\tau \in \mathcal{W}_{\widehat{\mathcal{S}}^+}^+) \tag{3.15}$$

$$\beta(\tilde{\tau}) = \tilde{\gamma}_{\tilde{\tau}} \qquad (\tilde{\gamma}_{\tilde{\tau}} \in \mathcal{W}_{\widehat{\mathcal{S}}^+}^-). \tag{3.16}$$

Proof. It is clear from (3.14) and Theorem 2.8.4 that α is a bijection of $\mathcal{S}$ onto $\mathcal{S}^+$, and from (3.15) and (3.16)—via (3.8) and (3.9)—that β is a bijection of $\mathcal{T}$ onto $\mathcal{W}_{\widehat{\mathcal{S}}+}$ satisfying $\beta(\mathcal{T}^+) = \mathcal{W}^+_{\widehat{\mathcal{S}}+}$ and $\beta(\mathcal{T}^-) = \mathcal{W}^-_{\widehat{\mathcal{S}}+}$. We must show that (α, β) is an embedding of $(\mathcal{S}, \mathcal{T})$ into $(\widehat{\mathcal{S}}^+, \mathcal{W}_{\widehat{\mathcal{S}}+})$ (cf. Definition 2.7.1). For any $S \in \mathcal{S}$ and $\tau \in \mathcal{T}$, we must consider four possible cases.

(i) $\tau \in \mathcal{T}^+$;
  (a) $\tau \notin \widehat{S}^+$, $\widehat{S}^+ \cup \{\tau\} \in \widehat{\mathcal{S}}^+$; by the definitions of α in (3.14) and β in (3.15) and the definition of γ_τ in (3.8), we have

$$\alpha(S)\beta(\tau) = \alpha(T) \iff \widehat{S}^+ \gamma_\tau = \widehat{T}^+ \iff \widehat{S}^+ + \{\tau\} = \widehat{T}^+ \iff S\tau = T.$$

  The last equivalence results from the definition of the positive content of a state in a medium.
  (b) $\tau \in \widehat{S}^+$ or $\widehat{S}^+ \cup \{\tau\} \notin \widehat{\mathcal{S}}^+$; then, omitting the justifications which are similar to those used in Case (i)(a), we have

$$\alpha(S)\beta(\tau) = \alpha(T) \iff \widehat{S}^+ \gamma_\tau = \widehat{T}^+ \iff \widehat{S}^+ = \widehat{T}^+ \iff S\tau = T.$$

Similar arguments establish the two remaining cases, that is:

(ii) $\tau \in \mathcal{T}^-$;
  (a) $\tau \in \widehat{S}^+$, $\widehat{S}^+ \setminus \{\tau\} \in \widehat{\mathcal{S}}^+$;
  (b) $\tau \notin \widehat{S}^+$ or $\widehat{S}^+ \setminus \{\tau\} \notin \widehat{\mathcal{S}}^+$.

We leave the proof of these two cases as Problem 3.4. □

An earlier version of this result which does not mention the 'sign' aspect of the isomorphism, was established by Ovchinnikov and Dukhovny (2000) (see also Ovchinnikov, 2006).

A number of important families of relations are wg-families and so lead to media via Theorem 3.3.1. We have encountered one of them, the strict partial order family, in Chapter 1. We discuss other important examples in Chapter 5.

In the case of an infinite oriented medium, Theorem 3.3.3 gives a representation in terms of a wg-family of infinite sets. However, if no orientation is given, a representation as a wg-family of finite sets is always possible. From Theorems 2.9.5 and 3.3.3, we derive:

3.3.4 Theorem. *Any medium $(\mathcal{S}, \mathcal{T})$ is isomorphic to the representing medium of a well-graded subfamily $\mathcal{F}$ of $\mathfrak{P}_{\mathrm{F}}(\mathcal{S})$, with $|\mathcal{F}| = |\mathcal{S}|$.*

As made clear by Example 2.10.1 and Remark 2.10.2, this result does not assume any bound on the cardinality of $\mathcal{S}$, $\mathcal{T}$, or the contents of the states.

Proof. Using Theorem 2.9.5, choose an orientation $\{\mathcal{T}^+, \mathcal{T}^-\}$ of $(\mathcal{S}, \mathcal{T})$ ensuring that the positive contents of all the states are finite sets. By Theorem 3.2.6, which holds for any oriented medium, $\widehat{\mathcal{S}}^+$ is a wg-family. Using Theorem 3.3.3, we can assert that $(\mathcal{S}, \mathcal{T})$ is sign-isomorphic to $(\widehat{\mathcal{S}}^+, \mathcal{W}_{\widehat{\mathcal{S}}+})$; thus $|\widehat{\mathcal{S}}^+| = |\mathcal{S}|$. □

This section, culminating with Theorems 3.3.1, 3.3.3 and 3.3.4, stresses the important role played by well-graded families in the study of media. In the rest of this chapter we pursue the investigation of this concept. While the power set $\mathfrak{P}(X)$ of a finite set X is obviously well-graded, this is no longer the case if X is infinite. However, we can always partition $\mathfrak{P}(X)$ into infinite, well-graded classes. In fact, such classes may be of arbitrary cardinality (for a large enough set X), leading thus to the study of large media.

3.4 Cluster Partitions and Media

In Example 2.10.1, we introduced the well-graded subfamily of $\mathbb{R}$ containing all the complements of finite subsets of $\mathbb{R}$. The intersection of this subfamily is clearly empty. There are other well-graded subfamilies of $\mathbb{R}$, such as $\mathfrak{P}_F(\mathbb{R})$, or the subfamily

$$\{S_X \,|\, S_X = \mathbb{Q} \setminus X, \text{ with } X \in \mathfrak{P}_F(\mathbb{Q})\}.$$

In fact, there are uncountably many well-graded subfamilies of $\mathfrak{P}(\mathbb{R})$ (see Theorem 3.4.4). By Theorem 3.3.1, each of these well-graded families represents an oriented medium. This prompts a systematic investigation of the collection of the well-graded subfamilies of a set.

Except when indicated otherwise, we deal in this section with a set X of arbitrary cardinality, and a subfamily $\mathcal{F}$ of $\mathfrak{P}(X)$ which is graded in the sense of Definition 3.2.1. Thus, by Lemma 3.2.2, $\mathcal{F}$ is discriminative if $\cap \mathcal{F} = \varnothing$. We recall that $\mathcal{G}_\mathcal{F}$ denotes the grading collection of $\mathcal{F}$ and that the pair $(\mathcal{F}, \mathcal{G}_\mathcal{F})$ is a token system satisfying Axiom [Mb] (see Lemma 3.2.3(iii)). By Theorem 3.3.1, Axiom [Ma] also holds if and only if $(\mathcal{F}, \mathcal{G}_\mathcal{F})$ is well-graded. Without loss of generality, we suppose that $\cup \mathcal{F} = X$.

Our study of the graded subfamilies of $\mathfrak{P}(X)$ inducing media via Theorem 3.3.1 will rely on some geometrical concepts.

3.4.1 Definition. Let $\bowtie$ be a binary relation on $\mathfrak{P}(X)$ defined by

$$P \bowtie Q \iff (P \triangle Q) \in \mathfrak{P}_F(X).$$

We also write

$$\langle S \rangle = \{V \in \mathfrak{P}(X) \,|\, V \bowtie S\} \qquad (S \in \mathfrak{P}(X)), \tag{3.17}$$

$$\mathfrak{P}(X)_\bowtie = \{\langle S \rangle \,|\, S \in \mathfrak{P}(X)\}. \tag{3.18}$$

For any $S \in \mathfrak{P}(X)$, we refer to $\langle S \rangle$ as a *cluster (subfamily)* of $\mathfrak{P}(X)$, and to $\mathfrak{P}(X)_\bowtie$ as the *cluster partition* of $\mathfrak{P}(X)$ (a term justified by the theorem below). There is no ambiguity in the phrase 'a cluster family $\mathcal{F}$' without explicit mention of a ground set X because we can define $X = \cup \mathcal{F}$. We have then $\mathcal{F} = \langle S \rangle$ for some $S \in \mathcal{F}$, with $\langle S \rangle$ defined by (3.17).

3.4.2 Remark. Our use of the term 'cluster' is consistent with its usage in statistics, in which it refers to "a maximal collection of suitably similar objects drawn from a larger collection of objets" (from the entry "Graph-theoretical cluster analysis", Matula, 1983, in the Encyclopedia of Statistical Sciences, Vol. 3, p. 511). In particular, the function $s(P,Q) = e^{-d(P,Q)}$ on $\mathfrak{P}(\mathcal{X}) \times \mathfrak{P}(\mathcal{X})$, with d taking its values in the extended real line is a 'similarity measure' in the sense of cluster analysis (cf. Harary, 1969).

3.4.3 Theorem. *The following statements hold for any nonempty set $\mathcal{X}$.*
   (i) *The relation $\bowtie$ is an equivalence relation on $\mathfrak{P}(\mathcal{X})$ and $\mathfrak{P}(\mathcal{X})_\bowtie$ is the partition of $\mathfrak{P}(\mathcal{X})$ induced by $\bowtie$;*
   (ii) *$\langle \varnothing \rangle$ is the class of all the finite subsets of $\mathcal{X}$;*
   (iii) *$\langle S \rangle$ is a wg-family for any $S \in \mathfrak{P}(\mathcal{X})$.*

Much more can be said about the cluster subfamilies $\langle S \rangle$ of $\mathfrak{P}(\mathcal{X})$ (see for example Theorems 3.4.5, 3.4.6, and especially 3.4.11).

Proof. (i) The transitivity of $\bowtie$ results from Lemma 3.1.1. Since the relation $\bowtie$ is symmetric and reflexive by definition, it is an equivalence relation and its induced partition is clearly $\mathfrak{P}(\mathcal{X})_\bowtie$ as defined by (3.18).
   (ii) We have obviously $S \in \langle \varnothing \rangle$ if and only if $S \in \mathfrak{P}_F(\mathcal{X})$.
   (iii) Take any two distinct sets P and Q in $\langle S \rangle$. Then, by Lemma 3.1.1, we have $d(P,Q) = n$ for some $n \in \mathbb{N}$. So, there exists a sequence $S_0 = P, S_1, \ldots, S_n = Q$, in $\mathfrak{P}(\mathcal{X})$ with $d(S_i, S_{i+1}) = 1$ for $0 \leq i \leq n-1$. It suffices to show that all the sets S_i are in $\langle S \rangle$. Since $d(P, S_1) = 1$, we have $P \bowtie S_1$, which together with $S \bowtie P$ and the transitivity of $\bowtie$ established in (i) yields $S \bowtie S_1$ and so $S_1 \in \langle S \rangle$. An induction completes the argument, establishing the wellgradedness of $\langle S \rangle$. It is clear that $\cap \langle S \rangle = \varnothing$. □

3.4.4 Remarks. (a) The set $\mathfrak{P}(\mathbb{R})_\bowtie$ is uncountable. Indeed, consider the uncountable collection of families $\{\langle]-\infty, \alpha]\rangle \,|\, \alpha \in \mathbb{R} \setminus \mathbb{Q}\} \subset \mathfrak{P}(\mathbb{R})_\bowtie$. If $\alpha \neq \beta$, then $]-\infty, \alpha] \triangle]-\infty, \beta]$ is uncountable. Thus, a different cluster of $\mathfrak{P}(\mathbb{R})_\bowtie$ is defined by each $\alpha \in \mathbb{R} \setminus \mathbb{Q}$.
   (b) Accordingly, by Theorem 3.4.3(iii), uncountably many clusters of $\mathfrak{P}(\mathbb{R})$ are wg-families.

3.4.5 Theorem. *Take any S in $\mathfrak{P}(\mathcal{X})$, and denote[3] by d the restriction to $\langle S \rangle$ of the symmetric difference distance on $\mathfrak{P}(\mathcal{X})$. The following two statements hold:*
   (i) *$(\langle S \rangle, d)$ is a metric space;*
   (ii) *for any R, P and Q in $\langle S \rangle$, we have*

$$d(P,R) + d(R,Q) = d(P,Q) \iff R \in [P \cap Q, P \cup Q].$$

---

[3] For simplicity, by abuse of notation.

Proof. (i) Since, $d(P,Q) = |P \triangle Q|$ is a nonnegative real number for any P and Q in $\langle S \rangle$, we have $d(P,Q) \geq 0$ with $d(P,Q) = 0$ if and only if $P = Q$, and $d(P,Q) = d(Q,P)$ (cf. 1.8.1). The triangle inequality follows from Lemma 3.1.1. So does (ii). □

Thus, the equivalence relation $\bowtie$ partitions the family $\mathfrak{P}(\mathcal{X})$ into disjoint metric spaces $(\langle S \rangle, d)$, with each class $\langle S \rangle$ forming a cluster.

3.4.6 Theorem. *Any two metric spaces $(\langle S \rangle, d)$ and $(\langle S' \rangle, d)$ are isometric.*

Proof. If $\mathcal{X}$ is a finite set, then $\langle \varnothing \rangle$ is the only equivalence class; so $\langle S \rangle = \langle S' \rangle = \langle \varnothing \rangle$ and the conclusion of the theorem follows trivially. So we may assume that $\mathcal{X}$ is an infinite set and $S' = \varnothing$. Define

$$\alpha : \langle S \rangle \to \langle \varnothing \rangle : Z \mapsto Z \triangle S \qquad (Z \in \langle S \rangle). \tag{3.19}$$

By the identity

$$(Z \triangle S) \triangle (Z' \triangle S) = Z \triangle Z', \tag{3.20}$$

α is an isometric embedding of $\langle S \rangle$ into $\langle \varnothing \rangle$.

For $Y \in \langle \varnothing \rangle$, let $Z = S \triangle Y$. Then

$$\alpha(Z) = (S \triangle Y) \triangle S = Y, \tag{3.21}$$

so α is a bijection. □

By definition of a wg-family, there always exists a tight path between any two sets (cf. 3.1.5). This concept evokes that of a concise message producing a state from another in a token system. The following theorem specifies the connection between the two concepts.

3.4.7 Theorem. *Let $(\mathcal{F}, \mathcal{G}_\mathcal{F})$ be a token system with $\mathcal{X} = \cup \mathcal{F}$ and let P and Q be two distinct sets in the family $\mathcal{F}$. A message $\boldsymbol{m} = \gamma_1 \gamma_2 \cdots \gamma_n$ producing Q from P is concise if and only if*

$$P_0 = P, \ P_1 = P_0 \gamma_1, \ \ldots, \ P_n = P_{n-1} \gamma_n = Q \tag{3.22}$$

is a tight path between P and Q.

Proof. (Necessity.) We use induction on $n = \ell(\boldsymbol{m})$. Suppose that $\boldsymbol{m} = \gamma_1$. Since γ_1 is effective and either $\gamma_1 = \gamma_x$ or $\gamma_1 = \tilde{\gamma}_x$ for some $x \in \mathcal{X}$, we have either $Q = P + \{x\}$ or $P = Q + \{x\}$. Therefore $d(P,Q) = 1$. Now, let us assume that the statement holds for all concise messages $\boldsymbol{p}$ with $\ell(\boldsymbol{p}) = n-1$ and let $\boldsymbol{m} = \gamma_1 \gamma_2 \cdots \gamma_n$ be a concise message. Clearly, $\boldsymbol{m}_1 = \gamma_2 \cdots \gamma_n$ is a concise message producing Q from $P_1 = P\gamma_1$ and $\ell(\boldsymbol{m}_1) = n-1$. Suppose $\gamma_1 = \gamma_x$ for some $x \in \mathcal{X}$. Since $\boldsymbol{m}$ is stepwise effective and consistent, $x \in Q \setminus P$. Therefore, with $P_1 = P\gamma_1 = P + \{x\}$, we have

$$P_1 \in [P \cap Q, P \cup Q]. \tag{3.23}$$

Now suppose $\gamma_1 = \tilde{\gamma}_x$ for some $x \in \mathcal{X}$. Because $\boldsymbol{m}$ is stepwise effective and consistent, we have $x \in P \setminus Q$. With $P_1 = P \setminus \{x\}$, we also get (3.23). Thus, by Lemma 3.1.1, Formula (3.2), we obtain $d(P, Q) = d(P, P_1) + d(P_1, Q)$. By the induction hypothesis, (3.22) is a tight path from X to Y.

(Sufficiency.) Let $P_0 = P, P_1 = P_0\gamma_1, \ldots, P_n = P_{n-1}\gamma_n = Q$ be a tight path from P to Q for some message $\boldsymbol{m} = \gamma_1 \cdots \gamma_n$. By definition, $\boldsymbol{m}$ is stepwise effective. To prove consistency, we use again induction on n. The statement is trivially true for $n = 1$. Suppose it holds for all messages of length less than n and let $\boldsymbol{m}$ be a message of length n. Suppose $\boldsymbol{m}$ is inconsistent. By the induction hypothesis, this can occur only if either $\gamma_1 = \gamma_x$, $\gamma_n = \tilde{\gamma}_x$ or $\gamma_1 = \tilde{\gamma}_x$, $\gamma_n = \gamma_x$ for some $x \in \mathcal{X}$. In the former case, $x \in P_1$, $x \notin P \cup Q$. In the latter case, we have $x \notin P_1$ with $x \in P \cup Q$. In both cases, we obtain $P_1 \notin [P \cap Q, P \cup Q]$, a contradiction. □

From Theorem 3.4.7, we immediately derive:

3.4.8 Corollary. *If $(\mathcal{F}, \mathcal{G}_\mathcal{F})$ is a medium, then $\mathcal{F} \subseteq \langle S \rangle$ for some $S \in \mathfrak{P}(\mathcal{X})$.*

3.4.9 Remark. We will revisit the concept of a cluster family in Chapter 4, where we consider a special case of Theorem 3.4.8. We prove there (cf. Theorem 4.3.12), that the situation of a medium $(\mathcal{F}, \mathcal{G}_\mathcal{F})$ with the wg-family $\mathcal{F}$ being equal to a cluster (rather than a subset of one, as in Theorem 3.4.8) can arise if and only the medium is 'complete' in the sense defined in 4.3.1.

We know (cf. Problem 2.11 in Chapter 2) that the reduction of a medium is not necessarily a medium. The following results holds, however.

3.4.10 Theorem. *(i) For any $S \in \mathfrak{P}(\mathcal{X})$ the token system $(\langle S \rangle, \mathcal{W}_{\langle S \rangle})$ is a medium.*

(ii) The reduction of the medium $(\langle S \rangle, \mathcal{W}_{\langle S \rangle})$ to a family $\mathcal{F} \subseteq \langle S \rangle$ is a medium if and only if $\mathcal{F}$ is a wg-family.

Proof. (i) For a given $S \in \mathfrak{P}(\mathcal{X})$, the cluster family $\langle S \rangle$ is a wg-family of subsets of $\mathcal{X}$ (cf. Theorem 3.4.3(ii)). Hence, by Theorem 3.3.1, $(\langle S \rangle, \mathcal{W}_{\langle S \rangle})$ is a medium.

(ii) By Definition 2.7.2, the reduction of $(\langle S \rangle, \mathcal{W}_{\langle S \rangle})$ to $\mathcal{F}$ is the token system $(\mathcal{F}, \mathcal{G}_\mathcal{F})$. The statement of the theorem follows from Theorem 3.3.1. □

3.4.11 Theorem. *For any two $S', S \in \mathfrak{P}(\mathcal{X})$, the media $(\langle S' \rangle, \mathcal{W}_{\langle S' \rangle})$ and $(\langle S \rangle, \mathcal{W}_{\langle S \rangle})$ are isomorphic. In particular, any medium $(\langle S \rangle, \mathcal{W}_{\langle S \rangle})$ is isomorphic to the medium $(\mathfrak{P}_F(\mathcal{X}), \mathcal{W}_{\mathfrak{P}_F(\mathcal{X})})$ of all finite subsets of $\mathcal{X}$.*

Proof. It suffices to consider the case when $S' = \varnothing$. We define as follows the isomorphism (α, β) of $(\langle \varnothing \rangle, \mathcal{W}_{\langle \varnothing \rangle})$ onto $(\langle S \rangle, \mathcal{W}_{\langle S \rangle})$:

3.4 Cluster Partitions and Media

$$\alpha : \mathfrak{P}_F(X) \to \langle S \rangle \; : \; P \mapsto P \triangle S,$$
$$\beta : W_{\mathfrak{P}_F(X)} \to W_{\langle S \rangle} \; : \; \gamma \mapsto \beta(\gamma),$$
$$\beta(\gamma) : \langle S \rangle \mapsto \langle S \rangle \; : \; T \mapsto T\beta(\gamma)$$

with

$$T\beta(\gamma) = \begin{cases} T \setminus \{x\} & \text{if } (x \in S \text{ and } \gamma = \gamma_x) \text{ or } (x \notin S \text{ and } \gamma = \tilde{\gamma}_x), \\ T \cup \{x\} & \text{if } (x \notin S \text{ and } \gamma = \gamma_x) \text{ or } (x \in S \text{ and } \gamma = \tilde{\gamma}_x). \end{cases} \quad (3.24)$$

Clearly, α and β are bijections. We prove the implication

$$P\gamma = Q \implies \alpha(P)\beta(\gamma) = \alpha(Q).$$

For each of the four possibilities of (3.24), we have the two alternative $x \in P$ and $x \notin P$, so we have eight cases to consider.

Case 1. $\gamma = \gamma_x$, $x \in S$.
  (a) $x \in P$. Then, $Q = P$ and
  $$\alpha(P)\beta(\gamma) = (P \triangle S) \setminus \{x\} = P \triangle S = Q \triangle S.$$
  (b) $x \notin P$. Then, $Q = P \cup \{x\}$ and
  $$\alpha(P)\beta(\gamma) = (P \triangle S) \setminus \{x\} = P \triangle (S \setminus \{x\}) = Q \triangle S.$$

Case 2. $\gamma = \gamma_x$, $x \notin S$.
  (a) $x \in P$. Then, $Q = P$ and
  $$\alpha(P)\beta(\gamma) = (P \triangle S) \cup \{x\} = P \triangle S = Q \triangle S.$$
  (b) $x \notin P$. Then $Q = P \cup \{x\}$ and
  $$\alpha(P)\beta(\gamma) = (P \triangle S) \cup \{x\} = Q \triangle S.$$

Case 3. $\gamma = \tilde{\gamma}_x$, $x \in S$.
  (a) $x \in P$. Then, $Q = P \setminus \{x\}$ and
  $$\alpha(P)\beta(\gamma) = (P \triangle S) \cup \{x\} = Q \triangle S.$$
  (b) $x \notin P$. Then, $Q = P$ and
  $$\alpha(P)\beta(\gamma) = (P \triangle S) \cup \{x\} = P \triangle S = Q \triangle S.$$

Case 4. $\gamma = \tilde{\gamma}_x$, $x \notin S$.
  (a) $x \in P$. Then, $Q = P \setminus \{x\}$ and
  $$\alpha(P)\beta(\gamma) = (P \triangle S) \setminus \{x\} = Q \triangle S.$$
  (b) $x \notin P$. Then, $Q = P$ and
  $$\alpha(P)\beta(\gamma) = (P \triangle S) \setminus \{x\} = P \triangle S = Q \triangle S.$$

The converse implication: $\alpha(P)\beta(\gamma) = \alpha(Q)$ only if $P\gamma = Q$, is established along the same lines and is left as Problem 3.8. □

The following result follows readily from Corollary 3.4.8, and Theorems 3.4.10 and 3.4.11.

3.4.12 Theorem. *The representing medium $(\mathcal{F}, \mathbf{W}_\mathcal{F})$ of any wg-family $\mathcal{F}$ is isomorphic to a submedium of the medium $(\mathfrak{P}_F(\mathcal{X}), \mathbf{W}_{\mathfrak{P}_F(\mathcal{X})})$ of all finite subsets of $\mathcal{X} = \cup \mathcal{F}$.*

3.4.13 Remark. Theorem 3.4.12 provides another route to the representation Theorem 3.3.4. Notice that a medium that is both infinite and oriented may have infinite positive contents of all its states. Consider, for instance, the wg-family of subsets of $\mathbb{Z}$ in the form $]-\infty, n]$ (cf. Problem 3.20(ii)).

3.5 An Application to Clustered Linear Orders

The representation theorems of this chapter, combined with the concept of a cluster partition, provide powerful tools for uncovering and classify media. We illustrate applications of these tools here by constructing a medium representation of the family of all the strict linear orders (cf. 1.8.3) contained in a cluster $\langle L_0 \rangle$, where L_0 is a particular strict linear order on a ground set $\mathcal{X}$. Such a representation proceeds by first constructing a wg-family, along the lines of Theorem 3.3.1[4].

This example is interesting for two reasons. One is that the family of all strict linear orders on a set is never well-graded; rather, in the finite case for example, it is 2-graded (Falmagne, 1997). This fact offers a more direct way of constructing a medium representation of a set of strict linear orders, in which the tokens consists of transpositions of pairs of adjacent elements in a strict linear order. These two media are in fact isomorphic (for a given cluster). We consider this topic at the end of this section.

The second reason is that, while by Theorem 3.4.11 any two clusters $\langle S_1 \rangle$ and $\langle S_2 \rangle$ define isomorphic media, such an isomorphism does not necessarily hold for media induced by families of strict linear orders in different clusters. In other words, if L_1 and L_2 are strict linear order taken in different clusters, and $\mathcal{L}_1$ and $\mathcal{L}_2$ are the set of strict linear order at finite distances of L_1 and L_2, respectively, then the media induced by $\mathcal{L}_1$ and $\mathcal{L}_2$ along the lines sketched above are not necessarily isomorphic (see Problem 3.17).

We prepare our construction by a few lemmas. We recall that x is said to cover y in some strict linear order L if yLx and there is no z such that $yLzLx$ (cf. 1.8.3, and Problem 1.8 in Chapter 1)[5].

3.5.1 Lemma. *If L is a strict linear order, then $(L \setminus \{yx\}) \cup \{xy\}$ is also a strict linear order if and only if x covers y in L.*

---

[4] For a different approach involving a countable set of strict linear orders, see Problem 3.9 and Ovchinnikov (2006).

[5] There are strict linear orders in which there are no x, y such that x covers y. In other words, the covering relation is empty. Consider the set $\mathbb{R}$ of all real numbers equipped with its natural strict linear order (see Example 3.5.4 (a)).

Proof. (Necessity.) Suppose that $L' = (L\setminus\{yx\})\cup\{xy\}$ is a strict partial order. If $yLzLx$ for some z, then also $zL'x$ and $yL'z$, yielding $yL'x$ by transitivity. So, both $yL'x$ and $xL'y$, contradicting the asymmetry of L'.

(Sufficiency.) Suppose that x covers y in L. As L' is asymmetric and connected by definition, we only have to verify its transitivity. Assume that $uL'vL'w$. We have to verify three cases.

1. $uv = yx$, so uLv with $vL'w$ yielding vLw because $w \neq x$ and $w \neq y$; we obtain uLw by the transitivity of L, and so $uL'w$.
2. $vw = yx$, so vLw with $uL'v$ yielding uLv because $u \neq x$ and $v \neq y$; we obtain again uLw and thus $uL'w$.
3. uLv, $uv \neq yx$ and vLw, $vw \neq yx$; then uLw by the transitivity of L, and $uw \neq yx$ because x covers y in L; so $uL'w$.

This completes the proof. □

3.5.2 Lemma. *Suppose that L and L' are two distinct strict linear orders on a set $\mathfrak{X}$, with $L \triangle L'$ finite. Then the two following conditions hold:*

(i) there exists $xy \in L \setminus L'$ such that y covers x in L; in fact, we have a finite sequence $x_1, \ldots, x_n$ in $\mathfrak{X}$ such that x_{i+1} covers x_i in L for $1 \leq i \leq n-1$ and we have

$$x_1 L x_2 L \ldots L x_{n-1} L x_n \quad \text{and} \quad x_n L' x_{n-1} L' \ldots L' x_2 L' x_1, \quad (3.25)$$
$$L \triangle L' = \{x_i x_j \in \mathfrak{X} \mid 1 \leq i \leq n, 1 \leq j \leq n, i \neq j\}; \quad (3.26)$$

(ii) either both L and L' well-order $\mathfrak{X}$, or neither of them does.

Proof. (i) Because L and L' are connected and asymmetric,
$$xy \in L \triangle L' \Rightarrow (xLy \Leftrightarrow yL'x).$$
As $L \triangle L'$ is finite, so is $Y = \{x \in \mathfrak{X} \mid xy \in L \triangle L', y \in \mathfrak{X}\}$; say $|Y| = n$. The restriction of L to Y is a strict linear order on a finite set, whose elements can be labeled $x_1, \ldots, x_n$ in such a way that (3.25) holds with x_{i+1} covering x_i in L, $1 \leq i \leq n-1$, and (3.26) follows from the labelling of the elements in Y.

(ii) Using (i), we get that if L is a well-ordering of $\mathfrak{X}$, so must be L'. We leave as Problem 3.12 the construction of two strict linear orders some set $\mathcal{Z}$, at finite distance from each other, neither of them well-ordering $\mathcal{Z}$. □

A particular construction will be used, which deserves a name.

3.5.3 Definition. Let $\mathfrak{P}(\mathcal{Y})_{\bowtie}$ be the *cluster partition* of a set $\mathcal{Y}$, and let $\mathcal{F}$ be a family of subsets of $\mathcal{Y}$. For any $S \in \mathcal{F}$, we call $\langle S \rangle \cap \mathcal{F}$ the *trace* of $\mathcal{F}$ on the cluster $\langle S \rangle$. In particular, if $\mathcal{Y} = \mathfrak{X} \times \mathfrak{X}$ for some set $\mathfrak{X}$, and $\mathcal{L}$ is the set of all linear orders on $\mathfrak{X}$, then for any $L \in \mathcal{L}$, we call $\langle L \rangle \cap \mathcal{L}$ a *linear trace*. It is clear from Lemma 3.5.2(ii) that if L is a well-ordering of $\mathcal{Y}$ then the trace of $\mathcal{L}$ on $\langle L \rangle$ only contains well-orderings. Such a trace is called a *well-ordering trace*.

3.5.4 Examples. (a) A linear trace may be reduced to a single linear order. Take the set $\mathbb{R}$ of real numbers, and let $\mathfrak{P}(\mathbb{R} \times \mathbb{R})_{\bowtie}$ the cluster partition of the family of all the relations on $\mathbb{R}$. Denote by $<$ the natural order of the reals and by $\mathcal{L}$ the set of all strict linear orders on $\mathbb{R}$. The trace of $\mathcal{L}$ on $\langle < \rangle$ contains a single linear order, namely $<$.

(b) On the other hand, the trace of $\mathcal{L}$ on $\langle W \rangle$, where W is a well-ordering of $\mathbb{R}$ is uncountable (see Problem 3.13).

(c) Different well-orderings on the same set may define different well-ordering traces. Consider the set $\mathbb{N}$ of natural numbers with its natural ordering $<$ and the ordering

$$1 <' 3 <' \ldots <' 2n+1 <' \ldots <' 2 <' 4 <' \ldots <' 2n <' \ldots$$

(Thus, the odd numbers precede all the even numbers in $<'$.) Both $<$ and $<'$ are well-orderings of $\mathbb{N}$. Write $\mathcal{L}_{\mathbb{N}}$ for the set of all strict linear orders on $\mathbb{N}$. The traces of $\mathcal{L}_{\mathbb{N}}$ on $\langle < \rangle$ and $\langle <' \rangle$ are disjoint. These traces are infinite, and equipollent (see Problems 3.10 and 3.11 in this connection).

We shall also need the following fact.

3.5.5 Lemma. *Let P, Q and R be antisymmetric, connected binary relations on the same set. Then*

$$P \cap R = Q \cap R \iff P = Q.$$

Proof. Suppose $P \cap R = Q \cap R$ and let $xy \in P$. If $xy \in R$, then $xy \in Q$. Otherwise, $yx \in R$. Since $yx \notin P$, we have $yx \notin Q$ implying $xy \in Q$. Thus $P \subseteq Q$. By symmetry, $P = Q$. □

As the main theorem of this section, we have:

3.5.6 Theorem. *Let $\mathcal{L}$ be the set of all strict linear orders on a set $\mathcal{X}$, and take any L_0 in $\mathcal{L}$. Suppose that $\mathcal{L}_0 = \langle L_0 \rangle \cap \mathcal{L}$, where $\langle L_0 \rangle$ is a cluster of $\mathfrak{P}(\mathcal{X} \times \mathcal{X})$ and $\mathcal{L}_0$ is the linear trace of $\mathcal{L}$ on $\langle L_0 \rangle$, contains more than one linear order, and define the function*

$$\alpha : \mathcal{L}_0 \to \langle L_0 \rangle : L \mapsto L \cap L_0. \tag{3.27}$$

Then, α is a 1-1 mapping of the set of linear orders $\mathcal{L}_0$ onto a wg-family of partial orders, all included in L_0.

In particular, if $\mathcal{X}$ is a finite set, then $\mathcal{L}_0 = \mathcal{L}$ and the function α defined by (3.27) is a 1-1 mapping of the set $\mathcal{L}$ of all the strict linear orderings on $\mathcal{X}$ onto a well-graded family of partial orders included in L_0. As we have seen in Example 3.5.4 (a), $\mathcal{L}_0$ may be reduced to a single strict linear order and the function α is then trivial. But nontrivial cases are easy to come by (see Examples 3.5.4 (b) and (c)).

Proof. From Lemma 3.5.5, we know that α is a one-to-one mapping from $\mathcal{L}_0$ onto the set of partial orders $\alpha(\mathcal{L}_0) \subseteq L_0$. Let P, P' be two distinct partial orders in $\alpha(\mathcal{L}_0)$ and L, L' be the corresponding strict linear orders. Using Lemma 3.5.2(i), we can assert the existence of a pair $xy \in L$ such that y covers x and $xy \notin L'$. By Lemma 3.5.1, L'' defined by

$$L'' = (L \setminus \{xy\}) \cup \{yx\} \in \mathcal{L}_0$$

is a linear order. Then

$$P'' = L'' \cap L_0 = ((L \cap L_0) \setminus (L_0 \cap \{xy\})) \cup (L_0 \cap \{yx\})$$
$$= \begin{cases} P \setminus \{xy\}, & xy \in L_0, \\ P \cup \{yx\}, & xy \notin L_0, \end{cases}$$

where $xy \in P$ if $xy \in L_0$ and $yx \notin P$ if $xy \notin L_0$. Hence, $P'' \neq P$ and $d(P, P'') = 1$. Clearly, $L \cap L' \subseteq L'' \subseteq L \cup L'$. Therefore

$$P \cap P' = L \cap L' \cap L_0 \subseteq P'' = L'' \cap L_0 \subseteq (L \cup L') \cap L_0 = P \cup P',$$

that is, $P'' \in [P \cap P', P \cup P']$. Invoking Theorem 3.4.5(ii), we obtain

$$d(P, P') = d(P, P'') + d(P'', P') = 1 + d(P'', P')$$

and the result follows by induction. □

As announced, we finally obtain the medium representation:

3.5.7 Theorem. *Let $\mathcal{X}, \mathcal{L}, L_0, \mathcal{L}_0$ and α be as in Theorem 3.5.6. The pair $(\alpha(\mathcal{L}_0), \mathcal{W}_{\alpha(\mathcal{L}_0)})$, with the set of tokens $\mathcal{W}_{\alpha(\mathcal{L}_0)}$ as in 3.3.2, is a medium with orientation $(\mathcal{W}^+_{\alpha(\mathcal{L}_0)}, \mathcal{W}^-_{\alpha(\mathcal{L}_0)})$ and tokens defined, for any $P = L \cap L_0 \in \alpha(\mathcal{L}_0)$ and $xy \in L_0$, by*

$$\rho_{xy} : P \mapsto P\rho_{xy} = \begin{cases} P + \{xy\}, & \text{if } P + \{xy\} \in \alpha(\mathcal{L}_0), \\ P, & \text{otherwise}, \end{cases} \quad (3.28)$$

for $\rho_{xy} \in \mathcal{W}^+_{\alpha(\mathcal{L}_0)}$, and

$$\tilde{\rho}_{xy} : P \mapsto P\tilde{\rho}_{xy} = \begin{cases} P \setminus \{xy\}, & \text{if } P \setminus \{xy\} \in \alpha(\mathcal{L}_0), \\ P, & \text{otherwise}, \end{cases} \quad (3.29)$$

for $\tilde{\rho}_{xy} \in \mathcal{W}^-_{\alpha(\mathcal{L}_0)}$.

Indeed, since $\alpha(\mathcal{L}_0)$ is a well-graded family of subsets of L_0, we can apply Theorem 3.3.1 and take $\alpha(\mathcal{L}_0)$ to be the set of states of the medium $(\alpha(\mathcal{L}_0), \mathcal{W}_{\alpha(\mathcal{L}_0)})$ (with $\mathcal{W}_{\alpha(\mathcal{L}_0)} = \mathcal{G}_{\alpha(\mathcal{L}_0)}$, cf. Definition 3.3.2), equipped with the orientation $(\mathcal{W}^+_{\alpha(\mathcal{L}_0)}, \mathcal{W}^-_{\alpha(\mathcal{L}_0)})$ and tokens defined by (3.28) and (3.29).

Simple examples show that $\alpha(\mathcal{L}_0)$ is a proper subset of the set of all strict partial orders contained in L_0. This subset is characterized in the following proposition.

3.5.8 Theorem. *Let L be a strict linear order on a set $\mathcal{Z}$ and $P \subseteq L$ be a strict partial order. Suppose also that $P = L \cap L'$ for some relation L' on $\mathcal{Z}$. Then $P' = L \setminus P$ is a strict partial order if and only if L' is a strict linear order.*

Proof. (Sufficiency.) Suppose $P = L \cap L'$. It suffices to prove that $P' = L \setminus P$ is transitive. Let $xy, yz \in P'$; then $xz \in L$. Suppose $xz \notin P'$. Then $xz \in P$, implying $xz \in L'$. Since $xy, yz \in P'$, we have $xy \notin L'$ and $yz \notin L'$, yielding $yx, zy \in L'$, which implies $zx \in L'$, a contradiction.

(Necessity.) Suppose now that $P \subseteq L$ and $P' = L \setminus P$ are strict partial orders. It is easily verified (cf. Problem 3.19) that

$$P = L \cap L' \iff L' = P \cup P'^{-1}. \tag{3.30}$$

We have to show that L' is a strict linear order. It is asymmetric by its definition in the r.h.s. of (3.30). To establish its connectedness, notice that the set $\{P \cup P'^{-1}, P^{-1}, P'\}$ is a partition of $(\mathcal{Z} \times \mathcal{Z}) \setminus \mathcal{I}_\mathcal{Z}$, where $\mathcal{I}_\mathcal{Z}$ is the identity on $\mathcal{Z}$. Thus $xy \notin L'$ implies $xP^{-1}y$ or $xP'y$; so, yPx or $yP'^{-1}x$, and in either case $yL'x$.

It remains to show that L' is transitive. Suppose that $xL'yL'z$. It suffices to consider only two cases:

(i) xPy, $yP'^{-1}z$. Since $P^{-1} \cap P'^{-1} = \varnothing$, we must have $x \neq z$, with moreover $xz \in \{P \cup P'^{-1}, P^{-1}, P'\}$. If $xP \cup P'^{-1}z$, then $xL'z$. Suppose that $xP^{-1}z$; thus, zPx, yielding zPy since xPy. We obtain $yP^{-1}z$, contradicting $yP'^{-1}z$ (because $P^{-1} \cap P'^{-1} = \varnothing$). Finally, $xP'z$ together with $yP'^{-1}z$ yields $xP'zP'y$, and so $xP'y$, contradicting xPy (again $P^{-1} \cap P'^{-1} = \varnothing$).

(ii) $xP'^{-1}y$, yPz, with as above $x \neq z$. Again, we must have $xz \in \{P \cup P'^{-1}, P^{-1}, P'\}$. Arguments similar to those used in Case (i) give $xP \cup P'^{-1}z$, and thus $xL'z$, as the only valid possibility. □

We illustrate Theorem 3.5.7 by an example in the finite case; we have thus $\mathcal{L}_0 = \mathcal{L} \subseteq \langle L_0 \rangle = \langle \varnothing \rangle$ (see also Example 1.3.1).

3.5.9 Example. Let $\mathcal{X} = \{1, 2, 3\}$. For simplicity, we recode the $3! = 6$ strict linear orders on $\mathcal{X}$ as 3-tuples:

$$L_0 \mapsto 123, \quad L_1 \mapsto 213, \quad L_2 \mapsto 231,$$
$$L_3 \mapsto 321, \quad L_4 \mapsto 312, \quad L_5 \mapsto 132.$$

These triples are represented by the six vertices in the diagram[6] of Figure 3.1 (which is also, after an appropriate relabeling of the vertices, the adjacency graph of the medium).

---

[6] One can compare this diagram with the diagram shown in Figure 5 in Falmagne (1997).

3.5 An Application to Clustered Linear Orders

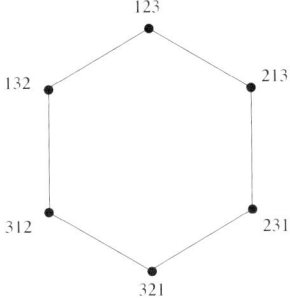

Figure 3.1. The diagram of $\mathcal{L}$ for $|\mathcal{X}| = 3$.

The strict partial orders forming the wg-family $\mathfrak{a}(\mathcal{L}_0)$ are included in L_0:

$$L_0 = \{12, 13, 23\}, \quad L_1 \cap L_0 = \{13, 23\}, \quad L_2 \cap L_0 = \{23\},$$
$$L_3 \cap L_0 = \emptyset, \quad L_4 \cap L_0 = \{12\}, \quad L_5 \cap L_0 = \{12, 13\}.$$

Note that the graph in Figure 3.1 is the graph of the permutohedron Π_2 (see, for instance Ziegler, 1995). The graph depicted in Figure 1.3.1 is the graph of the permutohedron Π_3.

3.5.10 An Unoriented Representation of Linear Orders. We construct here a medium representation of $\mathcal{L}_0$ which is is not oriented. In fact, no natural representation is available. This representation can nevertheless be shown to be isomorphic to the one spelled out in Theorem 3.5.7 (Problem 3.18).

Let us rewrite (3.28) explicitly as

$$(L \cap L_0)\rho_{xy} = \begin{cases} (L \cap L_0) + \{xy\}, & \text{if } (L \cap L_0) + \{xy\} \in \mathfrak{a}(\mathcal{L}_0), \\ L \cap L_0, & \text{otherwise}, \end{cases}$$
$$= \begin{cases} (L + \{xy\}) \cap L_0, & \text{if } (L + \{xy\}) \cap L_0 = L' \cap L_0, \\ L \cap L_0, & \text{otherwise}, \end{cases}$$

where L' is some well-ordering of $\mathcal{X}$. Since $yx \notin L_0$, we have

$$(L + \{xy\}) \cap L_0 = ((L \setminus \{yx\}) + \{xy\}) \cap L_0 = L' \cap L_0. \quad (3.31)$$

As x covers y in L, Lemma 3.5.1 implies that $L' = (L \setminus \{yx\}) + \{xy\}$ is a strict linear order. Equation (3.31) suggests a different kind of token directly transforming a strict linear order in $\mathcal{L}_0$ into another. We define for $xy \in L_0$

$$\tau_{xy} : \mathcal{L}_0 \to \mathcal{L}_0 : L \mapsto L\tau_{xy}$$
$$= \begin{cases} (L \setminus \{yx\}) + \{xy\} & \text{if } x \text{ covers } y \text{ in } L, \\ L & \text{otherwise}. \end{cases} \quad (3.32)$$

Notice that, for $xy \in L_0$ and with ρ_{xy} defined by (3.28),

$$(L \cap L_0)\rho_{xy} = L\tau_{xy} \cap L_0. \tag{3.33}$$

A similar argument relying on (3.29) shows that, again for $xy \in L_0$,

$$(L \cap L_0)\tilde{\rho}_{xy} = L\tau_{yx} \cap L_0 = L\tilde{\tau}_{xy} \cap L_0. \tag{3.34}$$

We obtained the set of tokens $\mathcal{T} = \{\tau_{xy}, \tilde{\tau}_{xy}\}_{xy \in L_0}$ by 'pulling back' tokens from the set $\mathcal{W}_{\alpha(\mathcal{L}_0)}$. The medium $(\mathcal{L}_0, \mathcal{T})$ is isomorphic to the medium $(\alpha(\mathcal{L}_0), \mathcal{W}_{\alpha(\mathcal{L}_0)})$ (Problem 3.18). It is the 'linear medium' introduced in Falmagne (1997).

3.6 A General Procedure

The steps used above to construct the linear medium $(\mathcal{L}_0, \mathcal{T})$ from the medium $(\alpha(\mathcal{L}_0), \mathcal{W}_{\alpha(\mathcal{L}_0)})$ exemplify a general procedure, based on the concept of media isomorphism, for constructing new media from old ones. Suppose that $\mathcal{X}$ is a set of features endowed with some particular 'transformation' structure $\mathfrak{S}$ which is not necessarily that of wg-family of sets. Suppose also that there is a medium $(\mathcal{S}, \mathcal{T})$ which models the set $\mathcal{X}$ and its structure $\mathfrak{S}$ in the sense that there is a bijection $\alpha : \mathcal{X} \to \mathcal{S}$ and that the tokens from $\mathcal{T}$ may be regarded as representing, in a natural way, the transformation occurring in $(\mathcal{X}, \mathfrak{S})$. Let us define a family $\mathcal{T}'$ of functions $\tau' : \mathcal{X} \to \mathcal{X}$ by 'pulling back' tokens from $\mathcal{T}$:

$$\tau' = \alpha \circ \tau \circ \alpha^{-1},$$

and let $\beta : \tau' \mapsto \tau$. We have the following proposition:

3.6.1 Theorem. *The pair (α, β) is an isomorphism from the token system $(\mathcal{X}, \mathcal{T}')$ onto the medium $(\mathcal{S}, \mathcal{T})$. Accordingly*[7], *the token system $(\mathcal{X}, \mathcal{T}')$ is a medium.*

We shall use this construction in our definition of a medium on the family of all the weak orders on a finite set in Chapter 9.

Problems

3.1 Which parts of Lemma 3.1.4, if any, hold for a k-graded path? Prove your answer.

---

[7] See Problem 2.10 in Chapter 2.

3.2 Let $\mathcal{F}$ be a family of subsets of a set $\mathcal{X} = \cup \mathcal{F}$. Verify that the family $\{x^* \,|\, x \in \mathcal{X}\}$ is a partition of $\mathcal{X}$ (cf. Definition 3.1.5). Prove that the two conditions: (i) $\mathcal{F}$ is well-graded, and (ii) $\mathcal{F}$ is discriminative, are independent. From Lemma 3.2.5, we know that any wg-family $\mathcal{F}$ satisfying $\cap \mathcal{F} = \varnothing$ is discriminative. Show by a counterexample that this result does hold in general for k-graded families, with $k \geq 2$. In this connection, ponder Theorem 3.1.7, however.

3.3 In which of the examples considered in Chapter 1 is the content family 2-graded (cf. Definition 3.1.5)?

3.4 Prove the two remaining cases in Theorem 3.3.3.

3.5 Prove Lemma 3.1.1. Especially, verify Formula (3.2).

3.6 In the study of the well-graded families of subsets of a set $\mathcal{X}$, why is there no loss of generality in formulating the hypotheses $\cap \mathcal{F} = \varnothing$ and $\cup \mathcal{F} = \mathcal{X}$?

3.7 Verify the identity (3.20) and also the second equation in (3.21).

3.8 Verify the converse implication $\alpha(P)\beta(\gamma) = \alpha(Q)$ only if $P\gamma = Q$ in the proof of Theorem 3.4.11.

3.9 Let $<$ be the natural linear order on $\mathbb{N}$. A relation L on $\mathbb{N}$ is a *locally finite linear order* if there exists $n \in \mathbb{N}$ such that

$$pLq \iff p < q, \qquad (p, q > n).$$

We write then $L_n = L$ and $L_0 = <$. Prove that the family $\{L_n \,|\, n = 0, 1, \ldots\}$ of locally finite linear orders on $\mathbb{N}$ is well-graded (cf. Ovchinnikov, 2006).

3.10 Prove that the two well-ordering traces of Example 3.5.4 are equipollent.

3.11 (Continuation.) Generalizing this example, show that there are countably many well-ordering traces of $\mathbb{N}$ on the clusters of $\mathfrak{P}(\mathbb{N} \times \mathbb{N})$.

3.12 Find a set $\mathcal{Z}$ and two strict linear orders on $\mathcal{Z}$ at finite distance from each other, neither of them well-orders $\mathcal{Z}$.

3.13 Prove that the trace of the set $\mathcal{L}$ of all strict linear orders on $\mathbb{R}$ on $\langle W \rangle$, where W is a well-ordering of $\mathbb{R}$ is uncountable.

3.14 For any three P, Q and S in $\mathfrak{P}(\mathbb{R})$, define the ternary relation

$$\langle PSQ \rangle \Leftrightarrow S \in [P,Q], \tag{3.35}$$

with $[P,Q]$ as in (3.3). Consider the following axioms formalizing a 'collinear betweenness' relation for triples of points (cf. Prenowitz and Jordan, 1965):

[B1] (SYMMETRY PROPERTY) $\langle PSQ \rangle \Rightarrow \langle QSP \rangle$.

[B2] (ANTICYCLIC PROPERTY) $\langle PSQ \rangle \Rightarrow \neg \langle QPS \rangle$.

[B3] (LINEAR COHERENCE) P, Q and S are collinear if and only if one of the following three hold: $\langle PSQ \rangle$ or $\langle PQS \rangle$ or $\langle SPQ \rangle$.

[B4] (SEPARATION PROPERTY) Suppose that T colline and is distinct from P, Q and S. Then $\langle PTQ \rangle$ implies $\langle QTS \rangle$ or $\langle PTS \rangle$, but not both.

[B5] (EXISTENCE) If $P \neq Q$, there exists S, T and W such that $\langle SPQ \rangle$, $\langle PTQ \rangle$ and $\langle PQW \rangle$.

Which of these axioms is satisfied by the relation $\langle \ldots \rangle$ defined by (3.35)? Prove your response by an argument or a counterexample. (Omit the 'collinearity' for the moment.)

3.15 (Continuation.) Answer the same question, but with a different definition of the ternary relation $\langle \ldots \rangle$, namely: Let P, S and Q be states in a token system. Define $\langle PSQ \rangle$ if there is a concise message $\boldsymbol{m} = \boldsymbol{nq}$ producing Q from P, such that $\boldsymbol{n}$ is a concise message producing S from P.

3.16 (Continuation.) Could you define 'collinearity' (for sets or for states in a token system) so that the ternary relation $\langle \ldots \rangle$ satisfies all the axioms [B1]-[B5], with either of the two definitions?

3.17 Show by a counterexample that the media induced, via Theorems 3.5.6 and 3.5.7, from linear traces in different clusters are not necessarily isomorphic.

3.18 Prove that the medium $(\mathcal{L}_0, \mathcal{T})$ constructed in 3.5.10 is isomorphic to the medium $(\alpha(\mathcal{L}_0), \mathcal{W}_{\alpha(\mathcal{L}_0)})$ of Theorem 3.5.7.

3.19 Verify the equivalence (3.30) in the proof of Theorem 3.5.8.

3.20 Let $\mathcal{J}$ be a nondegenerate (bounded or unbounded) interval in $\mathbb{Z}$ and let $\mathcal{T}_\mathcal{J}$ be the set of transformations of $\mathcal{J}$ defined by

$$\tau_i : x \mapsto x\tau_i = \begin{cases} x+1, & \text{if } x = i, \\ x, & \text{otherwise,} \end{cases}$$

and

$$\tilde{\tau}_i : x \mapsto x\tilde{\tau}_i = \begin{cases} x-1, & \text{if } x = i+1, \\ x, & \text{otherwise,} \end{cases}$$

for $\{i, i+1\} \subseteq \mathcal{J}$ and $x \in \mathcal{J}$.

(i) Prove that $(\mathcal{J}, \mathcal{T}_\mathcal{J})$ is a medium.
(ii) Prove that the medium $(\mathcal{J}, \mathcal{T}_\mathcal{J})$ is isomorphic to the medium $(\mathcal{F}, \mathcal{W}_\mathcal{F})$ where $\mathcal{F}$ is the wg-family of sets in the form $(-\infty, k] \cap \mathcal{J}$.

3.21 Prove that the set $\mathcal{L}_F$ of all strict linear orders on a finite set X is 2-graded.

3.22 (Continuation.) For every pair $xy \in X \times X$ of distinct elements, define a function

$$\tau_{xy} : \mathcal{L}_F \to \mathcal{L}_F : L \mapsto L\tau_{xy}$$

by the equation

$$L\tau_{xy} = \begin{cases} (L \setminus \{yx\}) \cup \{xy\} & \text{if } x \text{ covers } y \text{ in } L, \\ L & \text{otherwise.} \end{cases}$$

Let $\mathcal{Q}$ be the collection of all the functions τ_{xy}, with $xy \in X \times X$, $x \neq y$. Prove that the pair $(\mathcal{L}_F, \mathcal{Q})$ is a medium.

3.23 This problem and the next one investigate the possibility that the structure of a well-graded family (and so of the corresponding medium) could be captured by a relation guiding the organization and the compatibility of the tokens. Let X be a set, let R be a binary relation on X, and let $\mathcal{B}$ be a family of subsets of X. Define the operation $\triangleright : 2^{\mathcal{P}(X)} \to 2^{\mathcal{P}(X \times X)}$ by the equation

$$\mathcal{B} \triangleright R = \mathcal{B} + \{Y \subseteq X \mid \exists Z \in \mathcal{B}, \exists zx \in Z \times (X \setminus Z), Y = Z + \{x\} \text{ and } zRx\}$$

and then recursively $\mathcal{B}_0 = \mathcal{B}$, and $\mathcal{B}_{n+1} = \mathcal{B}_n \triangleright R$, for $n \in \mathbb{N}_0$.

The R-closure of $\mathcal{B}$ is the family $\mathcal{B}^R = \cup_{n=0}^\infty \mathcal{B}_n$. Under which necessary and sufficient conditions on a family $\mathcal{F}$ of subsets of a set X does there exists a relation R on X and a subset $\mathcal{B} \subseteq \mathcal{F}$ such that $\mathcal{F} = \mathcal{B}^R$, that is, $\mathcal{F}$ is the R-closure of $\mathcal{B}$?

3.24 (Continuation.) Investigate a similar problem with $R \subseteq 2^X \times X$.

4

Closed Media and ∪-Closed Families

Several examples of structures representable by media discussed earlier enjoy a closedness[1] property. For instance, the family of all partial orders on a set is closed under intersections (cf. Theorem 5.3.5). We define this concept for media and investigate its properties.

4.1 Closed Media

We begin by recalling standard concepts of closedness in set theory.

4.1.1 Definition. A family of sets $\mathcal{F}$ is *closed under unions*, or *∪-closed*, (resp. *closed under intersections*, or *∩-closed*) if for any nonempty[2] $\mathcal{G} \subseteq \mathcal{F}$ we have $\cup \mathcal{G} \in \mathcal{F}$ (resp. $\cap \mathcal{G} \in \mathcal{F}$). We say that a family $\mathcal{F}$ is *closed under finite unions* or $\cup_F$-*closed* if $\cup \mathcal{G} \in \mathcal{F}$ for any finite, nonempty subfamily $\mathcal{G} \subseteq \mathcal{F}$. Similar terminology and notation are used for finite intersections.

A family of sets $\mathcal{F}$ is *closed under subsets* if, for any nonempty set $S \in \mathcal{F}$, and any subset $T \subset S$, T also belongs to $\mathcal{F}$. Families of finite sets that are closed under subsets are sometimes called *independence systems* or *abstract simplicial complexes* (cf. Korte et al., 1991)

Thus, the family of all partial orders on a set is ∩-closed, as recalled above. The next definition captures a related concept of closedness in the framework of a medium.

4.1.2 Definition. An oriented medium $(\mathcal{S}, \mathcal{T})$ is *u-closed* (resp. *i-closed*) if for any state S and any two distinct positive (resp. negative) tokens τ, τ' both effective for S, we have

$$(S\tau = V, S\tau' = W) \implies V\tau' = W\tau. \qquad (4.1)$$

---

[1] We reserve the term 'closure' for the operation, as in 'transitive closure.'
[2] For some authors, the subfamily $\mathcal{G}$ may be empty, with $\cup \varnothing = \varnothing$. So, a ∪-closed family automatically contains the empty set. We do not use this convention here.

Results concerning u-closed media can be reformulated for i-closed media in a straightforward manner. In the sequel, we simplify our language and write that a medium is *closed* if it is an oriented medium that is u-closed. A medium that is not closed is said to be *open*. Clearly, a medium can be closed under one orientation without being closed under some other orientation. When a medium $(\mathcal{F}, \mathcal{W}_\mathcal{F})$ is defined, as in Theorem 3.3.1, from a wg-family $\mathcal{F}$, then $\mathcal{F}$ is closed under finite unions—or $\cup_F$-closed—if and only if $(\mathcal{F}, \mathcal{W}_\mathcal{F})$ is closed in the sense of Definition 4.1.2 (see Corollary 4.1.10).

4.1.3 Theorem. *In an oriented medium $(\mathcal{S}, \mathcal{T})$, the two conditions below are equivalent:*

(i) *$(\mathcal{S}, \mathcal{T})$ is closed.*

(ii) *Let $\boldsymbol{m} = \tau_1 \ldots \tau_n$ be any positive concise message from some state S, with $S_1 = S$, and $S_{i+1} = S_i \tau_i$ for $1 \leq i < n$. If a positive token $\tau \notin \mathcal{C}(\boldsymbol{m})$ is effective for some state S_i, $1 \leq i < n$, it is also effective for any state S_j, $i < j \leq n$.*

The proof is left as Problem 4.1. As a hint of a forthcoming result (Theorem 4.2.2), compare Condition (ii) with Axiom [K2] of a learning space on page 10. Another useful consequence of the closedness condition is given below.

4.1.4 Theorem. *Let $\boldsymbol{n} = \tau_1 \ldots \tau_k \tau_{k+1} \ldots \tau_n$ be a concise message from some state S in a closed medium, with τ_k negative and τ_{k+1} positive. Then $S\tau_1 \cdots \tau_k \tau_{k+1} \cdots \tau_n = S\tau_1 \cdots \tau_{k+1} \tau_k \cdots \tau_n$.*

In other words, the tokens τ_k and τ_{k+1} in the original message $\boldsymbol{n}$ can be transposed without changing the state produced.

Proof. Let $T = S\tau_1 \cdots \tau_k$. Then, there are distinct states W and W' such that $T\widetilde{\tau}_k = W = S\tau_1 \cdots \tau_{k-1}$ and $T\tau_{k+1} = W'$. Since both $\widetilde{\tau}_k$ and τ_{k+1} are positive and the medium is closed, we get $W\tau_{k+1} = W'\widetilde{\tau}_k$, and thus also, successively

$$S\tau_1 \cdots \tau_{k-1} \tau_{k+1} \tau_k = W\tau_{k+1}\tau_k = W'\widetilde{\tau}_k \tau_k = W'$$
$$= T\tau_{k+1} = S\tau_1 \cdots \tau_k \tau_{k+1}.$$

□

4.1.5 Definition. Suppose that $\boldsymbol{n} = \boldsymbol{mpm'}$, with $\boldsymbol{m}$ and $\boldsymbol{m'}$ two possibly ineffective messages, and $\boldsymbol{p}$ an effective one. Then $\boldsymbol{p}$ is a *segment* of $\boldsymbol{n}$. If $\boldsymbol{m}$ is empty, then (as $\boldsymbol{m}$ can be omitted) $\boldsymbol{p}$ is an *initial segment* or *prefix* of $\boldsymbol{n}$. Similarly, if $\boldsymbol{m'}$ is empty, then $\boldsymbol{p}$ is an *terminal segment* or *suffix* of $\boldsymbol{n}$. With respect to some orientation, a segment is said to be *positive* (resp. *negative*) if it contains only positive (resp. negative) tokens.

4.1.6 Definition. In an oriented medium, a concise message m producing a state V from a state S is called *canonical* if it satisfies one of the following three cases:

[1] m is positive;
[2] m is negative;
[3] $m = nn'$, with n a positive prefix and n' a negative suffix.

In Case [3], the canonical message $m = nn'$ is said to be *mixed*.

4.1.7 Theorem. *For any two distinct states S and V in a closed medium, there is a canonical message producing V from S.*

Proof. By [Ma], there is a concise message $p = \tau_1 \ldots \tau_n$ producing V from S. Suppose that p is not canonical. Then there must be an index i such that τ_i is negative and τ_{i+1} positive. By Theorem 4.1.4, the tokens τ_i and τ_{i+1} can be transposed without changing the state produced. The result follows. □

4.1.8 Theorem. *In an oriented medium, suppose that a state V is produced from a state S by mixed canonical message $n = mm'$, with $Sm = T$ and m a positive prefix of n; then, $\widehat{S}^+ \cup \widehat{V}^+ = \widehat{T}^+$.*

Proof. By Theorem 2.8.5, $\widehat{S}^+ \subset \widehat{T}^+$. We have also $\widehat{V}^+ \subset \widehat{T}^+$ because $\widetilde{m'}$ is positive, concise for V, and produces T from V. This implies that $\widehat{S}^+ \cup \widehat{V}^+ \subseteq \widehat{T}^+$. We get

$$\widehat{T}^+ \setminus (\widehat{S}^+ \cup \widehat{V}^+) = (\widehat{T}^+ \setminus \widehat{S}^+) \cap (\widehat{T}^+ \setminus \widehat{V}^+) = \mathcal{C}(m) \cap \mathcal{C}(\widetilde{m'}) = \varnothing,$$

the last equation holding because mm' is concise. Thus, $\widehat{T}^+ \subseteq \widehat{S}^+ \cup \widehat{V}^+$, and the result obtains. □

4.1.9 Theorem. *For any two states S and V in a closed medium $(\mathcal{S}, \mathcal{T})$, there is a unique state T whose positive content is the union of the positive contents of S and V. Consequently, the family $\widehat{\mathcal{S}}^+$ of all the positive contents is closed under finite unions, and the family $\widehat{\mathcal{T}}^-$ of all the negative contents is closed under finite intersections.*

Proof. For any two states S and V, there is a canonical message p producing V from S. Suppose that p is positive, then clearly $\widehat{S}^+ \subset \widehat{V}^+$, which gives $\widehat{S}^+ \cup \widehat{V}^+ = \widehat{V}^+ \in \widehat{\mathcal{T}}^+$, and $T = V$. If p is negative, then $\widetilde{p}$ is positive, and produces S from V, with a similar result. The case where p is a mixed canonical message is an immediate consequence of Theorem 4.1.8. The set T is unique because any state is defined by its content. The statement concerning the negative contents follows because, for any state W, we have $\widehat{W}^- = \mathcal{T} \setminus \widehat{W}^+$. Thus with S, V and T as above, we obtain

$$\widehat{S^-} \cap \widehat{V^-} = \overline{\widehat{S^+} \cap \widehat{V^+}} = \overline{\widehat{S^+} \cup \widehat{V^+}} = \overline{\widehat{T^+}} = \widehat{T^-}.$$ □

From Theorem 4.1.9 and the definitions, we get immediately:

4.1.10 Corollary. *Suppose that a medium $(\mathcal{F}, \mathsf{W}_{\mathcal{F}})$ has been defined from a wg-family $\mathcal{F}$ of subsets of some finite set, in the sense of Definition 3.3.2. Then, any set S in the family $\mathcal{F}$ is in a 1-1 correspondence $x \mapsto \gamma_x$ with the positive content $\widehat{S}^+$ of the corresponding state S of the medium $(\mathcal{F}, \mathsf{W}_{\mathcal{F}})$. Moreover, the family $\mathcal{F}$ is $\cup_F$-closed if and only if the medium $(\mathcal{F}, \mathsf{W}_{\mathcal{F}})$ is also closed.*

4.1.11 Theorem. *In a finite closed medium $(\mathcal{S}, \mathcal{T})$, there is a unique state Λ which is produced only by positive messages. Consequently, its positive content is identical to its content. In fact, we have $\widehat{\Lambda}^+ = \widehat{\Lambda} = \mathcal{T}^+$, and so $|\widehat{\Lambda}| = |\mathcal{T}|/2$.*

Proof. As the collection $\mathcal{S}$ of states is finite, we can show by induction that there exists a state whose positive content is equal to $\cup_{S \in \mathcal{S}} \widehat{S}^+$. Because a state is defined by its content, this state is unique. As it contains all the positive tokens, it cannot contain any negative one: by Theorem 2.4.3, we have $|\widehat{\Lambda}| = |\mathcal{T}|/2$. □

4.1.12 Definition. A state S in a medium with orientation $\{\mathcal{T}^+, \mathcal{T}^-\}$ is called the *apex* of the medium for that orientation, or more briefly the *apex* of $\{\mathcal{T}^+, \mathcal{T}^-\}$, if $\widehat{S} = \mathcal{T}^+$. Thus, the state Λ introduced in Theorem 4.1.11 is the apex of the closed medium. By Theorem 2.4.5, any oriented medium can have at most one apex.

Each state S of a medium can be the apex for a particular orientation by assuming that the set of positive tokens of that orientation is equal to $\widehat{S}$. This class of orientations has interesting properties.

4.1.13 Definition. Let $\mathcal{M} = (\mathcal{S}, \mathcal{T})$ be a medium with orientation $\{\mathcal{T}^+, \mathcal{T}^-\}$. For any state S in $\mathcal{S}$ we say that $\mathcal{M}$ is *apex oriented* (for S) if $\mathcal{T}^+ = \widehat{S}$. In such a case, we may write $\mathcal{T}_S^+ = \mathcal{T}^+$ and $\mathcal{T}_S^- = \mathcal{T}^-$.

By Theorem 4.1.11, any finite closed oriented medium is apex-oriented. We now investigate media for which all apex orientations form closed media, anticipating our investigation in Theorem 4.3.3 of media for which all orientations form closed media.

4.1.14 Definition. Let u, v, w and x be vertices in a graph G. Then x is a *median* for the triple (u, v, w) if x belongs to a shortest path in G between each pair of vertices in the triple. The graph G is a *median graph* if every triple of vertices has a unique median.

Finite median graphs have been extensively studied; see, e.g., Imrich et al. (1999) or Imrich and Klavžar (2000). Alternative characterizations of them are known; for example, they are exactly the 'retracts' of hypercubes (Bandelt, 1984). Every finite median graph is a partial cube (Mulder, 1980), from which it will follow by Theorem 7.1.10 that it is the graph of a medium. However, we will not use this result in what follows.

4.1.15 Example. Let G be a path. Then, if (u, v, w) is any triple of distinct vertices, the smallest connected subgraph of G that contains all three vertices in the triple is itself a path, with two of the triple as its endpoints. The third vertex of the triple is the median. Thus, any triple has a unique median and G is a median graph.

4.1.16 Lemma. *Let G be the graph of a medium $\mathcal{M}$, and let (S, T, U) be a triple of vertices in G that has a median V. Then a token τ belongs to $\widehat{V}$ if and only if it belongs to at least two of $\widehat{S}, \widehat{T}$, and $\widehat{U}$.*

Proof. Suppose that a token $\tau \in \widehat{V}$ does not belong to $\widehat{T}$ nor $\widehat{U}$. Then any path from T to U via V would correspond to a message $\boldsymbol{mn}$ producing U from T with $T\boldsymbol{m} = V$, $V\boldsymbol{n} = U$, $\tau \in \mathcal{C}(\boldsymbol{m})$ and $\tilde{\tau} \in \boldsymbol{n}$. So, $\boldsymbol{mn}$ would not be concise. Accordingly, V could not be on a shortest path from T to U. By symmetry, the same reasoning applies to the other cases in which fewer than two of the contents $\widehat{S}, \widehat{T}$, and $\widehat{U}$ contain τ.

Conversely, suppose that $\tau \in (\widehat{S} \cup \widehat{T}) \setminus U$. Since V belongs to a shortest path in G between S and T, we necessarily have $\boldsymbol{m} = \boldsymbol{np}$ for some concise messages $\boldsymbol{m}, \boldsymbol{n}$ and $\boldsymbol{p}$, with one of the two cases: (i) $S\boldsymbol{n} = V$ and $S\boldsymbol{m} = T$; or (ii) $T\boldsymbol{n} = V$ and $T\boldsymbol{m} = S$. Since $\tau \in \widehat{S} \cup \widehat{T}$, we get $\tau \in V$ in either case. The other possible situations arising from the hypothesis that the token τ belong to at least two of $\widehat{S}, \widehat{T}$, and $\widehat{U}$ are dealt with by similar arguments. □

4.1.17 Theorem. *Let $\mathcal{M}$ be a finite medium and let G be the graph of $\mathcal{M}$. Then all apex orientations of $\mathcal{M}$ give rise to closed media if and only if G is a median graph.*

Proof. (Sufficiency.) Assume that G is a median graph and let $\{\mathcal{T}_T^+, \mathcal{T}_T^-\}$ be the apex orientation for some state T. Suppose that the positive tokens τ and ρ are both effective for some state S. Since G is a median graph, we may find a median U of the triple $(S\tau, S\rho, T)$. (Lemma 5.4.3(iii) and Theorem 2.4.4 imply that the vertices in that triple must be distinct.) We will show that this entails $U = S\rho\tau = S\tau\rho$. We must have $S \neq U \neq S\tau$ because U must be on a shortest path from $S\rho$ to T. Similarly, we must have $U \neq S\rho$, as U must be on a shortest path from $S\tau$ to T. As we have eliminated all three vertices on the shortest path $S\rho$–S–$S\tau$ from $S\rho$ to $S\tau$, the vertex U must be the middle vertex on a different shortest path from $S\rho$ to $S\tau$. We conclude that $U = S\rho\tau = S\tau\rho$ must be true. Thus, the condition defining the closure of a medium is satisfied and the medium is closed.

(Necessity.) Assume that all apex orientations of $\mathcal{M}$ form closed media. To show that G is a median graph, we must prove that any triple of (S, T, U) of vertices has a median. To do so, we consider the apex orientation for T. Invoking Theorem 4.1.7, let $\boldsymbol{m}$ be a canonical message from S to U according to this orientation and let $\boldsymbol{n}$ be the positive prefix of $\boldsymbol{m}$. Then $S\boldsymbol{n}$ is the desired median. The uniqueness of this median follows from Lemma 4.1.16. Thus, any such triple has a unique median, and so G is a median graph. □

We finish this section with a result that is more naturally stated in terms of i-closed families than u-closed families. This will be relevant for the discussion of weak orders in Section 9.4.

4.1.18 Theorem. *Let $\mathcal{F}$ be an independence system. Then $\mathcal{F}$ is a wg-family and the associated oriented medium $(\mathcal{F}, \mathcal{W}_\mathcal{F})$ is i-closed.*

We leave the proof as an exercise (Problem 4.7).

4.2 Learning Spaces and Closed Media

This important representation of certain closed media has been defined in Chapter 1. We recall that, as mentioned there, a learning space is called an antimatroid in combinatorics (Edelman and Jamison, 1985; Welsh, 1995; Björner et al., 1999). We begin our discussion with the following result due to Cosyn and Uzun (2005).

4.2.1 Theorem. *The two following conditions are equivalent for a finite family $\mathcal{F}$ of subsets:*

(i) *$\mathcal{F}$ is a learning space (in the sense of Definition 1.6.1);*
(ii) *$\mathcal{F}$ contains the empty set and is well-graded and ∪-closed.*

Proof. Note that both conditions imply the finiteness of the set $Q = \cup \mathcal{F}$. Indeed, in both cases, we have $\varnothing, Q \in \mathcal{F}$. So, the finiteness of Q results from $\varnothing \subset Q$ and either [K1] for Condition (i) or wellgradedness for Condition (ii).

(ii) ⇒ (i). Let $\mathcal{F}$ be a ∪-closed wg-family. It is clear that Axiom [K1] is satisfied since it involves special cases of the situations covered by the wellgradedness condition. Suppose that $K \subset L$, with K, $K \cup \{q\}$ and L in $\mathcal{F}$ with $q \notin L$. Because $\mathcal{F}$ is ∪-closed, we have $(K \cup \{q\}) \cup L = L \cup \{q\} \in \mathcal{F}$; so, [K2] holds.

(i) ⇒ (ii). We begin by proving that $\mathcal{F}$ is ∪-closed. Take any distinct, nonempty L and K in $\mathcal{F}$. Applying [K1] to $\varnothing \subset L$, we obtain a sequence $q_1, \ldots, q_n$ in L forming a chain

$$\varnothing \subset \{q_1\} \subset \cdots \subset \{q_1, \ldots, q_i\} \subset \cdots \subset \{q_1, \ldots, q_n\} = L, \qquad (4.2)$$
$$\text{with} \quad \{q_1, \ldots, q_i\} \in \mathcal{F} \quad (1 \leq i \leq n). \qquad (4.3)$$

But we also have $\varnothing \subset K$, which by $\varnothing \cup \{q_1\} \in \mathcal{F}$ yields $K \cup \{q_1\} \in \mathcal{F}$, using [K2]. By induction, we derive $K \cup L \in \mathcal{F}$ from [K2] and (4.2), (4.3). Since Q is finite, $\mathcal{F}$ is $\cup$-closed.

Turning to the wellgradedness, we take any two distinct K and L in $\mathcal{F}$. As shown above, we must have $K \cup L \in \mathcal{F}$. Axioms [K1] implies the existence of two chains of subsets of $\mathcal{F}$

$$K \subset K \cup \{q_1\} \subset \cdots \subset K \cup \{q_1, \ldots, q_n\} = K \cup L,$$
$$L \subset L \cup \{v_1\} \subset \cdots \subset L \cup \{v_1, \ldots, v_m\} = K \cup L,$$

with $n = |L \setminus K|$ and $m = |K \setminus L|$, yielding $m + n = |(L \setminus K) + (K \setminus L)|$.

The theorem follows from defining the sequence

$$\begin{aligned}
K_0 &= K, \\
K_i &= K \cup \{q_1, \ldots, q_i\}, & (1 \le i \le n) \\
K_{n+j} &= (K \cup L) \setminus \{v_m, \ldots, v_{m-j+1}\} & (1 \le j \le m),
\end{aligned}$$

so $K_{m+n} = L$. $\square$

4.2.2 Theorem. *The two following propositions are equivalent for an oriented medium* $\mathcal{M} = (\mathcal{S}, \mathcal{T})$:

(i) $\mathcal{M}$ *is a closed rooted medium.*
(ii) *The collection* $\widehat{\mathcal{S}}^+$ *of positive contents of the states of* $\mathcal{M}$ *is a learning space in the sense of Definition 1.6.1.*

Proof. (i) $\Rightarrow$ (ii). We have to show that both $\mathcal{T}^+$ and $\varnothing$ are in $\widehat{\mathcal{S}}^+$, and that Axioms [K1] and [K2] hold. The fact that $\mathcal{T}^+ \in \widehat{\mathcal{S}}^+$ is established by Theorem 4.1.11. Since $\mathcal{M}$ is rooted, $\varnothing \in \widehat{\mathcal{S}}^+$ results from Theorem 2.9.2. From Theorems 3.2.6 and 4.1.9, we know that $\widehat{\mathcal{S}}^+$ is a wg-family that is $\cup$-closed. Accordingly, by Theorem 4.2.1, the collection $\widehat{\mathcal{S}}^+$ of positive contents must be a learning space.

(ii) $\Rightarrow$ (i). By Theorem 2.8.4, any state in an oriented medium is defined by its positive content. Observe that if the positive tokens τ^+ and μ^+ are effective for some state S of $\mathcal{M}$, then

$$\widehat{(S\tau^+)}^+ = \widehat{S}^+ + \{\tau^+\} \quad \text{and} \quad \widehat{(S\mu^+)}^+ = \widehat{S}^+ + \{\mu^+\}.$$

Because $\widehat{S}^+$ is $\cup$-closed,

$$\widehat{V}^+ = \widehat{S}^+ + \{\tau^+\} + \{\mu^+\}$$

is the positive content of some state V, with necessarily $V = S\tau\mu = S\mu\tau$. So, $\mathcal{M}$ is closed. We have $\varnothing \in \widehat{S}^+$ by definition of a learning space as a particular kind of knowledge structure. Thus $\varnothing$ is the positive content of some state R that is a root of $\mathcal{M}$, by Theorem 2.9.2. $\square$

We turn to media that are called 'complete' because they are endowed with a total symmetry (cf. Theorems 4.3.3 and 4.3.6).

4.3 Complete Media

4.3.1 Definition. A medium is *complete* if for any state S and any token τ, either τ or $\tilde{\tau}$ is effective for S.

4.3.2 Theorem. *Let $(\mathcal{S}, \mathcal{T})$ be a complete medium. For any state S and any two distinct tokens τ, τ' both effective for S, we have*

$$(S\tau = V,\ S\tau' = W) \quad \Longrightarrow \quad V\tau' = W\tau.$$

Proof. By the completeness of $(\mathcal{S}, \mathcal{T})$, either τ or $\tilde{\tau}$ is effective for W. Suppose that $W\tilde{\tau} = T \neq W$. Then $T \neq V$ (otherwise, we would have a return for S of odd length). Let $\boldsymbol{m}$ be a concise message producing V from T. By [Mb], the message $\tau'\tilde{\tau}\boldsymbol{m}\tilde{\tau}$ must be vacuous, which is impossible because $\boldsymbol{m}$ is concise. It follows that τ is effective for W. By symmetry, τ' is effective for V. Suppose that $P = S\tau\tau' \neq Q = S\tau'\tau$. Let $\boldsymbol{n}$ be a concise message producing Q from P. Thus, $\boldsymbol{p} = \tau\tau'\boldsymbol{n}\tilde{\tau}\tilde{\tau}'$ is a return message for S. By Axiom [Mb], this is impossible if $\boldsymbol{n}$ is concise. □

4.3.3 Theorem. *A medium is complete if and only if it is closed under any orientation.*

Proof. (Necessity.) By Theorem 4.3.2, a complete medium is closed under any orientation.

(Sufficiency.) Suppose that $(\mathcal{S}, \mathcal{T})$ is closed under any orientation. Take any $S \in \mathcal{S}$ and any $\tau \in \mathcal{T}$. Since τ is a token, there is a state V such that τ is effective for V. Let $\boldsymbol{m} = \tau_1 \ldots \tau_n$ be a concise message producing S from V. We consider two possible cases:

Case 1: $\tau \notin \mathcal{C}(\boldsymbol{m})$. By Theorem 2.9.3, there exists an orientation making V the root of $(\mathcal{S}, \mathcal{T})$. Since V is a root, the token τ and the message $\boldsymbol{m}$ are both positive. By Theorem 4.1.3, τ is effective for S.

Case 2: $\tau \in \mathcal{C}(\boldsymbol{m})$. If $\tau = \tau_n$, then $\tilde{\tau}$ is effective for S. Suppose that $\tau = \tau_i$ for $i < n$, and let $W = V\tau_1 \ldots \tau_i$. Then $\tilde{\tau}$ is effective for W and does not occur in the concise message $\boldsymbol{n} = \tau_{i+1} \ldots \tau_n$ producing S from W. By Theorem 2.9.3, there exists an orientation making W the root of $(\mathcal{S}, \mathcal{T})$. Since W is a root, the token $\tilde{\tau}$ and the message $\boldsymbol{n}$ are both positive. By Theorem 4.1.3, $\tilde{\tau}$ is effective for S. □

4.3.4 Example. Consider a collection $\mathcal{H}$ of n hyperplanes $x_i = 0$, $1 \leq i \leq n$, in $\mathbb{R}^n$. Following the steps outlined in 1.2.1, we construct a medium which is complete in the sense of Definition 4.3.1. Each state is a region in $\mathbb{R}^n$ defined by a system of inequalities $x_i P_i 0$, $1 \leq i \leq n$, where each P_i stands for either inequality $<$ or $>$ of the reals. There are 2^n states and any state has facets defined by all n hyperplanes in $\mathcal{H}$.

4.3.5 Example. As an application of Theorem 3.3.1, we consider the representing closed medium $(\mathfrak{P}_F(\mathcal{Z}), \mathcal{W}_{\mathfrak{P}_F(\mathcal{Z})})$ of the wg-family of all the finite subsets of a set $\mathcal{Z}$ (cf. Definition 3.3.2). This medium has a natural orientation and is clearly complete (see Problem 4.2). If $\mathcal{Z}$ is a finite set, then the medium $(\mathfrak{P}_F(\mathcal{Z}), \mathcal{W}_{\mathfrak{P}_F(\mathcal{Z})})$ is isomorphic to the medium of Example 4.3.4, with $n = |\mathcal{Z}|$.

4.3.6 Theorem. *A medium $(\mathcal{S}, \mathcal{T})$ is complete if and only if for some orientation $\{\mathcal{T}^+, \mathcal{T}^-\}$, we have*

$$\mathfrak{P}_F(\mathcal{T}^+) = \widehat{\mathcal{S}}^+ = \{\widehat{S}^+ \mid S \in \mathcal{S}\}, \tag{4.4}$$

that is, every finite subset of positive tokens is the positive content of some state. Moreover, any medium isomorphic to a complete medium is complete.

Proof. (Sufficiency.) Let $\{\mathcal{T}^+, \mathcal{T}^-\}$ be some orientation and suppose that (4.4) holds. Take any state $S \in \mathcal{S}$ and any token $\tau^+ \in \mathcal{T}^+$. By Theorem 2.4.3, we have either $\tau^+ \in \widehat{S}$ or $\tau^- \in \widehat{S}$. Suppose that $\tau^+ \in \widehat{S}^+$. By Eq. (4.4), there exists some state T with $\widehat{T}^+ = \widehat{S}^+ \setminus \{\tau^+\}$. Since $(\mathcal{S}, \mathcal{T})$ is a medium, there is by [Ma] a concise message from S to T. This message can only be the single token τ^-. Thus, τ^- is effective for S. In the case $\tau^- \in \widehat{S}$, a similar argument establishes that there is a state T with $\widehat{S}^+ \cup \{\tau^+\} = \widehat{T}^+$, with τ^+ effective for S. Thus, $(\mathcal{S}, \mathcal{T})$ is complete by Definition 4.3.1.

(Necessity.) Suppose that $(\mathcal{S}, \mathcal{T})$ is complete and let T_0 be one of its state. By Theorem 2.9.3, there exists an orientation making T_0 the root of $(\mathcal{S}, \mathcal{T})$. Theorem 2.9.2 implies that $\widehat{T}_0^+ = \varnothing$. Because $(\mathcal{S}, \mathcal{T})$ is complete, every positive token τ is effective on T_0. Thus, for every $\tau \in \mathcal{T}^+$, $\{\tau\}$ is the positive content of a state. By Theorems 4.3.3 and 4.1.9, the family $\widehat{\mathcal{S}}^+$ of all the positive contents is $\cup_F$-closed. Accordingly, we must have $\widehat{\mathcal{S}}^+ = \mathfrak{P}_F(\mathcal{T}^+)$.

The last proposition is obvious. □

4.3.7 Theorem. *If $(\mathcal{S}, \mathcal{T})$ is an oriented medium, then its sign-isomorphic representing medium $(\widehat{\mathcal{S}}^+, \mathcal{W}_{\widehat{\mathcal{S}}^+})$ is a submedium of the complete medium $(\mathfrak{P}(\widehat{\mathcal{S}}^+), \mathcal{W}_{\mathfrak{P}(\widehat{\mathcal{S}}^+)})$.*

This derives immediately from Theorem 3.3.3. As a preparation for the next section, we have the following definition and results relating star media (cf. Definition 2.9.4) and completeness.

4.3.8 Definition. A rooted medium $(\mathcal{S}, \mathcal{T})$ is an *extended star* or *E-star* if it has a star submedium $(\mathcal{Q}, \mathcal{V})$ with the same root satisfying $|\mathcal{V}| = |\mathcal{T}|$. An E-star $\mathcal{N} = (\mathcal{Q}, \mathcal{V})$ that is a submedium of some oriented, not necessarily rooted, medium $\mathcal{M}$ is called a *E-star* of $\mathcal{M}$ if there is no E-star submedium $\mathcal{N}' = (\mathcal{Q}', \mathcal{V}')$ of $\mathcal{M}$ having the same root as $\mathcal{N}$ and satisfying $\mathcal{Q} \subset \mathcal{Q}'$ and $|\mathcal{V}'| = |\mathcal{V}|$. It is clear that any oriented medium has at least one E-star.

82   4 Closed Media and ∪-Closed Families

In the display of an oriented medium as a digraph, we will often indicate only the edges corresponding to the positive tokens. We may sometimes refer to such a digraph as a *positive digraph*.

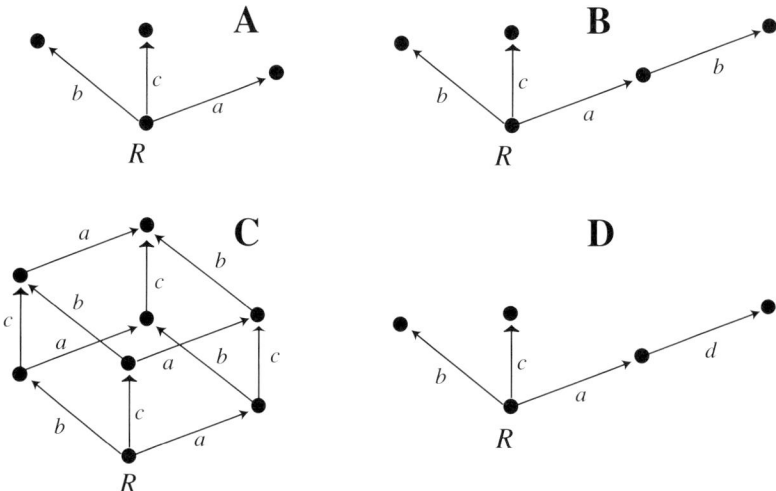

Figure 4.1. The digraphs A, B and C represent E-stars, but the digraph D does not. The digraph in C is that of a complete medium.

The labeling of the edges is critical in a digraph representation of an E-star. For example, A, B, C and D in Figure 4.1 are positive digraphs of rooted media (with only their positive token represented). Each of A, B, and C represent E-stars, but the digraph in D does not: its star submedium has 2×3 tokens while the full rooted medium has 2×4 tokens, contradicting the definition of an E-star.

4.3.9 Theorem. *Any closed E-star is a complete medium.*

Proof. Let $\mathcal{N} = (\mathcal{Q}, \mathcal{V})$ be a closed E-star with $\mathcal{V}^+ = \{\tau_1, \ldots, \tau_n\}$ and a root denoted by R. By definition of the root of an E-star, we must have $\widehat{R}^+ = \emptyset$ and $\widehat{R\tau_i}^+ = \{\tau_i\}$ for $1 \leq i \leq n$. Since $\mathcal{N}$ is closed, Theorem 4.1.9 implies that $\widehat{\mathcal{Q}}^+$ is closed under finite unions; so $\widehat{\mathcal{Q}}^+ = \mathfrak{P}_F(\mathcal{V}^+)$. Applying Theorem 4.3.6, we conclude that $\mathcal{N}$ is complete. □

4.3.10 Lemma. *Suppose that S is a state of a medium $(\mathcal{S}, \mathcal{T})$, and that $S\tau \neq S \neq S\tau'$ for some tokens τ and τ', with moreover either $S\tau\tau' = S\tau$ or $S\tau'\tau = S\tau'$. Then there exists a open E-star $(\mathcal{Q}, \mathcal{T}_\mathcal{Q})$ of $(\mathcal{S}, \mathcal{T})$ such that $S \in \mathcal{Q}$ with $\tau_\mathcal{Q}$ and $\tau'_\mathcal{Q}$ the reductions of τ and τ' to $\mathcal{Q}$, respectively.*

This lemma is useful in proving the theorem below. We leave the two proofs as Problem 4.9.

4.3.11 Theorem. *A medium is closed if and only if all its E-stars are closed.*

Finally, we return, from the standpoint of the completeness of media, to the cluster subfamilies studied in Chapter 3. We learned from Theorem 3.4.6 that, for a fixed set $\mathcal{X}$, all the clusters $\langle S \rangle$ of the power set $\mathfrak{P}(\mathcal{X})$ form isometric metric spaces $(\langle S \rangle, d)$ (where d stands for the restriction to $\langle S \rangle$ of the symmetric difference distance on $\mathfrak{P}(\mathcal{X})$). We also know that all the media $(\langle S \rangle, \mathcal{W}_{\langle S \rangle})$ are isomorphic; in particular, they are isomorphic to $(\mathfrak{P}_F(\mathcal{X}), \mathcal{W}_{\mathfrak{P}_F(\mathcal{X})})$ (cf. Theorems 3.4.10(i) and 3.4.11). In addition, it is easy to see that any medium $(\langle S \rangle, \mathcal{W}_{\langle S \rangle})$ is complete. The converse is also true as the following theorem asserts.

4.3.12 Theorem. *Let $\mathcal{F}$ be a wg-family of subsets of a set $\mathcal{X}$. The representing medium $(\mathcal{F}, \mathcal{W}_\mathcal{F})$ is complete if and only if $\mathcal{F} = \langle S \rangle$ for some $S \in \mathfrak{P}(\mathcal{X})$.*

Proof. We only need to establish the necessity. Suppose thus that $(\mathcal{F}, \mathcal{W}_\mathcal{F})$ is a complete medium. By Theorem 3.4.8, we have $\mathcal{F} \subseteq \langle S \rangle$ for some $S \in \mathfrak{P}(\mathcal{X})$. For a given $P \in \langle S \rangle$, let $k = d(P, S)$. We prove that $P \in \mathcal{F}$ by induction on k. If $k = 1$, then either $P = S + \{x\}$ or $P = S \setminus \{x\}$ for some $x \in \mathcal{X}$ and $P \neq S$. In the former case, $x \notin S$ implying that $\hat{\gamma}_x$ is not effective on S. By completeness, $S\gamma_x = P$. Thus $P \in \mathcal{F}$. Similarly, if $P = S \setminus \{x\}$, then $x \in S$ and $S\tilde{\gamma}_x = P$. Suppose $Q \in \mathcal{F}$ for all $Q \in \langle S \rangle$ such that $d(S, Q) = k$ and let P be an element in $\langle S \rangle$ such that $d(S, P) = k + 1$. Then there exists $R \in \mathcal{F}$ such that $d(S, R) = k$ and $d(R, P) = 1$. Because $\langle R \rangle = \langle S \rangle$, it follows from the above argument that $P \in \mathcal{F}$. □

4.4 Summarizing a Closed Medium

In certain cases, a structure can be economically summarized by a well chosen part it. A familiar case is the Hasse diagram $\check{P}$ of a partial order P on a finite set: the transitive closure of $\check{P}$ is equal to P (cf. 1.8.3, Eq. (1.8)). We face a similar situation in the case of certain empirical large rooted media, whose number of states may occasionally be of the order of millions[3]. An appealing idea is to represents such a medium by a much smaller rooted submedium whose 'closure' would reconstruct the large medium. The last two theorems suggest that the essential step towards the closure of a rooted medium is the closure of all its E-stars. This has to be done recursively since, typically, the closure of one E-star may create new ones.

While this idea is attractive, its application to the closure of some open medium is delicate because the most obvious potential technique is not feasible.

---

[3] Personal communication from Eric Cosyn concerning learning spaces (cf. Theorem 4.2.1 and Chapter 13).

4.4.1 A Counterexample. A priori, an appealing method is to go back to the definition of 'closed medium' and to enforce it recursively on some open medium $(\mathcal{S}, \mathcal{T})$, namely:

1. pick a state W such that, for two distinct tokens τ and μ
$$W\tau \neq W \neq W\mu \text{ with } W\tau\mu = W\tau, W\mu\tau = W\mu; \qquad (4.5)$$
2. add a new state S to $\mathcal{S}$ and recast the tokens τ and μ as the tokens $\tau' : \mathcal{S} \cup \{S\} \to \mathcal{S} \cup \{S\}$ and $\mu' : \mathcal{S} \cup \{S\} \to \mathcal{S} \cup \{S\}$ defined by:

$$\tau' : T \mapsto T\tau' = \begin{cases} T\tau & \text{if } S \neq T \neq W\mu \\ S & \text{if } T = W\mu \\ T & \text{otherwise} \end{cases} \qquad (4.6)$$

$$\mu' : T \mapsto T\mu' = \begin{cases} T\mu & \text{if } S \neq T \neq W\tau \\ S & \text{if } T = W\tau \\ T & \text{otherwise.} \end{cases} \qquad (4.7)$$

Figure 4.2 demonstrates that this method does not necessarily lead to construct a medium. In the open medium[4] pictured by the digraph A, there are four possible cases of some state W satisfying the condition of Eq. (4.5). The corresponding vertices are labeled $W_1, \ldots, W_4$ in that digraph. There are thus four ways of adding the state S. In each these cases, which are displayed by the digraphs B, C, D, and E, there is no consistent message from the new state S to one of the other state. Accordingly, the resulting token system fails Axiom [Ma] and so is not a medium.

The required construction is more elaborate and we shall follow a rather different route in the next section, based on the fact that any oriented medium can be faithfully represented by its family of positive contents. We just discuss here the much simpler case of the closure of an exemplary E-star.

4.4.2 Example. Consider the medium $\mathcal{M} = (\mathcal{S}, \mathcal{T})$ displayed in Figure 4.3 by the digraph of its positive tokens. It is a slight modification of the example of Figure 2.1 setting its root to be W and equipped with the orientation defined by $\mathcal{T}^+ = \{\tau_1, \tau_4, \tau_6\}$. The complete medium forming the closure $\mathcal{M}^\square$ of the medium $\mathcal{M}$ has the collection of positive contents (cf. Theorem 4.3.6)

$$\mathfrak{P}\{\tau_1, \tau_4, \tau_6\} = \{\varnothing, \{\tau_1\}, \{\tau_4\}, \{\tau_6\}, \{\tau_4, \tau_6\}, \{\tau_1, \tau_4\},$$
$$\{\tau_1, \tau_6\}, \{\tau_1, \tau_4, \tau_6\}\}$$

five of which are already in the E-star of the figure. We are thus missing three states. The expanded set of states is

$$\mathcal{S}^\square = \mathcal{S} + \{S_{\{\tau_1, \tau_4\}}, S_{\{\tau_1, \tau_6\}}, S_{\{\tau_1, \tau_4, \tau_6\}}\}.$$

---
[4] The fact that this particular medium is rooted has no bearing on our discussion.

4.4 Summarizing a Closed Medium   85

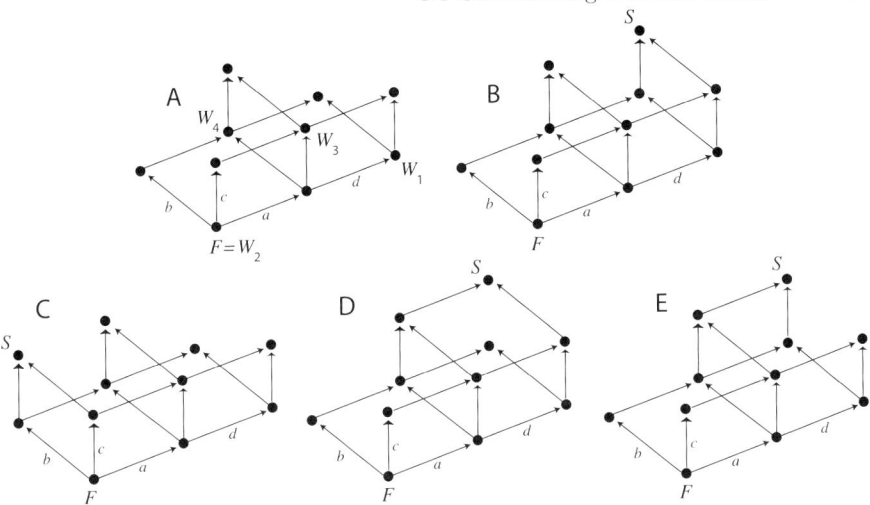

Figure 4.2. The positive digraph A represents an open medium, while B, C, D and E picture four failed attempts to construct an enlarged medium by adding a state S via the method of Eqs. (4.6)–(4.7).

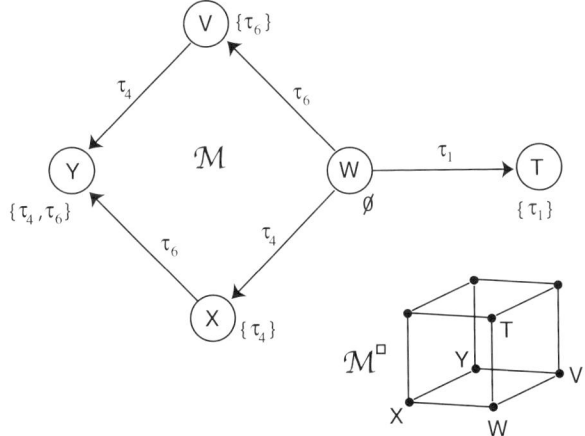

Figure 4.3. Positive digraph of an E-star $\mathcal{M} = (\mathcal{S}, \mathcal{T})$ with root W, $\mathcal{S} = \{V, W, X, Y, T\}$, and $\mathcal{T}^+ = \{\tau_1, \tau_4, \tau_6\}$. The graph of the closure $\mathcal{M}^\square$ of $\mathcal{M}$ is also indicated schematically.

We also have to expand the definition of the three pairs of tokens $(\tau_i, \tilde{\tau}_i)$, $i = 1, 4, 6$. To avoid repetition, we only give five examples illustrating all the important cases for the positive tokens. Denoting by $\tau_i^\square$, $\tilde{\tau}_i^\square$, $i = 1, 4, 6$ the 'expansions' of the tokens of $\mathcal{M}$, we get

$$T\tau_1^\square = T\tau_1 = T, \quad W\tau_1^\square = W\tau_1 = T, \quad X\tau_1^\square = S_{\{\tau_1,\tau_4\}}, \tag{4.8}$$
$$S_{\{\tau_1,\tau_4\}}\tau_1^\square = S_{\{\tau_1,\tau_4\}}, \quad S_{\{\tau_1,\tau_4\}}\tau_6^\square = S_{\{\tau_1,\tau_4,\tau_6\}}. \tag{4.9}$$

The neologism 'expansion' is justified. As illustrated by the last equation in (4.8), we cannot use 'extension' there because, as some loops have been removed, some of the images of states under τ_i are changed under $\tau_i^\square$. (For symmetrical reasons, we used 'reduction' rather than 'restriction' in our definition of a submedium—cf. 2.7.2.)

4.5 ∪-Closed Families and their Bases

As announced, we investigate here another avenue for summarizing a closed medium. For some of the material in this section, we follow Doignon and Falmagne (1999) except that we do not assume that the empty set necessarily belongs to the ∪-closed family. Any result adapted from their monograph is marked by '[DF99].' Some additional results in this section are from Eppstein et al. (2007).

4.5.1 Definition. The *(finite) span* of a family of sets $\mathcal{F}$ is the family $\mathcal{F}'$ containing any set which is the union of some (finite) subfamily[5] of $\mathcal{F}$. In such a case, we write $\mathbb{S}(\mathcal{F}) = \mathcal{F}'$ (resp. $\mathbb{S}_F(\mathcal{F}) = \mathcal{F}'$) and we say that $\mathcal{F}$ *spans* (resp. *(finitely) spans*) $\mathcal{F}'$. By definition $\mathbb{S}(\mathcal{F})$ and $\mathbb{S}_F(\mathcal{F})$ are thus ∪-closed and $\cup_F$-closed, respectively. A *base* of a ∪-closed family $\mathcal{H}$ is a minimal subfamily $\mathcal{J}$ of $\mathcal{H}$ spanning $\mathcal{H}$ (where 'minimal' is meant with respect to set inclusion: if $\mathbb{S}(\mathcal{F}) = \mathcal{H}$ for some $\mathcal{F} \subseteq \mathcal{J}$, then $\mathcal{F} = \mathcal{J}$.) Note that a base of a $\cup_F$-closed family need not be finite (cf. Problem 4.8).

4.5.2 Lemma. *The finite span of a wg-family is well-graded.*

Proof. Let $\mathbb{S}_F(\mathcal{F})$ be the finite span of some wg-family $\mathcal{F}$. Take any two distinct X, Y in $\mathbb{S}_F(\mathcal{F})$. Since $\mathbb{S}_F(\mathcal{F})$ is $\cup_F$-closed by definition, $X \cup Y$ is in $\mathbb{S}_F(\mathcal{F})$. Notice that $S \triangle T$ is a finite set for any $S, T \in \mathbb{S}_F(\mathcal{F})$. (This follows easily from the facts that $\mathcal{F}$ is a wg-family and that both S and T are finite unions of sets in $\mathcal{F}$.) We can thus write $d(X,Y) = d(X, X \cup Y) + d(X \cup Y, Y)$. Accordingly, it suffices to prove that there is in $\mathbb{S}_F(\mathcal{F})$ a tight path

$$X_1 = X, X_2, \ldots, X_n = X \cup Y, \tag{4.10}$$

with in fact $X_i \subset X_{i+1}$, $1 \leq i \leq n-1$. By definition of the finite span, there exists finite $\mathcal{G}, \mathcal{H} \subseteq \mathcal{F}$ such that $X = \cup \mathcal{G}$ and $Y = \cup \mathcal{H}$. Without loss of

---

[5] Contrary to the convention used by Doignon and Falmagne (1999), the empty subfamily of $\mathcal{F}$ is not allowed; so $\varnothing \in \mathbb{S}(\mathcal{F})$ only if $\varnothing \in \mathcal{F}$.

generality (exchanging the roles of X and Y if needed), we can assume that there exists some $H \in \mathcal{H}$ such that $H \setminus X \neq \varnothing$. Choose $G \in \mathcal{G}$ arbitrarily. By the well-gradedness of $\mathcal{F}$, there is a tight path $G_1 = G, \ldots, G_m = H$. Let k be the first index such that $G_k \setminus X \neq \varnothing$. (Such an index must exist because $H \setminus X \neq \varnothing$.) We necessarily have $|G_k \setminus X| = 1$. Defining $X_2 = (\cup \mathcal{G}) \cup G_k$, we obtain $X_1 = X \subset X_2 \subseteq X \cup Y$ with $|X_2 \setminus X_1| = 1$. An induction completes the proof. □

Note that a base of a ∪-closed wg-family need not be well-graded.

4.5.3 Example. The ∪-closed wg-family

$$\mathcal{F} = \{\varnothing, \{a\}, \{b\}, \{c\}, \{a,b\}, \{a,c\}, \{b,c\}, \{c,d\}, \{a,b,c\},$$
$$\{a,c,d\}, \{b,c,d\}, \{a,b,c,d\}, \{a,b,c,d,e\}\}. \quad (4.11)$$

has the base

$$\{\varnothing, \{a\}, \{b\}, \{c\}, \{c,d\}, \{a,b,c,d,e\}\},$$

which is not well-graded. Moreover, $\mathcal{F}$ has two minimal well-graded subfamilies spanning $\mathcal{F}$:

$$\{\varnothing, \{a\}, \{b\}, \{c\}, \{a,b\}, \{a,c\}, \{c,d\}, \{a,b,c\},$$
$$\{a,c,d\}, \{a,b,c,d\}, \{a,b,c,d,e\}\}, \quad (4.12)$$
$$\{\varnothing, \{a\}, \{b\}, \{c\}, \{a,b\}, \{b,c\}, \{c,d\}, \{a,b,c\},$$
$$\{b,c,d\}, \{a,b,c,d\}, \{a,b,c,d,e\}\}. \quad (4.13)$$

In the context of Theorems 3.3.1 and 3.3.3 we can of course consider the family represented by (4.11) as that of the positive contents of a closed medium. This example raises the question: how can we recognize or characterize a base of a wg-family? One answer is given as Theorem 4.5.13 for the $\cup_F$-closed case.

4.5.4 Example. The unique base of the set $\mathfrak{P}_F(\mathbb{Z})$ of all finite subsets of $\mathbb{Z}$ is the family

$$\{\{0\}, \{1\}, \{-1\}, \ldots, \{n\}, \{-n\}, \ldots\}.$$

4.5.5 Example. A base of a family which is both ∪-closed and ∩-closed (that is, closed under intersection) is not necessarily well-graded. Indeed, consider the family

$$\mathcal{G} = \{\varnothing, \{a\}, \{b\}, \{d\}, \{a,b\}, \{a,d\}, \{b,d\}, \{a,b,c\}, \{a,b,d\},$$
$$\{a,b,c,d\}, \{a,b,c,d,e\}\},$$

for which $\{\varnothing, \{a\}, \{b\}, \{d\}, \{a,b,c\}, \{a,b,c,d,e\}\}$ is a base.

4.5.6 Theorem. [DF99] *Let $\mathcal{B}$ be a base of a $\cup$-closed family $\mathcal{F}$. If $\mathcal{G}$ is any family spanning $\mathcal{F}$, then $\mathcal{B} \subseteq \mathcal{G}$. Accordingly, any $\cup$-closed family has at most one base.*

Proof. Let $\mathcal{B}$, $\mathcal{F}$ and $\mathcal{G}$ be as in the statement of the theorem and take any $X \in \mathcal{B}$. So $X \in \mathcal{F}$ and there exists $\mathcal{H} \subseteq \mathcal{G}$ such that $X = \cup \mathcal{H}$. Since $\mathcal{B}$ is a base of $\mathcal{F}$, there is for any $Y \in \mathcal{H}$ a subfamily $\mathcal{B}_Y$ of $\mathcal{B}$ such that $Y = \cup \mathcal{B}_Y$. We obtain
$$X = \cup \mathcal{H} = \cup \left(\cup_{Y \in \mathcal{H}} \mathcal{B}_Y \right) \in \mathcal{B}. \tag{4.14}$$
As $\mathcal{B}$ is a minimal family spanning $\mathcal{F}$, the set X cannot be the union of other sets in $\mathcal{B}$. So, the only possibility for (4.14) to be true is to have $\mathcal{H} = \{X\}$, yielding $X \in \mathcal{G}$. □

4.5.7 Theorem. [DF99] *Finite $\cup$-closed families have always a base.*

Proof. Consider the set $\mathfrak{S}(\mathcal{F})$ of all the families spanning some $\cup$-closed family $\mathcal{F}$, partially ordered by inclusion. If $\mathcal{F}$ is finite, that partial order has a minimal element, which by Theorem 4.5.6 is equal to $\cap \mathfrak{S}$. □

If a $\cup$-closed family $\mathcal{F}$ has a base, its element can be described precisely: for any X in $\mathcal{F}$, we can test whether or not X belongs to the base of $\mathcal{F}$.

4.5.8 Definition. For any x in $\cup \mathcal{F}$, where $\mathcal{F}$ is a $\cup$-closed family, an *atom at x* is a minimal set of $\mathcal{F}$ containing x (where 'minimal' is with respect to set inclusion). A set X in $\mathcal{F}$ is called an *atom*[6] if either $X = \varnothing \in \mathcal{F}$, or there is some $x \in \cup \mathcal{F}$ such that X is an atom at x.

The next two theorems recast results from Doignon and Falmagne (1999).

4.5.9 Theorem. [DF99] *A nonempty set X in a $\cup$-closed family $\mathcal{F}$ is an atom if and only if $X \in \mathcal{H}$ for any subfamily $\mathcal{H}$ of $\mathcal{F}$ such that $\cup \mathcal{H} = X$.*

We omit the proof (see Problem 4.11.)

4.5.10 Theorem. *Let $\mathcal{A}$ be the collection of all the atoms of a $\cup$-closed family $\mathcal{F}$, with $\mathcal{X} = \cup \mathcal{F}$.*

(i) Suppose that for every point of $\mathcal{X}$ there is an atom at x. Then $\mathcal{A}$ is the base of $\mathcal{F}$.

(ii) If $\mathcal{F}$ has a base $\mathcal{B}$, then $\mathcal{A} = \mathcal{B}$. In such a case, there is not necessarily an atom at every point of $\mathcal{X}$.

---

[6] Our meaning of the term 'atom' is different from its usage in lattice theory; cf. Birkhoff (1967), Davey and Priestley (1990). It also slightly differs from that in Doignon and Falmagne (1999) because we do not allow the empty union of a family (see Footnote 5).

Proof. Let $\mathcal{A}$ be the collection of all the atoms of $\mathcal{F}$. Suppose that there is an atom at every point of $\mathcal{X}$. We first assume that $\varnothing \notin \mathcal{F}$. Thus, all the atoms are nonempty. We claim that $\mathcal{A}$ must be the base of $\mathcal{F}$. Notice that, for any X in $\mathcal{F}$, the set $\mathcal{A}_X = \{Y \in \mathcal{A} \mid \exists x \in X, \ x \in Y \subseteq X\}$ exists because there is an atom at every point of $X \subseteq \mathcal{X}$. We have thus $\cup \mathcal{A}_X = X$ and so $\mathcal{A}$ spans $\mathcal{X}$. Let now $\mathcal{H}$ be another subfamily of $\mathcal{F}$ spanning $\mathcal{F}$. Take any $Z \in \mathcal{A}$. Since $\mathcal{H}$ spans $\mathcal{F}$, there must be a subfamily $\mathcal{G}$ of $\mathcal{H}$ such that $\cup \mathcal{G} = Z$. By Theorem 4.5.9, we must have $Z \in \mathcal{G} \subseteq \mathcal{H}$; this yields $\mathcal{A} \subseteq \mathcal{H}$. Thus $\mathcal{A}$ is a minimal family spanning $\mathcal{F}$ and so is its base. In the case $\varnothing \in \mathcal{F}$, we have $\varnothing \in \mathcal{A}$ by definition with $\cup\{\varnothing\} = \varnothing$, and so $\mathcal{A}$ is also the base.

It remains to show that if the base $\mathcal{B}$ of $\mathcal{F}$ exists, then every X in $\mathcal{B}$ must be an atom. It is clear that we must have $\varnothing \in \mathcal{B}$ if $\varnothing \in \mathcal{F}$. Suppose that some nonempty X in $\mathcal{B}$ is not an atom. Then, there must exist for every $x \in X$ a set $Y_x \subset X$ containing x. We have thus $\cup_{x \in X} Y_x = X$. We obtain the family $\mathcal{G} = (\mathcal{B} \setminus \{X\}) \cup (\{Y_x \mid x \in X\}$ spanning $\mathcal{F}$, with $\mathcal{G} \not\subseteq \mathcal{B}$, which is impossible by Theorem 4.5.6 and our hypothesis that $\mathcal{B}$ is the base of $\mathcal{F}$. This proves that every $X \in \mathcal{B}$ is an atom. The wg-family $\mathcal{H}$ in Counterexample 4.5.12 is ∪-closed and is its own base, but has no atom at the point 1. □

4.5.11 Remarks. (a) Various algorithmic issues are raised by the material in this section, such as: how do we construct the base of a ∪-closed family $\mathcal{F}$? How do we efficiently deploy either of these the bases to recover $\mathcal{F}$? An algorithm for constructing the base of a finite ∪-closed family was proposed by Dowling (1993b). The problem of recovering all the states given the base is discussed in Chapter 10.

(b) In Theorem 4.5.10, we are assuming that the base and the set of atoms exists, without assuming that the family is finite. In fact, there are infinite ∪-closed families having a base (cf. Problem 4.12). As shown by the counterexample below, there are also infinite wg-families with no base.

4.5.12 A Counterexample. Consider the infinite family $\mathcal{F} = \mathcal{G} + \mathcal{H}$, with

$$\mathcal{G} = \{G_n \mid G_n = \{\ldots, \frac{1}{n+1}, \frac{1}{n}\}, n > 1\} \tag{4.15}$$

$$\mathcal{H} = \{H_n \mid H_n = G_n + \{1\}, G_n \in \mathcal{G}\}. \tag{4.16}$$

The family $\mathcal{F}$ is ∪-closed and well-graded and there is no atom at 1. It is easily verified that this ∪-closed family $\mathcal{F}$ has no base (Problem 4.16). The ∪-closed family $\mathcal{H}$ is its own base and has not atom at 1 either.

As a consequence of Lemma 4.5.2, we have:

4.5.13 Theorem. *Let $\mathcal{F}$ be a $\cup_F$-closed family with base $\mathcal{B}$. Then $\mathcal{F}$ is a wg-family if and only if, for any two distinct sets K and L in $\mathcal{B}$, there is a tight path in $\mathcal{F}$ from K to $L \cup K$. Thus, if $\mathcal{B}$ contains the empty set, then $\mathcal{F}$ is well-graded if and only if there is a tight path from $\varnothing$ to K for any K in $\mathcal{B}$.*

In other words, the first statement asserts that $\mathcal{F}$ is well-graded if and only if $d(K, L)$ is finite for any two sets K and L in $\mathcal{B}$ and moreover, if $d(K, K \cup L) = n > 1$, there exists for all $1 \leq i \leq n-1$, a subfamily $\mathcal{A}(i) \subseteq \mathcal{B}$ such that $K \subset \cup \mathcal{A}(i) \subset K \cup L$ and $d(K, \cup \mathcal{A}(i)) = i$.

Proof. As $\mathcal{F}$ is $\cup_\mathrm{F}$-closed with base $\mathcal{B}$, the necessity is clear. To establish the sufficiency, we point out that the family $\mathcal{B}^*$ defined by

$$M \in \mathcal{B}^* \iff \begin{cases} M = \cup_\mathrm{F} \mathcal{A} \text{ for some } \mathcal{A} \subseteq \mathcal{B} \text{ such that} \\ K \subseteq \cup_\mathrm{F} \mathcal{A} \subseteq K \cup L \text{ for some } K, L \in \mathcal{B} \end{cases} \quad (4.17)$$

includes $\mathcal{B}$ since $K = \cup\{K\}$ and $K \subseteq \cup\{K\} \subseteq K \cup L$ for any K and L in $\mathcal{B}$. Moreover, it follows from the definition of $\mathcal{B}^*$ that for any finite subfamily $\mathcal{L}$ of $\mathcal{B}^*$, there is a finite subfamily $\mathcal{K}$ of $\mathcal{B}$ such that $\cup \mathcal{L} = \cup \mathcal{K}$. Thus, $\mathcal{B}^*$ finitely spans $\mathcal{F}$. We claim that $\mathcal{B}^*$ is well-graded, which implies by Lemma 4.5.2 that $\mathcal{F}$ is well-graded. The main line of our argument is similar to that used in the proof of Lemma 4.5.2.

Take any two distinct $V, W \in \mathcal{B}^*$. By definition of $\mathcal{B}^*$, we have $V = \cup_\mathrm{F} \mathcal{V}$ and $W = \cup_\mathrm{F} \mathcal{W}$ for some subfamilies $\mathcal{V}$ and $\mathcal{W}$ of $\mathcal{B}$. It follows easily from Condition (i) of the theorem that $d(V, V \cup W)$ is finite, say $d(V, V \cup W) = n$. We must show that there exists in $\mathcal{B}^*$ a tight path $V_0 = V, V_1, \ldots, V_n = V \cup W$ from V to $V \cup W$. Without loss of generality (exchanging the roles of V and W if needed), we can assume that there is some $H \in \mathcal{W}$ such that $H \setminus V \neq \varnothing$. Choose $G \in \mathcal{V}$ arbitrarily. Then $G \subset H \cup G \subseteq V \cup W$, with H and G in $\mathcal{B}$. By hypothesis, there is a tight path $G_0 = G, G_1, \ldots, G_m = G \cup H$ from G to $G \cup H$ in $\mathcal{F}$, with $G \subset G_i \subset G \cup H$ and $d(G, G_i) = i$ for $1 \leq i \leq m$. Let k be the first index such that $G_k \setminus V \neq \varnothing$. (Such an index must exist because $H \setminus V \neq \varnothing$.) We necessarily have $|G_k \setminus V| = 1$. Defining $V_1 = (\cup \mathcal{V}) \cup G_k$, we obtain $V_0 = V \subset V_1 \subseteq V \cup W$ with $|V_1 \setminus V_0| = 1$. An induction completes the proof. The second statement follows easily (see Problem 4.13). □

4.5.14 Remarks. (a) The set $\mathcal{B}^*$ constructed in the proof of Theorem 4.5.13 is not necessarily a minimal wg-family spanning $\mathcal{F}$. Indeed, the definition of $\mathcal{B}^*$ by (4.17) includes all the unions $\cup \mathcal{A}(i)$, while only some of them may be needed. An example of such a situation was provided by the wg-family of Example 4.5.3. In this case, each of (4.12) and (4.13) is a minimal wg-family including the base and spanning the wg-family $\mathcal{F}$ defined by (4.11). The set $\mathcal{B}^*$ in this case would be the union of the two families in (4.12) and (4.13), which is in fact equal to $\mathcal{F}$.

(b) The results concerning the base of a ∪-closed wg-family have a straightforward application to closed media. In view of the relationship between a closed medium and the ∪-closed wg-family of its positive contents described by Theorems 3.3.3, 4.1.9 and 4.2.2, any closed medium can be faithfully and economically summarized by the base $\mathcal{B}$ of its family of positive contents, whenever such a base exists (see Theorem 4.5.17). Note that when this base is graded in the sense of Definition 3.2.1 with a grading collection $\mathcal{G}_\mathcal{B}$, then Lemma 3.2.3(i) tells us that $(\mathcal{B}, \mathcal{G}_\mathcal{B})$ is a token system which can be constructed. The usefulness of such a translation is actually doubtful, as suggested by Example 4.5.3 and the corresponding digraphs of Figure 4.4. The closed medium is that corresponding to the wg-family $\mathcal{F}$ of Example 4.5.3 and is represented by the positive digraph A. The positive digraph B is that of the token system representing the base of the family of positive contents of $\mathcal{F}$; C and D represent the closed media induced by the two minimal wg-subfamilies of $\mathcal{F}$ spanning $\mathcal{F}$.

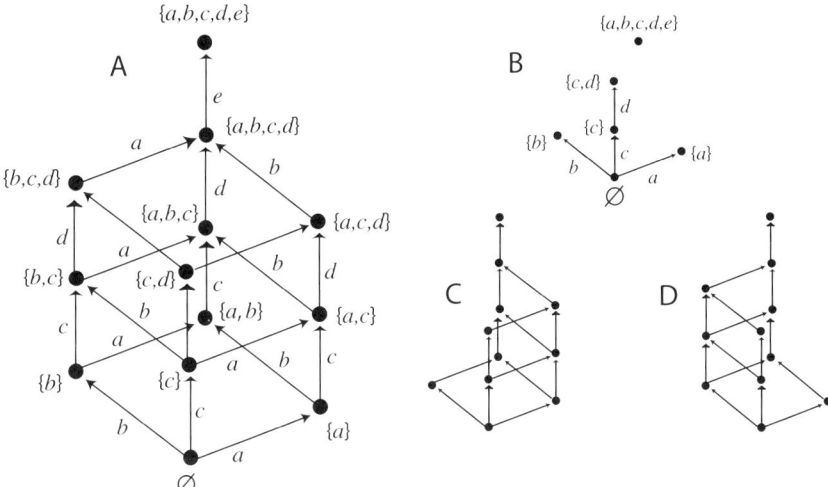

Figure 4.4. The positive digraph A is that of the representing closed medium of the ∪-closed family $\mathcal{F}$ defined by Eq. (4.11) in Example 4.5.3. The positive digraph B is that of the token system representing the base of the family of positive contents of $\mathcal{F}$ (notice that the vertex corresponding to the state with positive content $\{a, b, c, d, e\}$ has no link to the other vertices); C and D are those of the two open media corresponding to the two minimal wg-subfamilies (4.12) and (4.13) spanning the family $\mathcal{F}$ defined by (4.11).

(c) A different approach was taken by Koppen (1989) to capture the well-gradedness of a ∪-closed family $\mathcal{F}$, based on a property of the 'surmise system' of the family $\mathcal{F}$. The concept of a surmise system was introduced by Doignon and Falmagne as a pair $(\mathcal{X}, \sigma)$, where σ is a function mapping a ground set $\mathcal{X}$ into $\mathfrak{P}(\mathfrak{P}(\mathcal{X}))$ and satisfying certain conditions (Doignon and Falmagne, 1985, 1999). In particular, for any $x \in \mathcal{X}$, any $C \in \sigma(x)$ must contain x and must be minimal for set inclusion among the sets in $\sigma(x)$. In the context of learning spaces applied to education, the interpretation of a family $\sigma(x)$ is that from observing that a student has mastered problem x, one can infer that the student has also mastered all the problems in at least one subset C

in $\sigma(x)$. In other words, the sets in $\sigma(x)$ are the possible sets of prerequisites for the mastery of x.

Related mathematical results regarding can also be found in the literature of convex geometries (Edelman and Jamison, 1985; Van de Vel, 1993).

The following theorem paves the way for the application to closed media of the results concerning the existence of the base of a $\cup$-closed wg-family. Indeed, by Theorem 4.2.2, we know that the family of positive contents of a closed rooted medium is a learning space.

4.5.15 Theorem. *Any learning space has a base.*

Notice that no assumption of finiteness is made here.

Proof. Let $\mathcal{F}$ be a learning space, with $\cup\mathcal{F} = \mathcal{X}$. By Theorem 4.2.1, $\mathcal{F}$ contains the empty set and is well-graded. Take any $x \in \mathcal{X}$, and any set $S \in \mathcal{F}$ containing x. By the wellgradedness of $\mathcal{F}$, there exists at least one chain of sets $\varnothing \subset S_1, \ldots, S_n = S$ with $S_i \in \mathcal{F}$ and $|S_{i+1} \setminus S_i| = 1$ for $1 \leq i \leq n-1$. Note that all such chains have the same cardinality n. Let $\mathfrak{K} = \{\mathcal{K}_k \mid k \in \mathcal{J}\}$ be the collection of all such well-graded chains from $\varnothing$ to S, and let $\mathfrak{S} = \{S_{i,k} \mid 1 \leq i \leq n, k \in \mathcal{J}\}$ be the family of all the sets belonging to such chains; thus

$$\varnothing \subset S_{1,k} \subset \ldots \subset S_{i,k} \subset \ldots \subset S_{n,k} = S \qquad (k \in \mathcal{J}).$$

The collection $\mathfrak{K}$ is finite, since $|\mathfrak{K}| \leq n!$. Let j be the smallest index $1 \leq j \leq n$ such that $x \in S_{j,k}$ for some set $S_{j,k} \in \mathfrak{S}$. (There may be more than one such set in $\mathfrak{S}$.) The set $S_{j,k}$ is thus an atom at x. So, there is an atom at every point of $\mathcal{X}$. By Theorem 4.5.10, the set of all those atoms is the base of $\mathcal{F}$. □

4.5.16 Definition. By extension, we call the *base* of a closed medium the base of its family of positive contents.

4.5.17 Theorem. (i) *Any finite closed medium has a base.*
(ii) *Any rooted closed medium has a base.*
(iii) *Some infinite closed media have no base.*

Proof. We recall that, by Theorem 3.3.1, the family $\widehat{\mathcal{S}}^+$ of positive contents of an oriented medium $\mathcal{M} = (\mathcal{S}, \mathcal{T})$ is well-graded and satisfies both $\cup\widehat{\mathcal{S}}^+ = \mathcal{T}^+$ and $\cap\widehat{\mathcal{S}}^+ = \varnothing$. By Theorem 4.1.9 and Corollary 4.1.10, the medium $\mathcal{M}$ is closed if and only if $\widehat{\mathcal{S}}^+$ is closed under finite unions.

(i) This is clear by Theorem 4.5.7 and the fact that $\widehat{\mathcal{S}}^+$ is finite.

(ii) As Theorem 4.2.2 implies that $\widehat{\mathcal{S}}^+$ is a learning space, this case follows from Theorem 4.5.15.

(iii) Take the medium $\mathcal{M}$ induced by the family of positive contents defined by the equivalence

$$\tau_q \in \widehat{S}^+ \iff q \in (\mathcal{G} + \mathcal{H}), \qquad (4.18)$$

with $\mathcal{G}$ and $\mathcal{H}$ as in equations (4.15) and (4.16) of Counterexample 4.5.12. Then $\widehat{S}^+$ is ∪-closed and so $\mathcal{M}$ is closed. But $\widehat{S}^+$ has no atom at τ_1 and thus has no base by Theorem 4.5.10. The positive digraph of this medium is displayed in Figure 4.5. □

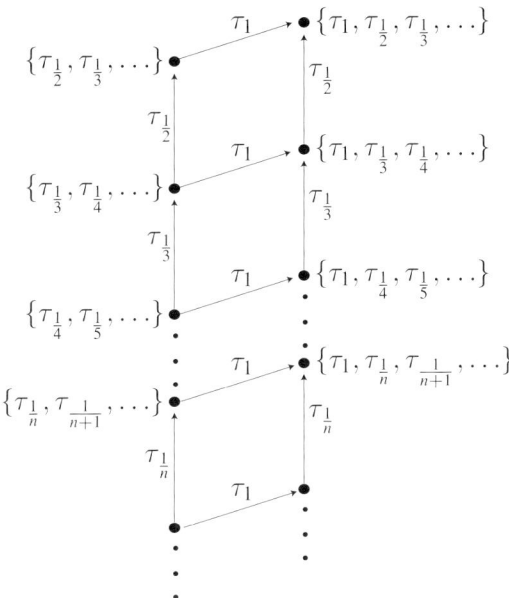

Figure 4.5. Positive digraph of the closed medium induced by the family of positive contents $\widehat{S}^+$ defined by Eq. (4.18); cf. Eqs. (4.15) and (4.16) of Counterexample 4.5.12. The positive content of the corresponding state is indicated near each vertex.

We end this section with a reformulation, in the context of a medium, of the well–known Union-Closed Sets Conjecture proposed by Frankl in 1979[7] but stated in print much later, for example as an open problem by D. Duffus (in Rival, 1985, p. 525), and in any event by Frankl himself (see Frankl, 1995, p. 1296).

4.5.18 Definition. For any medium $(S, \mathcal{T})$ and any $\tau \in \mathcal{T}$, we write

$$S_\tau = \{ S \in S \mid \tau \in \widehat{S} \}. \qquad (4.19)$$

4.5.19 Conjecture. Let $(S, \mathcal{T})$ be a closed medium with orientation $\{\mathcal{T}^+, \mathcal{T}^-\}$. Then there exists a token $\tau \in \mathcal{T}^+$ and a 1-1 function $\mathfrak{f}: S_{\widetilde{\tau}} \to S_\tau$.

---
[7] Frankl's original formulation involved sets closed under intersection, not union.

While proving Conjecture 4.5.19 would only establish a special case of the Union-Closed Sets Conjecture (which does not assume wellgradedness), it may achieve a useful step. There are numerous partial results of the Union-Closed Sets Conjecture. Some of them could be found, for example, in Sarvate and Renaud (1989, 1990); Lo Faro (1994a,b); Vaughan (2002), but the question is still open.

4.6 Projection of a Closed Medium

The concept of a projection of a medium $(\mathcal{S}, \mathcal{T})$ was introduced in Chapter 2 as a token system constructed from $(\mathcal{S}, \mathcal{T})$ by gathering the states in $\mathcal{S}$ into equivalence classes, and arranging for each of these classes to be a state of the new token system. The corresponding equivalence relation $\sim$ is obtained by choosing a symmetric subset $\mathcal{U} \subseteq \mathcal{T}$ of tokens (that is, $\tau \in \mathcal{U}$ if and only if $\tilde{\tau} \in \mathcal{U}$), and declaring that $S \sim V$ for two states of $\mathcal{S}$ if V is produced from S by a concise messages whose tokens are all in $\mathcal{T} \setminus \mathcal{U}$. The tokens in $\mathcal{U}$ are then redefined so as to act on the equivalence classes (see Definitions 2.11.2 and 2.11.4). We have seen in Theorem 2.11.6 that not only is the projection so defined a genuine medium—which we denoted by $(\mathcal{S}_{|\mathcal{U}}, \mathcal{T}_{|\mathcal{U}})$—but each of the equivalence classes other than singletons forms the set of states of a submedium of $(\mathcal{S}, \mathcal{T})$. Moreover, if $(\mathcal{S}, \mathcal{T})$ is oriented, then so is $(\mathcal{S}_{|\mathcal{U}}, \mathcal{T}_{|\mathcal{U}})$, and the orientation of $(\mathcal{S}_{|\mathcal{U}}, \mathcal{T}_{|\mathcal{U}})$ is in fact 'inherited' from $(\mathcal{S}, \mathcal{T})$. We show here that a similar hereditary property holds in the case where $(\mathcal{S}, \mathcal{T})$ is closed.

For convenience, we recall the definitions of the projection concepts given in 2.11.4. We start with a medium $(\mathcal{S}, \mathcal{T})$ and choose a nonempty symmetric subset $\mathcal{U}$ of $\mathcal{T}$. We then define the equivalence relation $\sim$ on $\mathcal{S}$ induced by $\mathcal{U}$ via the formula

$$S \sim T \iff \begin{cases} \text{either } S = T \\ \text{or } S\boldsymbol{m} = T \text{ and } \mathcal{C}(\boldsymbol{m}) \subseteq (\mathcal{T} \setminus \mathcal{U}) \\ \text{for some concise message } \boldsymbol{m}. \end{cases} \quad (4.20)$$

$$[S] = \{T \in \mathcal{S} \,|\, S \sim T\}, \qquad (S \in \mathcal{S});$$
$$\mathcal{S}_{|\mathcal{U}} = \{[S] \,|\, S \in \mathcal{S}\},$$
$$\tau_{|\mathcal{U}} : \mathcal{S}_{|\mathcal{U}} \to \mathcal{S}_{|\mathcal{U}} : [S] \mapsto [S]\tau_{|\mathcal{U}}, \qquad (\tau \in \mathcal{U});$$
with
$$[S]\tau_{|\mathcal{U}} = \begin{cases} [T] & \text{if } S \not\sim T \text{ and } \exists (Q, W) \in [S] \times [T], Q\tau = W, \\ [S] & \text{otherwise,} \end{cases} \quad (4.21)$$
$$\mathcal{T}_{|\mathcal{U}} = \{\tau_{|\mathcal{U}} \,|\, \tau \in \mathcal{U}\}.$$

The pair $(\mathcal{S}_{|\mathcal{U}}, \mathcal{T}_{|\mathcal{U}})$ is called the *projection* of $(\mathcal{S}, \mathcal{T})$ under $\mathcal{U}$.

4.6 Projection of a Closed Medium

4.6.1 Theorem. *Suppose that* $\mathcal{M} = (\mathcal{S}, \mathcal{T})$, $\mathcal{U}$ *and* $\mathcal{M}_{|\mathcal{U}} = (\mathcal{S}_{|\mathcal{U}}, \mathcal{T}_{|\mathcal{U}})$ *are as above, and have the matching orientations* $\{\mathcal{T}^+, \mathcal{T}^-\}$, $\{\mathcal{T}^+_{|\mathcal{U}}, \mathcal{T}^-_{|\mathcal{U}}\}$, *with*

$$\mathcal{T}^+_{|\mathcal{U}} = \{\tau_{|\mathcal{U}} \in \mathcal{T}_{|\mathcal{U}} \mid \tau \in \mathcal{T}^+\}, \qquad \mathcal{T}^-_{|\mathcal{U}} = \{\tau_{|\mathcal{U}} \in \mathcal{T}_{|\mathcal{U}} \mid \tau \in \mathcal{T}^-\}. \qquad (4.22)$$

If $\mathcal{M}$ *is closed, the following two statements are true.*

(i) $\mathcal{M}_{|\mathcal{U}}$ *is also closed and we have*

$$(S\tau = T, S\tau' = T', T \neq T') \implies ([T] \neq [T'], [T]\tau'_{|\mathcal{U}} = [T']\tau_{|\mathcal{U}})$$
$$(\tau, \tau' \in \mathcal{T}^+ \cap \mathcal{U},\ \tau_{|\mathcal{U}}, \tau'_{|\mathcal{U}} \in \mathcal{T}^+_{|\mathcal{U}}).$$

(ii) *If* $\mathcal{M}$ *is rooted, then* $\mathcal{M}_{|\mathcal{U}}$ *is also rooted. Moreover, if* R *is the root of* $\mathcal{M}$, *then* $[R]$ *is the root of* $\mathcal{M}_{|\mathcal{U}}$.

Proof. Note that by Theorem 2.11.6, $(\mathcal{S}_{|\mathcal{U}}, TU)$ is a medium.

(i) If $\tau_{|\mathcal{U}}, \tau'_{|\mathcal{U}} \in \mathcal{T}^+_{|\mathcal{U}}$, with $[S]\tau_{|\mathcal{U}} \neq [S]\tau'_{|\mathcal{U}}$, then $S\tau \neq S\tau'$ and $S\tau\tau' = S\tau'\tau$ by definition of $\mathcal{T}^+_{|\mathcal{U}}$ and $\mathcal{T}^-_{|\mathcal{U}}$ in (4.22) and because $\mathcal{M}$ is closed. This gives $[S\tau]\tau'_{|\mathcal{U}} = [S]\tau_{|\mathcal{U}}\tau'_{|\mathcal{U}} = [S\tau']\tau_{|\mathcal{U}} = [S]\tau'_{|\mathcal{U}}\tau_{|\mathcal{U}}$.

(ii) Since the root R of $\mathcal{M}$ is in $[R]$, the messages from $[R]$ producing all the other states of $(\mathcal{S}_{|\mathcal{U}}, \mathcal{T}_{|\mathcal{U}})$ are positive by (4.22). □

4.6.2 Example. The situation described in statement (ii) of the above theorem is illustrated by Figure 4.6, which displays a rooted medium $\mathcal{M}$ represented by its family of positive contents, which is a learning space on the set $\{A, B, C, D, E, F\}$ (cf. Theorem 4.2.2). The set of tokens of $\mathcal{M}$ is thus $\mathcal{T} = \{\gamma_A, \tilde{\gamma}_A, \ldots, \gamma_F, \tilde{\gamma}_F\}$. Defining $\mathcal{U} = \{\gamma_B, \tilde{\gamma}_B, \gamma_D, \tilde{\gamma}_D, \gamma_E, \tilde{\gamma}_E, \gamma_F, \tilde{\gamma}_F\}$ gives the projection $\mathcal{M}_{|\mathcal{U}}$ which is also represented by its own learning space.

4.6.3 Remark. Note that if a medium $(\mathcal{S}, \mathcal{T})$ is closed for some orientation $\{\mathcal{T}^+, \mathcal{T}^-\}$, then any submedium $(\mathcal{Q}, \mathcal{T}_{\mathcal{Q}})$, with $\mathcal{Q} \subset \mathcal{S}$, is closed for the orientation $\{\mathcal{T}^+_{\mathcal{Q}}, \mathcal{T}^-_{\mathcal{Q}}\}$ induced by $\{\mathcal{T}^+, \mathcal{T}^-\}$, that is, $\tau_{\mathcal{Q}} \in \mathcal{T}^+_{\mathcal{Q}}$ if and only if $\tau \in \mathcal{T}^+$ (see Problem 4.19). This fact obviously applies, in particular to all the submedia $([S], \mathcal{T}_{[S]})$ of Theorem 4.6.1 with $S \in \mathcal{S}$.

Example 4.6.2 was motivated by the concept of a learning space. Our next and last illustration, in Figure 4.7 and Example 4.6.4, is inspired by the technique of *triangulation* in computational geometry.

4.6.4 Example. A *triangulation* of a planar point set is a graph, drawn with non-crossing straight line segments as edges, having the points as its vertices, such that all interior faces of the drawing are triangles. Two triangles are connected by a *flip* if and only if they differ by the removal of one edge and the insertion of a single replacement edge. Eppstein (2007a) proved that the system of triangulations and flips of a point set forms the states and tokens

96   4 Closed Media and ∪-Closed Families

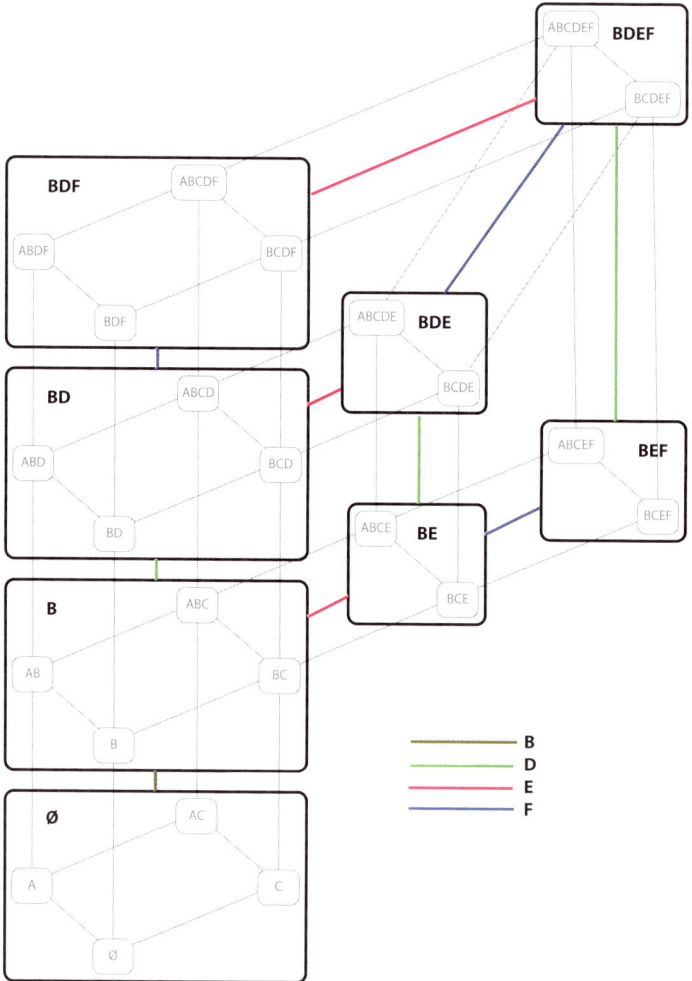

Figure 4.6. A rooted medium $\mathcal{M} = (\mathcal{S}, \mathcal{T})$ represented by its family of positive contents forming a learning space on the set $\{A, B, C, D, E, F\}$, together with one of its projection $\mathcal{M}_{|\mathcal{U}} = (\mathcal{S}_{|\mathcal{U}}, \mathcal{T}_{|\mathcal{U}})$ represented the same way, as a learning space on the set $\{B, D, E, F\}$.

of a medium, if and only if the point set does not contain any five points forming the vertices of an *empty pentagon*; that is, a pentagon that does not contain any other points from the set. In this medium, there are two tokens for each pair of edges that can be flipped; each such token performs the replacement of one edge in the pair for the other, in triangulations for which such a replacement is possible. The medium is closed when oriented by flips towards the *Delaunay triangulation*.

4.6 Projection of a Closed Medium   97

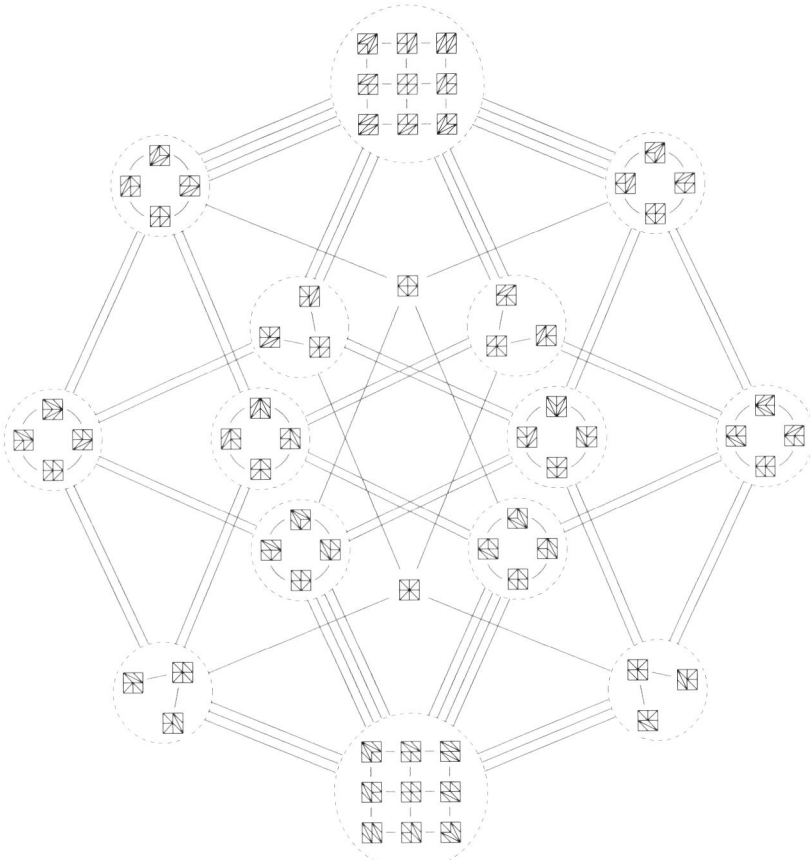

Figure 4.7. The medium of triangulations on a 3 × 3 grid of integer points in the plane. The states of the medium are grouped according to the set of length-$\sqrt{2}$ edges present in each triangulation, forming a projection onto a four-dimensional hypercube. The number of lines drawn between each pair of groups of states indicates the number of transitions of the original medium connecting the pair of groups. From Eppstein (2007a).

In particular, the integer lattice contains no empty pentagon, so we get a medium for any $m \times n$ *grid* of integer points; Figure 4.7 depicts the states and transitions of the resulting medium for a 3 × 3 grid of nine points. This figure depicts a projection of the medium in which the set $\mathcal{U}$ of tokens defining the projection are the flips that preserve the length of the flipped edge. Such a flip must replace one length-$\sqrt{2}$ diagonal of a unit square by the other diagonal of the same square. If $\mathcal{U}$ is the set of tokens corresponding to length-preserving flips, then the grouping in the figure depicts the projection of the medium for $\mathcal{U}$. In any $m \times n$ grid, and any triangulation of such a grid in which all edges have unit length or length $\sqrt{2}$, each length-preserving flip can be performed independently of each other such flip; thus, the number of states in the projection for this set of tokens is $2^{(m-1)(n-1)}$. In the depicted case of a 3×3 grid, the resulting projection is a 4-dimensional hypercube. The original medium has 64 states, shown in the figure as the small triangulations, while its projection has 16 states, shown as the groups of triangulations enclosed by dashed circles.

Problems

4.1 Prove Theorem 4.1.3.

4.2 Verify that the medium defined in Example 4.3.5 is complete. Let $\mathcal{Z}$ be a nonempty set. Give an example of a well-graded subfamily $\mathcal{F} \subset \mathfrak{P}_F(\mathcal{Z})$ (where $\mathfrak{P}_F(\mathcal{Z})$ is the set of all finite subsets of $\mathcal{Z}$) inducing a complete medium in the sense of (3.8) and (3.9) and Definition 3.3.2. What can we say about such a family?

4.3 A medium is said to be *taut* if, for any two tokens τ and τ' and any orderly circuit $\boldsymbol{m}\tau\tau'\boldsymbol{n}$, the segment $\tau\tau'$ must be part of an orderly circuit of length 4. Prove or disproved that any oriented taut medium having a root must be closed.

4.4 Disprove the following proposition by a counterexample: *If $\mathcal{N}$ is a closed submedium of a finite closed medium $\mathcal{M}$. If $\mathcal{N}$ and $\mathcal{M}$ have the same number of tokens, then $\mathcal{N} = \mathcal{M}$.*

4.5 Prove by a counterexample that the statement in the first sentence of Theorem 4.1.11 does not hold if the word 'finite' is removed.

4.6 Which conditions (necessary and/or sufficient) on the digraph of an oriented medium ensure that the family of all positive contents is closed under intersection?

4.7 Prove Theorem 4.1.18.

4.8 Give an example of a $\cup_F$-closed family having an infinite base.

4.9 Prove Lemma 4.3.10 and Theorem 4.3.11.

Problems for Chapter 4

4.10 Definition. A *quasimedium* is a token system $(S, \mathcal{T})$ satisfying Axiom [Mb], plus the following weaker version of Axiom [Ma]:

[Ma†] For any two distinct states S and V in S, there is a message transforming S into V.

Which part, if any, of Lemma 2.3.1 remains true? Prove your results and give a counterexample for any part of the lemma that no longer holds.

4.11 Prove Theorem 4.5.9.

4.12 Can an infinite $\cup$-closed wg-family have a base? Give an example or prove that this cannot happen.

4.13 Prove the last statement in Theorem 4.5.13.

4.14 Sketch an algorithm for verifying that the base of a finite $\cup$-closed family satisfies the condition of Theorem 4.5.13.

4.15 (Continuation.) Assuming that the base of a finite $\cup$-closed family does not satisfy the condition of Theorem 4.5.13, sketch an algorithm adding missing states in some optimal way (for example minimizing the number of new states), thus ensuring that the resulting $\cup$-closed family is well-graded.

4.16 (a) Verify the $\cup$-closed family of Counterexample 4.5.12 has no base and has no atom at the point 1. (b) Verify that the wg-family $\mathcal{H}$ of this example is $\cup$-closed is its own base, with again no atom at the point 1.

4.17 For a closed medium $(S, \mathcal{T})$, denote by $S_{[\tau_1,\ldots,\tau_n]}$ the subset of all states S such that τ_i is effective for S, for $1 \leq i \leq n$. Consider $S_{[\tau]}$ and $S_{[\tau']}$ with $\tau \neq \tau'$; described $S_{[\tau]} \cap S_{[\tau']}$ in each of the possible cases and try to derive some conclusions.

4.18 We can now reformulate Theorem 3.4.12 in the context of complete media. Prove the following: *The representing medium $(\mathcal{F}, \mathcal{W}_\mathcal{F})$ of any wg-family $\mathcal{F}$ is a submedium of a complete medium and is isomorphic to a submedium of the complete medium $(\mathfrak{P}_F(\mathcal{X}), \mathcal{W}_{\mathfrak{P}_F(\mathcal{X})})$ of all finite subsets of $\mathcal{X} = \cup \mathcal{F}$.*

4.19 Verify that if a medium $(S, \mathcal{T})$ is closed for some orientation $\{\mathcal{T}^+, \mathcal{T}^-\}$, then any submedium $(\mathcal{Q}, \mathcal{T}_\mathcal{Q})$, with $\mathcal{Q} \subset S$, is closed for the orientation $\{\mathcal{T}_\mathcal{Q}^+, \mathcal{T}_\mathcal{Q}^+\}$ induced by $\{\mathcal{T}^+, \mathcal{T}^-\}$, that is, $\tau_\mathcal{Q} \in \mathcal{T}_\mathcal{Q}^+$ if and only if $\tau \in \mathcal{T}^+$.

5
Well-Graded Families of Relations

We have seen in Chapter 3 that wg-families of sets offer a systematic way of representing oriented media in which the sets of the family stand for the states of the medium, and the addition or removal of a given point from any set stand for a token or its reverse. The representation results are formulated by Theorems 3.3.3 and 3.3.4. The purpose of this chapter is to establish that several well-known families of relations, such as the partial orders, the biorders, the interval orders and the semiorders are well-graded, and can thus be represented by media. We also discuss another family, the almost connected orders or ac-orders, which are well-graded if and only if the size of their ground set does not exceed four, offering thus a revealing counterexample.

The material in this chapter follows closely Doignon and Falmagne (1997) and, in the last section, Doble et al. (2001). For all these examples, we assume that the sets involved are finite.

5.0.1 Remark. The finiteness assumption simplifies our discussion, but is not critical. Following the pattern used in Chapter 3 in our treatment of a possibly infinite family of all the linear orders on a set that are at finite distance from one of them, we could consider the family $\mathcal{F}$ of all the semiorders (for example) on some arbitrary set $\mathcal{X}$ that are at finite distance from one of them, that is, lying in one class of the cluster partition $\mathfrak{P}(\mathcal{X} \times \mathcal{X})_{\bowtie}$ (cf. Definition 3.4.1). With some adaptation of the techniques of this chapter, we could prove that the family $\mathcal{F}$ is well-graded. The other families of relations discussed in this chapter could be dealt with similarly. However interesting such a development would be, it would go beyond what is intended here, which is to provide a collection of examples of well-known relations whose families define media. In any event, a few of the problems at the end of the chapter are devoted to exploring this avenue.

5 Well-Graded Families of Relations

5.1 Preparatory Material

In some of our proofs, we use relative product calculus and notation whenever they enable substantially shorter arguments. We recall some concepts encountered earlier but specified here in compact set-theoretical notation. We also introduces a couple of new properties.

5.1.1 Definition. Let $\mathcal{X}$ be of arbitrary cardinality, and let R be a relation on that set. We define

$$I_\mathcal{X} = R^0 = \{xx \,|\, x \in \mathcal{X}\} \quad \text{as the } \textit{identity} \text{ on } \mathcal{X}$$
$$R^{-1} = \{xy \,|\, yx \in R\} \quad \text{as the } \textit{converse} \text{ of } R$$
$$\bar{R} = \{xy \,|\, xy \notin R, \, x, y \in \mathcal{X}\} \quad \text{as the } \textit{complement} \text{ of } R$$
$$\text{(with respect to } \mathcal{X} \times \mathcal{X}\text{).}$$

When no ambiguity can arise regarding the ground set $\mathcal{X}$, we abbreviate $I_\mathcal{X}$ as I. As recalled above from 1.8.2, we have $R^0 = I$.

A relation R on $\mathcal{X}$ is

reflexive	if	$I \subseteq R$	(5.1)
irreflexive	if	$I \cap R = \varnothing$	(5.2)
symmetric	if	$R \subseteq R^{-1}$	(5.3)
asymmetric	if	$R \subseteq \bar{R}^{-1}$	(5.4)
antisymmetric	if	$R \cap R^{-1} \subseteq I$	(5.5)
transitive	if	$RR \subseteq R$	(5.6)
strongly connected	if	$\bar{R}^{-1} \subseteq R$	(5.7)
n-connected	if	$R^n \bar{R}^{-1} \subseteq R$.	(5.8)

Thus, R is 0-connected if and only if it is strongly connected ($\bar{R}^{-1} \subseteq R$), and R is 2-connected if and only if

$$xRy \wedge yRz \wedge \neg(wRz) \Rightarrow xRw \qquad (x, y, z, w \in \mathcal{X}).$$

Our last condition also involve a second set $\mathcal{Y}$ which is possibly (but not necessarily) distinct from $\mathcal{X}$. A relation $R \subseteq \mathcal{X} \times \mathcal{Y}$ is

a biorder	if	$R\bar{R}^{-1}R \subseteq R$.	(5.9)

We say then that R is a biorder from $\mathcal{X}$ to $\mathcal{Y}$. The biorder defining condition applies obviously also to the case in which $\mathcal{X} = \mathcal{Y}$.

Five types of relations are studied here from the standpoint of the wellgradedness of their families: the partial orders, the biorders, the interval orders, the semiorders, and the almost connected orders, or ac-orders. We define or redefine four of these concepts.

A relation R on a set $\mathcal{X}$ is

a *partial order*	if	it is transitive and asymmetric	(5.10)
a *strict weak order*	if	it is 1-connected and asymmetric	(5.11)
an *interval order*	if	it is irreflexive biorder	(5.12)
a *semiorder*	if	it is a 2-connected interval order	(5.13)
an *ac-order*	if	it is 2-connected and asymmetric.	(5.14)

5.1.2 Remark. The almost connected orders or ac-orders are especially interesting because they are well-graded only if $|\mathcal{X}| \leq 4$, but satisfy in general the weaker properties of being 'upgradable' and 'downgradable.' We discuss this concept in the last section.

We introduce an important tool for the study of wg-families.

5.2 Wellgradedness and the Fringes

5.2.1 Definition. Let $\mathcal{F}$ be a graded family of subsets of a set $\mathcal{X} = \cup \mathcal{F}$; that is, $|\mathcal{F}| \geq 2$ and for any $x \in \mathcal{X} \setminus \cap \mathcal{F}$, there is some Q and S in $\mathcal{F}$ satisfying $Q \triangle S = \{x\}$; cf. Definition 3.2.1. Writing $\overline{S} = \mathcal{X} \setminus S$ for any $S \in \mathcal{F}$, define the three sets

$$S^{\mathfrak{I}} = \{x \in S \mid S \setminus \{x\} \in \mathcal{F}\},$$
$$S^{\mathfrak{O}} = \{x \in \overline{S} \mid S + \{x\} \in \mathcal{F}\},$$
$$S^{\mathfrak{F}} = S^{\mathfrak{O}} \cup S^{\mathfrak{I}}.$$

The sets $S^{\mathfrak{I}}$, $S^{\mathfrak{O}}$, and $S^{\mathfrak{F}}$ are called the *inner fringe*, the *outer fringe*, and the *fringe* of the set $S \in \mathcal{F}$, respectively. Note that the fringes of a set are always defined with respect to a particular family of sets.

The following result is due to Doignon and Falmagne (1997). Except for minor details[1], the proof below is essentially that given in their monograph Doignon and Falmagne (1999, Theorem 2.8, p. 48), and is included here for completeness[2].

5.2.2 Theorem. *The five conditions listed below are equivalent for any graded family of sets $\mathcal{F}$. Conditions (ii) to (v) apply to any two sets S and T in $\mathcal{F}$.*

---

[1] Our definition of wellgradedness slightly differs from that used in Doignon and Falmagne (1999) but equivalent to it, as established in fact by their Theorem 2.8
[2] An adaptation of the concept of fringe to the states of a medium was already encountered in Chapter 2; see Definition 2.5.1 and Theorem 2.5.2.

(i) The family $\mathcal{F}$ is well-graded.
(ii) There exists a path $(S_j)_{0 \leq j \leq k}$ from $S = S_0$ to $T = S_k$ such that, for $0 \leq j \leq k-1$, we have
$$S_j \cap T \subseteq S_{j+1} \subseteq S_j \cup T. \tag{5.15}$$

(iii) $(S \triangle T) \cap S^{\mathfrak{F}} \neq \varnothing$.
(iv) If $S^{\mathfrak{I}} \subseteq T$ and $S^{\mathfrak{D}} \subseteq \overline{T}$, then $S = T$.
(v) If the four inclusions
$$S^{\mathfrak{I}} \subseteq T, \quad S^{\mathfrak{D}} \subseteq \overline{T}, \quad T^{\mathfrak{I}} \subseteq S, \text{ and } T^{\mathfrak{D}} \subseteq \overline{S}$$

hold, then $S = T$.

Proof. We prove (i) $\Rightarrow$ (ii) $\Rightarrow$ (iii) $\Rightarrow$ (iv) $\Rightarrow$ (v) $\Rightarrow$ (i).

(i) $\Rightarrow$ (ii). By definition of wellgradedness (3.1.2), there exists a tight path $S_0 = S, S_1, \ldots, S_k = T$ from S to T. Problem 5.2 requires the reader to prove that such a tight path verifies (5.15) for $0 \leq j \leq k-1$.

(ii) $\Rightarrow$ (iii). Take any two states S and T, and let $(S_j)_{0 \leq j \leq k}$ be the path described in Condition (ii). Then S and S_1 differ by exactly one point x, and we have moreover $S \cap T \subseteq S_1 \subseteq S \cup T$. Either the x belongs to S, or it belongs to T, but not both. Hence x belongs to $(S \triangle T) \cap T^{\mathfrak{F}}$.

(iii) $\Rightarrow$ (iv). Proceeding by contradiction, we take two sets S and T satisfying $S^{\mathfrak{I}} \subseteq T$ and $S^{\mathfrak{D}} \subseteq \overline{T}$, with some $x \in (S \triangle T) \cap S^{\mathfrak{F}}$. If $x \in S$, then $x \in S^{\mathfrak{I}} \subseteq T$, contradicting $x \in S \triangle T$. Hence $x \notin S$, but then $x \in T \cap S^{\mathfrak{D}}$, yielding $x \in S^{\mathfrak{D}} \subseteq \overline{T}$, a contradiction.

(iv) $\Rightarrow$ (v). Trivial.

(v) $\Rightarrow$ (i). Let S and T be two distinct sets in $\mathcal{F}$ with $d(S,T) = k > 0$. We establish (i) by constructing a tight path $(S_j)_{0 \leq j \leq k}$ from S and T. Since $S \neq T$, Condition (v) implies that there must be some point
$$x \in (S^{\mathfrak{I}} \setminus T) \cup (S^{\mathfrak{D}} \cap T) \cup (T^{\mathfrak{I}} \setminus S) \cup (T^{\mathfrak{D}} \cap S).$$

If $x \in S^{\mathfrak{I}} \setminus T$, we set $S_1 = S \setminus \{x\}$. In the other three cases, we set $S_1 = S \cup \{x\}$ or $S_{k-1} = T \setminus \{x\}$ or $S_{k-1} = T \cup \{x\}$. We obtain either $d(S_1, T) = k - 1$ (in the first two cases), or $d(S, S_{k-1}) = k - 1$ (in the last two cases). The result follows by induction. □

5.2.3 Remark. A consequence of Theorem 5.2.2 is that any set in a well-graded family $\mathcal{F}$ is defined by its fringes in the sense of the condition

[F] $(S^{\mathfrak{I}} = T^{\mathfrak{I}} \text{ and } S^{\mathfrak{D}} = T^{\mathfrak{D}}) \iff R = S$ $(S, T \in \mathcal{F})$.

However, Condition [F] does not imply wellgradedness (Problem 5.6).

On the other hand, if a family $\mathcal{F}$ is closed under finite intersection and also satisfies $\varnothing \in \mathcal{F}$, then it is well-graded if and only if any set in $\mathcal{F}$ is defined by its inner fringe in the sense of the condition

[IF] $\qquad S^{\mathcal{J}} = T^{\mathcal{J}} \iff S = T \qquad (S, T \in \mathcal{F})$.

The proof of this statement is left as Problem 5.9.

We have encountered statements similar to [F] and [IF] in Chapter 2, concerning the effective set and the productive sets of a state in a medium (cf. Definition 2.5.1 and Theorem 2.5.2). We recall that, by Theorem 3.2.6, the family $\widehat{S}^+$ of positive content of an oriented medium $(S, \mathcal{T})$ is well-graded. The links between the effective and the productive sets $S^{\mathcal{E}}$ and $S^{\mathcal{P}}$ of a state S and the fringes of $\widehat{S}^+$ with respect to $\widehat{S}^+$ are specified below.

5.2.4 Definition. For any subset of tokens Γ in a medium $(S, \mathcal{T})$ we write
$$\widetilde{\Gamma} = \{\tau \in \mathcal{T} \mid \tilde{\tau} \in \mathcal{T} \setminus \Gamma\}.$$

5.2.5 Theorem. *Let $(S, \mathcal{T})$ be an oriented medium, with orientation $\{\mathcal{T}^+, \mathcal{T}^-\}$ and family of positive contents $\widehat{S}^+$. Then, the following two equalities hold for any state S:*

$$S^{\mathcal{E}} = \left(\widehat{S}^+\right)^{\mathcal{O}} \cup \widetilde{\left(\widehat{S}^+\right)^{\mathcal{J}}} \qquad (5.16)$$

$$S^{\mathcal{P}} = \left(\widehat{S}^+\right)^{\mathcal{J}} \cup \widetilde{\left(\widehat{S}^+\right)^{\mathcal{O}}}. \qquad (5.17)$$

Proof. For any token τ and state S, we have successively,

$$\begin{aligned}
\tau \in S^{\mathcal{E}} &\iff (\tau \in \mathcal{T}^+ \text{ or } \tau \subset \mathcal{T}^-) \text{ and } (S\tau = T \neq S) &&\text{(for some } T \in \mathcal{S}) \\
&\iff \left(\tau \in \left(\widehat{S}^+\right)^{\mathcal{O}}\right) \text{ or } (\tau \in \mathcal{T}^-, S\tau = T \neq S) &&\text{(for some } T \in \mathcal{S}) \\
&\iff \left(\tau \in \left(\widehat{S}^+\right)^{\mathcal{O}}\right) \text{ or } (\tilde{\tau} \in \mathcal{T}^+, T\tilde{\tau} = S \neq T) &&\text{(for some } T \in \mathcal{S}) \\
&\iff \left(\tau \in \left(\widehat{S}^+\right)^{\mathcal{O}}\right) \text{ or } \left(\tilde{\tau} \in \left(\widehat{S}^+\right)^{\mathcal{J}}\right) \\
&\iff \left(\tau \in \left(\widehat{S}^+\right)^{\mathcal{O}}\right) \text{ or } \left(\tau \in \widetilde{\left(\widehat{S}^+\right)^{\mathcal{J}}}\right) \\
&\iff \tau \in \left(\left(\widehat{S}^+\right)^{\mathcal{O}} \cup \widetilde{\left(\widehat{S}^+\right)^{\mathcal{J}}}\right).
\end{aligned}$$

The argument establishing (5.17) is similar (Problem 5.4). □

5.3 Partial Orders

We use the results of the following four lemmas to show that the family of all the partial orders on a finite set is well-graded (see Theorem 5.3.5).

5.3.1 Lemma. *Let $\mathcal{F}$ be a family of subsets closed under finite intersection. Then $\mathcal{F}$ is well-graded if and only if for any two sets S, T in $\mathcal{F}$ with $S \subset T$ and $|T \setminus S| = k$, there exists a chain of sets $S_0 = S \subset S_1 \subset \ldots \subset S_k = T$ in $\mathcal{F}$ such that $|S_j \setminus S_{j-1}| = 1$ for $j = 1, 2, \ldots, k$.*

Proof. The necessity follows immediately from the definition of wellgradedness. The sufficiency is also easily established: if T and T' are in $\mathcal{F}$, then so is $T \cap T'$ and there exists two chains $S_0 = T \cap T' \subset S_1 \subset \ldots \subset S_k = T$ and $S'_0 = T \cap T' \subset S'_1 \subset \ldots \subset S'_m = T$ such that $|S_j \setminus S_{j-1}| = 1$ for $j = 1, 2, \ldots, k$ and $|S'_i \setminus S'_{i-1}| = 1$ for $i = 1, 2, \ldots, m$. Concatenating these two chains and relabelling the indices yields the sequence

$$T_0 = T = S_k, T_1 = S_{k-1}, \ldots, T_k = S_0 = T \cap T' = S'_0,$$
$$T_{k+1} = S'_1, \ldots, T_{k+m} = S'_m = T'$$

required by the definition of wellgradedness. □

We now specify the fringes in a family of all partial orders of a finite set.

5.3.2 Lemma. *The two fringes of a partial order P on a finite set are described by the formulas*

$$P^{\mathfrak{I}} = P \setminus PP \tag{5.18}$$
$$P^{\mathfrak{O}} = \bar{P} \setminus (I \cup \bar{P}P^{-1} \cup P^{-1}\bar{P}). \tag{5.19}$$

We omit the proof (Problem 5.5). Note that $P^{\mathfrak{I}}$ is the Hasse diagram of P. The pairs in the outer fringe $P^{\mathfrak{O}}$ of P are referred to as *critical pairs* or *nonforced pairs* by Trotter (1992).

In passing, we prove a variant of Lemma 3.5.1 which is of general interest and is relevant to Problem 5.7.

5.3.3 Lemma. *Suppose that L is a strict linear order. Then $L \setminus \{xy\}$ is a partial order if and only if x covers y in L.*

Proof. (Necessity.) Let L and $P = L \setminus \{xy\}$ be a strict linear order and a partial order, respectively. We cannot have both xPz and zPy, because then also xPy by the transitivity of P, contradicting the construction of P. So, x covers y in L.

(Sufficiency.) Let L be strict linear order in which x covers y. The relation $P = L \setminus \{xy\}$ is asymmetric by definition. To verify its transitivity, suppose that $uPvPw$. If $|\{x,y\} \cap \{u,v,p\}| \leq 1$, we get $uLvLw$, and uLw follows from the transitivity of L, which gives uPw because $uw \neq xy$. Suppose that $|\{x,y\} \cap \{u,v,p\}| = 2$. Three cases must be checked.

1. $uv = yx$, so uLv with vPw yielding vLw because $w \neq x$ and $w \neq y$; we obtain uLw by the transitivity of L, and so uPw.
2. $vw = yx$, so vLw with uPv yielding uLv because $u \neq x$ and $v \neq y$; we obtain again uLw and thus uPw.
3. uLv, $uv \neq yx$ and vLw, $vw \neq yx$; then uLw by the transitivity of L, and $uw \neq yx$ because x covers y in L; so uPw. □

5.3.4 Lemma. *Suppose that P and Q are two partial orders on the same set X satisfying $Q \subset P$ and $d(P,Q) = n \in \mathbb{N}$. Then there exists a chain $Q_0 = Q \subset Q_1 \subset \ldots \subset Q_n = P$ of partial orders such that $d(Q_i, Q_{i+1}) = 1$ for $0 \leq i \leq n-1$.*

Proof. Pick a pair $xy \in P^J$ and define $Q_{n-1} = P \setminus \{xy\}$; so, $d(P, Q_{n-1}) = 1$ and, since $Q \subseteq Q_{n-1} \subset Q_n = P$, we obtain $d(Q, Q_{n-1}) = n - 1$. The lemma follows by induction. □

5.3.5 Theorem. *The family $\mathcal{P}$ of all partial orders on a finite set X is closed under intersection and well-graded.*

Proof. It is well-known (and in any event, easily verified—see Problem 5.1) that the intersection of a family of partial orders is a partial order. This implies that the family $\mathcal{P}$ is closed under finite intersection. Applying first Lemma 5.3.4, and then Lemma 5.3.1, we conclude that $\mathcal{P}$ is well-graded. □

This result is due to Bogart (1973) (cf. also Kuzmin and Ovchinnikov, 1976).

5.4 Biorders and Interval Orders

The concept of a biorder first appeared in the literature in the form of a device for the measurement of human attitudes or aptitudes (Guttman, 1944). It was widely used in the behavioral sciences under the name 'Guttman scales.' The corresponding combinatoric object, called a 'Ferrers relation', was independently studied by Riguet (1951) (see also Cogis, 1982). The connection was established by Ducamp and Falmagne (1969), who relabelled them 'bi-quasi-series' in the context of ordinal measurement. They proved that, in the case of finite sets X and Y, a relation $R \subseteq X \times Y$ satisfies the biorder condition (5.9), that is, $R\bar{R}^{-1}R \subseteq R$, if and only if there exists functions $f : X \to \mathbb{R}$ and $g : Y \to \mathbb{R}$ such that

$$xRy \iff f(x) > g(y) \qquad (x \in X, y \in Y). \qquad (5.20)$$

Also independently, Fishburn (1970) introduced the interval orders which, as we have seen, are irreflexive biorders from X to X. The name 'biorder' was coined[3] by Doignon et al. (1984), who extended the representation (5.20) to the case of infinite sets $\mathcal{X}$ and $\mathcal{Y}$ satisfying a condition of order density. This term is justified by the fact that a biorder $R \subseteq \mathcal{X} \times \mathcal{Y}$ induces two strict weak orders; one on the set $\mathcal{X}$, defined by $R\bar{R}^{-1}$; and the other on the set $\mathcal{Y}$, defined by $\bar{R}^{-1}R$ (see Lemma 5.4.1(ii)). The wellgradedness of the family of all biorders between two finite sets $\mathcal{X}$ and $\mathcal{Y}$ was proved by Doignon and Falmagne (1997). A slightly more detailed version of their argument is given here. The corresponding result for interval orders is also stated[4]. We suppose throughout that any biorder under consideration belongs to the family of all the biorders from a finite set $\mathcal{X}$ to a finite set $\mathcal{Y}$.

We begin with some straightforward consequences of the definition of a biorder. We omit the simple proofs of Lemmas 5.4.1 and 5.4.3 and of Theorem 5.4.2 (see Problems 5.8, 5.10 and 5.11).

5.4.1 Lemma. *If R is a biorder, then:*

(i) $\bar{R}$ *and* R^{-1} *are biorders;*
(ii) $R\bar{R}^{-1}$ *and* $\bar{R}^{-1}R$ *are strict weak orders.*

Not surprisingly, the main tool of our proof lies in the two fringes of the biorders in the family. The next theorem specify those fringes.

5.4.2 Theorem. *The inner and outer fringes of a biorder R from $\mathcal{X}$ to $\mathcal{Y}$ are respectively defined by*

$$R^\mathcal{J} = R \setminus R\bar{R}^{-1}R, \quad \text{and} \quad R^\mathcal{O} = \bar{R} \setminus \bar{R}R^{-1}\bar{R}.$$

We immediately reformulate these two equalities in the next lemma.

5.4.3 Lemma. *For any biorder R, we have*

$$R \cap \overline{R^\mathcal{J}} = R\bar{R}^{-1}R, \tag{5.21}$$

$$\bar{R}^{-1} \cap \overline{R^\mathcal{O}} = \bar{R}R^{-1}\bar{R}. \tag{5.22}$$

To establish that the family of all biorders between two finite sets is well-graded, we prove that any two biorders in the family satisfies Condition (iv) of Theorem 5.2.2. The next theorem is the key step. A more intricate version of the same idea is also exploited in the next section to deal with the wellgradedness of semiorders (cf. Definition 5.5.3 and Theorem 5.5.4).

---

[3] As far as we can tell, the name 'biorder' is gaining acceptance.
[4] The wellgradedness of interval orders, which was presented as an obvious corollary in Doignon and Falmagne (1997), results in fact from a similar, but separate argument.

5.4.4 Lemma. *Any biorder R in the family of all biorders between two finite sets $\mathcal{X}$ and $\mathcal{Y}$ satisfies the condition*

$$R = \cup_{n=0}^{\infty}(R\bar{R}^{-1})^n R = \cup_{n=0}^{\infty}(R^{\mathtt{J}}(R^{\mathtt{O}})^{-1})^n R^{\mathtt{J}}. \tag{5.23}$$

Proof. We show that

$$R \subseteq \cup_{n=0}^{\infty}(R\bar{R}^{-1})^n R \subseteq \cup_{n=0}^{\infty}(R^{\mathtt{J}}(R^{\mathtt{O}})^{-1})^n R^{\mathtt{J}} \subseteq R. \tag{5.24}$$

The first inclusion follows from the term $n = 0$ in the r.h.s. To get the second inclusion, take any $xy \in \cup_{n=0}^{\infty}(R\bar{R}^{-1})^n R$. Thus, xy is in at least one of the terms $(R\bar{R}^{-1})^n R$ and may belong to more than one. Because $\mathcal{X}$ and $\mathcal{Y}$ are finite, we may suppose that

$$xy \in (R\bar{R}^{-1})^k R \tag{5.25}$$

with k maximal. This implies that each of the $k+1$ factors R of the product in (5.25) can be replaced by $R^{\mathtt{J}}$ with still $xy \in (R^{\mathtt{J}}\bar{R}^{-1})^k R^{\mathtt{J}}$. Indeed, if this were not true, there would be some z, either in $\mathcal{Y}$ with $xz \in R \cap \overline{R^{\mathtt{J}}} = R\bar{R}^{-1}R$ or in $\mathcal{X}$ with $xz \in R \cap \overline{R^{\mathtt{J}}} = R\bar{R}^{-1}R$, both equalities resulting from (5.21). This would imply that we could rewrite (5.25) with $(k+2)+1$ factors R, contradicting the maximality of k. A similar argument, based on (5.22), proves that we can replace each of the k factors $\bar{R}^{-1}$ in (5.25) by $(R^{\mathtt{O}})^{-1}$, and still have $xy \in (R^{\mathtt{J}}(R^{\mathtt{O}})^{-1})^k R^{\mathtt{J}}$. Thus, the second inclusion in (5.24) holds. To obtain the third inclusion, we notice that $R^{\mathtt{J}} \subseteq R$ and both $R^{\mathtt{J}}(R^{\mathtt{O}})^{-1}R^{\mathtt{J}} \subseteq R$ and $R(R^{\mathtt{O}})^{-1}R^{\mathtt{J}} \subseteq R$ by the biorder inclusion $R\bar{R}^{-1}R \subseteq R$. This gives $(R^{\mathtt{J}}(R^{\mathtt{O}})^{-1})^0 R^{\mathtt{J}} = R^{\mathtt{J}} \subseteq R$ for $n = 0$, and for any $n > 0$, using induction,

$$(R^{\mathtt{J}}(R^{\mathtt{O}})^{-1})^n R^{\mathtt{J}} = \underbrace{\overbrace{(R^{\mathtt{J}}(R^{\mathtt{O}})^{-1})(R^{\mathtt{J}}(R^{\mathtt{O}})^{-1})}^{\subseteq R} \ldots (R^{\mathtt{J}}(R^{\mathtt{O}})^{-1})}_{n \text{ products } (R^{\mathtt{J}}(R^{\mathtt{O}})^{-1})} R^{\mathtt{J}}$$

$$\subseteq \underbrace{\overbrace{(R(R^{\mathtt{O}})^{-1})(R^{\mathtt{J}}(R^{\mathtt{O}})^{-1})}^{\subseteq R} \ldots (R^{\mathtt{J}}(R^{\mathtt{O}})^{-1})}_{\text{one product } (R(R^{\mathtt{O}})^{-1}),\ n-2 \text{ products } (R^{\mathtt{J}}(R^{\mathtt{O}})^{-1})} R^{\mathtt{J}}$$

$$\ldots$$

$$\subseteq R(R^{\mathtt{O}})^{-1} R^{\mathtt{J}} \subseteq R.$$

We conclude that (5.23) holds. □

5.4.5 Lemma. *For any two biorders R and S from $\mathcal{X}$ to $\mathcal{Y}$, we have*

$$(R^{\mathtt{J}} \subseteq S \text{ and } R^{\mathtt{O}} \subseteq \bar{S}) \implies R = S.$$

Proof. Suppose that $R^{\mathcal{J}} \subseteq S \wedge R^{\mathcal{O}} \subseteq \bar{S}$. We have successively

$$\begin{aligned}
R &= \cup_{n=0}^{\infty} (R^{\mathcal{J}}(R^{\mathcal{O}})^{-1})^n R^{\mathcal{J}} &&\text{(by Lemma 5.4.4)} \\
&\subseteq \cup_{n=0}^{\infty} (S\bar{S}^{-1})^n S &&\text{(because } R^{\mathcal{J}} \subseteq S \text{ and } R^{\mathcal{O}} \subseteq \bar{S}) \\
&= S &&\text{(by Lemma 5.4.4).}
\end{aligned}$$

To establish the converse inclusion, note that both $\bar{R}$ and $\bar{S}$ are also biorders (Lemma 5.4.1(i)), and that $(\bar{R})^{\mathcal{J}} = R^{\mathcal{O}}$ and $(\bar{R})^{\mathcal{O}} = R^{\mathcal{J}}$ (Problem 5.8(iii)). Our hypothesis can be restated as $(\bar{R})^{\mathcal{J}} \subseteq \bar{S}$ and $(\bar{R})^{\mathcal{O}} \subseteq \overline{(\bar{S})}$. Using the argument of the first part of this proof yields $\bar{R} \subseteq \bar{S}$, so $S \subseteq R$. □

5.4.6 Theorem. *The family $\mathcal{F}$ of all the biorders R from a finite set $\mathcal{X}$ to a finite set $\mathcal{Y}$ is well-graded.*

Proof. Notice that $\mathcal{F}$ is a graded family. Indeed, for any $xy \in \mathcal{X} \times \mathcal{Y}$, both $\mathcal{X} \times \mathcal{Y}$ and $\mathcal{X} \times \mathcal{Y} \setminus \{xy\}$ are biorders. The result follows readily from Lemma 5.4.5 and the implication (iv) $\Rightarrow$ (i) of Theorem 5.2.2. □

5.4.7 Theorem. *The family of all interval orders on a finite set $\mathcal{X}$ is well-graded.*

Although every interval order on a set $\mathcal{X}$ is a biorder from $\mathcal{X}$ to $\mathcal{X}$, Theorem 5.4.7 cannot be regarded as a corollary of Theorem 5.4.6 because a biorder from a set to itself may be reflexive: consider the biorder $\mathcal{X} \times \mathcal{X}$. However, a careful check shows that the argument used to establish the wellgradedness of biorders also applies to the interval orders. The verification is left to the reader (Problem 5.13).

5.5 Semiorders

From Definition 5.1.1, (5.13), we know that a relation R on a set $\mathcal{X}$ is a semiorder if the following three conditions are satisfied:

$$R \cap I = \varnothing \qquad (R \text{ is irreflexive}) \qquad (5.26)$$
$$R\bar{R}^{-1}R \subseteq R \qquad (R \text{ is a biorder}) \qquad (5.27)$$
$$RR\bar{R}^{-1} \subseteq R \qquad (R \text{ is 2-connected}). \qquad (5.28)$$

The concept of a semiorder is due to by Luce (1956) who, inspired by the notion of 'just-noticeable-difference' in psychophysics, introduced it as a form of linear order tempered by a (fixed) threshold. The standard numerical representation for semiorder is consistent with this intuition. It was shown by Scott and Suppes (1958) that if $\mathcal{X}$ is a finite set, then R is a semiorder if and only if there exists a function $f: \mathcal{X} \to \mathbb{R}$ such that

$$xRy \iff f(x) > f(y) + 1. \tag{5.29}$$

Note that (5.29) is a special case of (5.20) (Problem 5.14).

The family of all the semiorders on a finite set is well-graded. Our argument establishing this fact is similar to that used in the previous section to prove the wellgradedness of a family of biorders, but relies on a more intricate machinery. We begin by noting some useful equivalences between formulas.

5.5.1 Lemma. *For any relation R, we have:*

$$R\bar{R}^{-1}R \subseteq R \iff \bar{R}R^{-1}R \subseteq \bar{R} \tag{5.30}$$

and

$$RR\bar{R}^{-1} \subseteq R \iff \bar{R}^{-1}RR \subseteq R \tag{5.31}$$
$$\iff \bar{R}\bar{R}R^{-1} \subseteq \bar{R} \tag{5.32}$$
$$\iff R^{-1}R\bar{R} \subseteq \bar{R}. \tag{5.33}$$

We omit the simple proofs (Problem 5.15). Next, we specify the fringes of a semiorder.

5.5.2 Theorem. *The family of all semiorders on a finite set X is graded. Moreover, the inner and outer fringes of a semiorder R on X are defined by the equalities*

$$R^{\mathcal{I}} = R \setminus \left(R\bar{R}^{-1}R \cup \bar{R}^{-1}RR \cup RR\bar{R}^{-1} \right) \tag{5.34}$$

and

$$R^{\mathcal{O}} = \bar{R} \setminus \left(I \cup \bar{R}R^{-1}\bar{R} \cup R^{-1}\bar{R}\bar{R} \cup \bar{R}\bar{R}R^{-1} \right). \tag{5.35}$$

Problem 5.16 requires the reader to verify these facts.

Our next definition introduces a compact notation for two particular types of unions of products of the two relations R and $\bar{R}^{-1}$. These devices will permit us to deal efficiently with a wide variety of products of the type $RR\bar{R}^{-1}R$, $\bar{R}^{-1}RR\bar{R}^{-1}$, etc.

5.5.3 Definition. *For any product $r = R_1 \cdots R_n$, $n \in \mathbb{N}$ of relations on the same set, we denote by $\mathfrak{c}_R(r)$ the number of occurrences of the relation R in the product r; thus $\mathfrak{c}_R(RR\bar{R}^{-1}R) = 3$ and $\mathfrak{c}_{\bar{R}^{-1}}(RR\bar{R}^{-1}R) = 1$. We then define, for any two relation R and S on the same set, the collection of products*

$$\mathcal{R}_{RS} = \{r \mid r = R_1 \cdots R_n,\ n \in \mathbb{N},\ R_i = R \text{ or } R_i = S,\ 1 \leq i \leq n\},$$

and the two unions

$$R \trianglerighteq S = \bigcup \{r \in \mathcal{R}_{RS} \mid \mathfrak{c}_R(r) \geq \mathfrak{c}_S(r)\}$$
$$R \triangleright S = \bigcup \{r \in \mathcal{R}_{RS} \mid \mathfrak{c}_R(r) > \mathfrak{c}_S(r)\}.$$

Thus, the difference between $R \rhd S$ and $R \vartriangleright S$ is that the products included in the later have strictly more R's than S's. For example, $R \rhd \bar{R}^{-1}$ is the union of the products

$$R, \, RR, \ldots \qquad (\mathfrak{c}_{\bar{R}^{-1}}(r) = 0)$$
$$RR\bar{R}^{-1}, \, R\bar{R}^{-1}R, \, \bar{R}^{-1}RR, \, RRR\bar{R}^{-1}, \, RR\bar{R}^{-1}R, \ldots \qquad (\mathfrak{c}_{\bar{R}^{-1}}(r) = 1)$$
$$RRR\bar{R}^{-1}\bar{R}^{-1}, \, RR\bar{R}^{-1}R\bar{R}^{-1}, \ldots, RRRR\bar{R}^{-1}\bar{R}^{-1}, \, \ldots \qquad (\mathfrak{c}_{\bar{R}^{-1}}(r) = 2)$$
$$\ldots$$

The following result is our key step.

5.5.4 Theorem. *For any semiorder R on a finite set X, we have*

$$R = R \rhd \bar{R}^{-1} = R^{\mathtt{J}} \rhd (\bar{R}^{\mathtt{O}})^{-1}. \tag{5.36}$$

We establish this theorem via the intermediate result below.

5.5.5 Lemma. *The relation $R \rhd \bar{R}^{-1}$ is irreflexive if R is a semiorder.*

Proof. We prove by induction on $n+m$ that, for all $n \geq m$, $n \geq 1$, we have

$$r = R_1 \cdots R_{n+m} \subseteq \bar{I} \quad (r \in \mathcal{R}_{R\bar{R}^{-1}}, \, \mathfrak{c}_R(r) = n \geq \mathfrak{c}_{\bar{R}^{-1}} = m, \, n \geq 1).$$

Since $R \rhd \bar{R}^{-1}$ is the union of all such products, this proves the lemma. Because R is irreflexive, we have $r = R \subseteq \bar{I}$, thus for $\mathfrak{c}_R(r) + \mathfrak{c}_{\bar{R}^{-1}}(r) = 1 + 0 = 1$. The inclusion $r \subseteq \bar{I}$ also holds for $\mathfrak{c}_R(r) + \mathfrak{c}_{\bar{R}^{-1}}(r) = 1 + 1 = 2$ and $\mathfrak{c}_R(r) + \mathfrak{c}_{\bar{R}^{-1}}(r) = 2 + 0 = 2$ because both $R\bar{R}^{-1}$ and $\bar{R}^{-1}R$ are irreflexive (Problem 5.8(ii)), and $RR \subseteq R$. Suppose that $n+m = k$. If $n = k$ and $m = 0$, we get $r = R^k \subseteq R$, which is irreflexive. In all other cases, the expression of the product r must contain a partial product having one of the six forms:

(1) $RR\bar{R}^{-1}$, \qquad (2) $R\bar{R}^{-1}R$, \qquad (3) $\bar{R}^{-1}RR$,
(4) $R\bar{R}^{-1}\bar{R}^{-1}$, \qquad (5) $\bar{R}^{-1}R\bar{R}^{-1}$, \qquad (6) $\bar{R}^{-1}\bar{R}^{-1}R$.

Using Lemma 5.5.1, we see that the first three products can be replaced by R, and the last three producs by $\bar{R}^{-1}$. This results in replacing r with $\mathfrak{c}_R(r) + \mathfrak{c}_{\bar{R}^{-1}}(r) = n+m = k$ by r' with $\mathfrak{c}_R(r') + \mathfrak{c}_{\bar{R}^{-1}}(r') = n' + m' = k - 2$. An induction argument establishes the lemma. $\square$

PROOF OF THEOREM 5.5.4. We prove that

$$R \subseteq R \rhd \bar{R}^{-1} \subseteq R^{\mathtt{J}} \rhd (\bar{R}^{\mathtt{O}})^{-1} \subseteq R. \tag{5.37}$$

OBSERVATION. If a product r in the union $R \rhd \bar{R}^{-1}$ has a subproduct m containing some pair zz, say $r = n\,z\,m\,z\,p$, then removing that subproduct m from r yields some other product $r' = np \subseteq R \rhd \bar{R}^{-1}$ such that

$$\mathfrak{c}_R(\boldsymbol{r'}) - \mathfrak{c}_{\bar{R}^{-1}}(\boldsymbol{r'}) > \mathfrak{c}_R(\boldsymbol{r}) - \mathfrak{c}_{\bar{R}^{-1}}(\boldsymbol{r}). \tag{5.38}$$

This is a consequence of Lemma 5.5.5: removing the subproduct $\boldsymbol{m}$ must increase the difference between the number of R's and the number of $\bar{R}^{-1}$'s.

The first inclusion in (5.37) holds by the definition of $\triangleright$. Suppose that $xy \in R \triangleright \bar{R}^{-1}$. This implies that $x\,\boldsymbol{r}\,y$ for some product $\boldsymbol{r} \subseteq R \triangleright \bar{R}^{-1}$. Obviously, there may be more than one such product. Consider the set $\mathcal{V}$ of all sequences $x = x_0, \ldots, x_k = y$ for some $k \in \mathbb{N}$, such that $x_i\,(R \cup \bar{R}^{-1})\,x_{i+1}$ for $0 \le i \le k-1$. Suppose that $x_i = x_j$ for some $0 \le i < j \le k$, say $x_i R_i x_{i+1} R_{i+1} \cdots R_j x_j$, with $R_{i+h} = R$ or $R_{i+h} = \bar{R}^{-1}$ for $0 \le h \le j-i$. By the above Observation, removing $R_i R_{i+1} \cdots R_j = \boldsymbol{m}$ from the product $\boldsymbol{r}$ leaves a product $\boldsymbol{r'}$ satisfying (5.38). Because there is a finite number of sequences in $\mathcal{V}$ having no repeated elements, there is at least one product $\boldsymbol{q} \subseteq R \triangleright \bar{R}^{-1}$ such that $x\,\boldsymbol{q}\,y$ and $\mathfrak{c}_R(\boldsymbol{q})$ is maximal. Take one such product $\boldsymbol{q}$. We claim that any occurrence of R in the product $\boldsymbol{q}$ can be replaced by R^{J}. Indeed, if this were not true for a particular instance of R in $\boldsymbol{q}$, we could, in view of Eq. (5.34) in Theorem 5.5.2, replace that R by one of $R\bar{R}^{-1}R$, $\bar{R}^{-1}RR$ or $RR\bar{R}^{-1}$, thus yielding in each of these three cases a product $\boldsymbol{q'}$ still satisfying $x\,\boldsymbol{q'}\,y$, with $\mathfrak{c}_R(\boldsymbol{q'}) - \mathfrak{c}_{\bar{R}^{-1}}(\boldsymbol{q'}) = \mathfrak{c}_R(\boldsymbol{q}) - \mathfrak{c}_{\bar{R}^{-1}}(\boldsymbol{q})$, but also

$$\mathfrak{c}_R(\boldsymbol{q'}) = \mathfrak{c}_R(\boldsymbol{q}) + 1,$$

contradicting the maximality of $\mathfrak{c}_R(\boldsymbol{q})$. A similar argument, based on choosing $\boldsymbol{q}$ so that $\mathfrak{c}_{\bar{R}^{-1}}(\boldsymbol{q})$ is maximal and $x\,\boldsymbol{q}\,y$ holds, shows that each $\bar{R}^{-1}$ in $\boldsymbol{q}$ can be replaced by $(\bar{R}^{\mathsf{O}})^{-1}$. We have thus $R \triangleright \bar{R}^{-1} \subseteq R^{\mathsf{J}} \triangleright (\bar{R}^{\mathsf{O}})^{-1}$. As $R^{\mathsf{J}} \subseteq R$ and $(\bar{R}^{\mathsf{O}})^{-1} \subseteq \bar{R}^{-1}$, we also have $R^{\mathsf{J}} \triangleright (\bar{R}^{\mathsf{O}})^{-1} \subseteq R \triangleright \bar{R}^{-1}$, which gives $R^{\mathsf{J}} \triangleright (\bar{R}^{\mathsf{O}})^{-1} = R \triangleright \bar{R}^{-1}$. It remains to show that $R \triangleright \bar{R}^{-1} \subseteq R$. Suppose that $xy \in (R \triangleright \bar{R}^{-1})$; thus $x\,\boldsymbol{r}\,y$ for some product $\boldsymbol{r} \subseteq R \triangleright \bar{R}^{-1}$. Since $\mathfrak{c}_R(\boldsymbol{r}) > \mathfrak{c}_{\bar{R}^{-1}}(\boldsymbol{r}) \le 0$, we have either $\boldsymbol{r} = R^n \subseteq R$ for ≥ 1, in which case $xy \in R$, or the product $\boldsymbol{r}$ must have a subproduct having one of the three forms:

$$R R \bar{R}^{-1}, \quad R \bar{R}^{-1} R, \text{ and } \bar{R}^{-1} R R, \tag{5.39}$$

each of which is included in R. This subproduct may be removed from the expression of $\boldsymbol{r}$, yielding a product $\boldsymbol{r'}$ with still $x\,\boldsymbol{r'}\,y$ and

$$\mathfrak{c}_R(\boldsymbol{r'}) - \mathfrak{c}_{\bar{R}^{-1}}(\boldsymbol{r'}) = \mathfrak{c}_R(\boldsymbol{r}) - \mathfrak{c}_{\bar{R}^{-1}}(\boldsymbol{r})$$
$$\mathfrak{c}_R(\boldsymbol{r'}) = \mathfrak{c}_R(\boldsymbol{r}) - 1\,.$$

Applying induction, we can reduce $\boldsymbol{r}$ ultimately to one of the forms in (5.39), which yields $\boldsymbol{r} \subseteq R$, and so $xy \in R$. □

5.5.6 Lemma. *Let R and S be two semiorders on a finite set X; then*

$$(R^{\mathsf{J}} \subseteq S \wedge R^{\mathsf{O}} \subseteq \bar{S} \wedge S^{\mathsf{J}} \subseteq R \wedge S^{\mathsf{O}} \subseteq \bar{R}) \quad \Longrightarrow \quad R = S.$$

Proof. Use Theorem 5.5.4. □

5 Well-Graded Families of Relations

5.5.7 Theorem. *The family of all semiorders on a finite set is well-graded.*

Proof. Use the implication (v) $\Rightarrow$ (i) of Theorem 5.2.2 together with Lemma 5.5.6. □

5.5.8 Remark. Results related to our Lemma 5.5.6 have been obtained by Pirlot (1991) who introduces the concept of a 'reduced semiorder' as a semiorder S on a set X satisfying the condition

$$(\forall z \in X, \ xSz \Leftrightarrow ySz \text{ and } zSx \Leftrightarrow zSy) \iff x = y.$$

He defines the concept of 'noses' and 'hollows', which correspond, for reduced semiorders, to the pairs in the outer and inner fringes in our sense, respectively. He shows that a reduced semiorder is defined by its noses and hollows. Theorem 5.5.2, which[5] is due to Doignon and Falmagne (1997), solves a problem raised by Pirlot in the conclusions of his paper.

5.6 Almost Connected Orders

We recall that a relation R on a set X is an ac-order if it is asymmetric and 2-connected, that is, R satisfies the condition

$$R^2 \bar{R}^{-1} \subseteq R. \tag{5.40}$$

Evidently, the term 'almost connected' was chosen because (5.40) is a natural generalization of a connectedness condition[6]. Other names are used for Condition (5.40), such as 'semitransitivity', by Chipman (1971), Fishburn (1997) and Fishburn and Trotter (1999). Monjardet (1978) refers to ac-orders as 'S-relations', and Fishburn (1985) calls them 'partial semiorders.' Examples and counterexamples of ac-orders are displayed in Figure 5.1.

5.6.1 Remarks. This section summarizes the results of Doble et al. (2001), from which Figures 5.1, 5.2 and 5.3 are adapted. Example (a) is in fact borrowed from Fishburn (1985). It is a semiorder, a special case of an ac-order. Example (b) is a strict weak order. Examples (c) and (d) are ac-orders which are neither semiorders nor strict weak orders. Each of Examples (f) and (g) satisfies exactly one of the two axioms of ac-orders: Counterexample (f) fails asymmetry, and Counterexample (g) 2-connectedness. Neither (e) nor (h) is an ac-order.

---

[5] Along with most of the results presented so far in this chapter.
[6] For example, a relation R is strongly connected if $R^0 \bar{R}^{-1} \subseteq R$ holds. It is a strict weak order if it is asymmetric and satisfies $R\bar{R}^{-1} \subseteq R$, another connectedness condition.

Examples

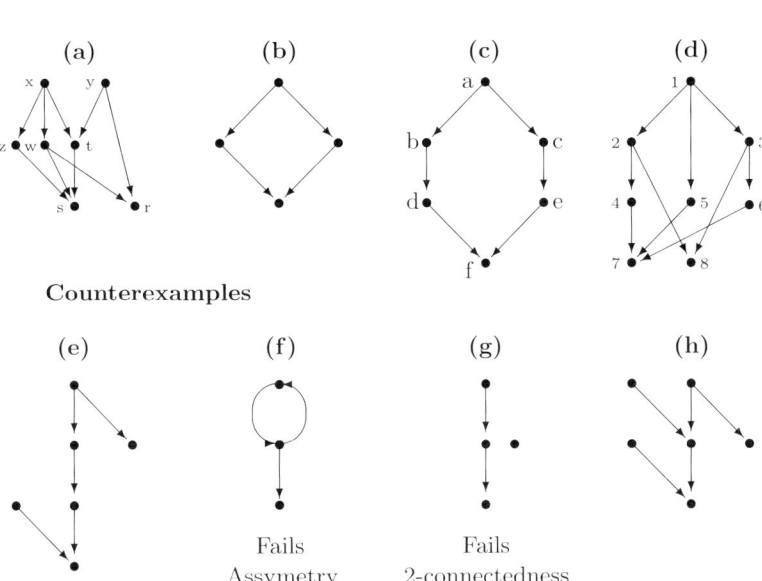

Figure 5.1. Examples and counterexamples of ac-orders.

Even though the attention given to ac-orders is relatively recent, noteworthy results have been obtained. For example, Fishburn (1985) showed that the product of two ac-orders is an ac-order (Problem 5.18). Fishburn and Trotter (1999) note that the (order) dimension of finite ac-orders (in the sense of Dushnik and Miller, 1941) is unbounded, and that so are the semiorder dimension and the interval order dimension. For other recent results, see also Skandera (2001), Trenk (1998) and Gimbel and Trenk (1998).

Only those results from Doble et al. (2001) that are relevant to wellgradedness will be detailed here. Among other facts established in that paper, we mention the following: if $W \subset R \subset W'$, where R is any relation and (W, W') is a covering pair of strict weak orders on the same set, that is, there is no strict weak order W'' satisfying $W \subset W'' \subset W'$, then R is an ac-order. A partial converse also holds. Other results of that paper are covered in the problem section.

We omit the proof of the next lemma (Problem 5.17).

5.6.2 Lemma. *Any ac-order R is is transitive and irreflexive. Moreover, the product $R\bar{R}^{-1}$ is irreflexive.*

We turn to the fringes. As customary, we write I for the identity on the ground set $\mathfrak{X}$.

5.6.3 Theorem. *The inner and outer fringes of an ac-order R in the family $\mathcal{A}$ of all the ac-orders on a finite set $\mathcal{X}$ are defined by the two equations*

$$R^{\mathfrak{I}} = R \setminus (R R \bar{R}^{-1} \cup \bar{R}^{-1} R R), \tag{5.41}$$

$$R^{\mathfrak{O}} = \bar{R} \setminus (I \cup \bar{R} \bar{R} R^{-1} \cup R^{-1} \bar{R} \bar{R}). \tag{5.42}$$

Proof. Suppose that $x(R^{\mathfrak{I}} \cap RR\bar{R}^{-1})y$. By definition of the inner fringe, we have $R^{\mathfrak{I}} \subseteq R$, and $S = R \setminus \{xy\}$ is an ac-order; moreover, $xRzRw\bar{R}^{-1}y$ for some $z, w \in \mathcal{X}$. We claim that

$$xSzSw\bar{S}^{-1}y \tag{5.43}$$

must hold because neither z nor w may be in $\{x, y\}$. Indeed, we have $z \notin \{x, y\}$ since S is irreflexive and does not contain xy by definition. As for $w \notin \{x, y\}$, this must be true since $w = x$ would contradict the transitivity and irreflexivity of R together applied to $xRRx$, and $w = y$ would yield $yR\bar{R}^{-1}y$, contradicting the irreflexivity of $R\bar{R}^{-1}$ for an ac-order R. Hence, (5.43) holds, which implies xSy because S is an ac-order, but contradicts $S = R \setminus \{xy\}$. The proof that $R^{\mathfrak{I}} \cap \bar{R}^{-1}RR = \varnothing$ is obtained by a similar contradiction. We have thus $R^{\mathfrak{I}} \subseteq R \setminus (RR\bar{R}^{-1} \cup \bar{R}^{-1}RR)$.

To prove the reverse inclusion, it suffices to show that for any

$$zw \in R \setminus (RR\bar{R}^{-1} \cup \bar{R}^{-1}RR), \tag{5.44}$$

the relation $T = R \setminus \{zw\}$ is an ac-order. Note that $z \neq w$ and that T is asymmetric because R is. Suppose that $TT\bar{T}^{-1} \not\subseteq T$; we have thus $xTsTt\bar{T}^{-1}y$ and $xy \notin T$ for some $x, s, t, y \in \mathcal{X}$, which implies $xRsRt$ and either $yt = zw$ and $xy \notin R$, or $xy = zw$. In the first case, we get $y\bar{R}^{-1}xRsRt$, yielding $z\bar{R}^{-1}xRsRw$, contradicting (5.44). The second case gives $x\bar{R}^{-1}y$ because $x\bar{T}^{-1}y$; so $xRsR\bar{R}^{-1}y$, that is $zRsR\bar{R}^{-1}w$, again a contradiction of (5.44).

Turning to the outer fringe, we suppose that $xy \in R^{\mathfrak{O}}$. Thus, $S = R + \{xy\}$ is an ac-order. Thus $\neg(xIy)$ because ac-orders are irreflexive. We must still prove that $xy \notin (\bar{R}\bar{R}R^{-1} \cup R^{-1}\bar{R}\bar{R})$. Suppose that $x\bar{R}\bar{R}R^{-1}y$; thus, $x\bar{R}s\bar{R}tR^{-1}y$ for some $s, t \in \mathcal{X}$. We get xSy, together with ySt and $s\bar{S}^{-1}x$ (because yRt and $x\bar{R}^{-1}s$). This gives $s\bar{S}^{-1}xSySt$, yielding sSt because S is an ac-order, but contradicting $s\bar{R}t$. Similarly, $xR^{-1}\bar{R}\bar{R}y$ gives $xR^{-1}s\bar{R}t\bar{R}y$ for some $s, t \subset \mathcal{X}$, which yields $sSxSy\bar{S}^{-1}t$, and thus sSt, again contradicting $s\bar{R}t$. We conclude that $R^{\mathfrak{O}} \subseteq \bar{R} \setminus (I \cup \bar{R}\bar{R}R^{-1} \cup R^{-1}\bar{R}\bar{R})$.

Suppose now that

$$xy \in \bar{R} \setminus (I \cup \bar{R}\bar{R}R^{-1} \cup R^{-1}\bar{R}\bar{R}). \tag{5.45}$$

We have to prove that $S = R + \{xy\}$ is an ac-order, that is, verify asymmetry and 2-connectedness. Because R is asymmetric, the only way S could fail asymmetry arises with yRx. Note that $\neg(xIy)$, and so $x \neq y$. Because R is irreflexive, we get $x\bar{R}x$. This leads to $x\bar{R}\bar{R}R^{-1}y$, giving xRy, which together with yRx contradict the asymmetry of R. We conclude that S is asymmetric. It remains to show that $S = R + \{xy\}$ is 2-connected. Given that R is 2-connected, this could fail in only two cases:

1. $sSxSy\bar{S}^{-1}t$ and $s\bar{S}t$; but then sRx, $x\bar{R}y$, $t\bar{R}y$ and $s\bar{R}t$, yielding $xR^{-1}s\bar{R}t\bar{R}y$, contradicting $xy \notin R^{-1}\bar{R}\bar{R}$ assumed by (5.45).
2. $xSyStS^{-1}s$ and $x\bar{S}s$; but then $x\bar{R}y$, yRt, $s\bar{R}t$ and $x\bar{R}s$, yielding $x\bar{R}s\bar{R}tR^{-1}y$, contradicting $xy \notin \bar{R}\bar{R}R^{-1}$ assumed by (5.45).

Thus $\bar{R} \setminus (I \cup \bar{R}\bar{R}R^{-1} \cup R^{-1}\bar{R}\bar{R}) \subseteq R^{\mathrm{o}}$ and so (5.42) holds, completing the proof of the theorem. □

One additional concept is needed.

5.6.4 Definition. For any relation R on a set $\mathcal{X}$, we denote by $\|_R$ the *incomparability* relation of R on $\mathcal{X}$, that is

$$x \| y \iff (x\bar{R}y \text{ and } y\bar{R}x) \qquad (x, y \in \mathcal{X}).$$

We may simply write $\|$ when the context makes it clear that $\|_R$ is intended. We recall that $\breve{\|}$ denotes the transitive closure of the incomparability relation $\|$ on $\mathcal{X}$ (cf. 1.8.3).

5.6.5 Theorem. *The family $\mathcal{A}$ of all the ac-orders on a finite set $\mathcal{X}$ is well-graded if and only if $|\mathcal{X}| \leq 4$.*

Proof. If $|\mathcal{X}| \leq 3$, the ac-orders on $\mathcal{X}$ are confounded with the partial orders on that set and the result follows from Theorem 5.3.5. In the case $|\mathcal{X}| = 4$, we rely on the implication (v) $\Leftrightarrow$ (i) in Theorem 5.2.2 to prove the wellgradedness of $\mathcal{A}$. Take any R and S in $\mathcal{A}$ and suppose that $R^{\mathrm{J}} \subseteq S$, $R^{\mathrm{o}} \subseteq \bar{S}$, $S^{\mathrm{J}} \subseteq R$, and $S^{\mathrm{o}} \subseteq \bar{R}$. If $R^{\mathrm{J}} = R$ and $S^{\mathrm{J}} = S$, we get $R \subseteq S$ and $S \subseteq R$; thus, $R = S$. So, we can assume that $R^{\mathrm{J}} \subset R$. An examination of the possible cases leads to conclude that there are only two (see Figure 5.2):

$$x \| y R z R w R^{-1} x \quad \text{(Case 1)} \quad \text{and} \quad xR^{-1}yRzRw \| x \quad \text{(Case 2)}.$$

Case 1 gives $R^{\mathrm{J}} = \{yz, zw\}$, and because $R^{\mathrm{J}} \subseteq S$, we get

$$ySzSw. \tag{5.46}$$

Since $R^{\mathrm{o}} = \{yx, xz\}$ and $R^{\mathrm{o}} \subseteq \bar{S}$, we have $y\bar{S}x$ and $x\bar{S}z$. Now, $y\bar{S}x$ together with (5.46) gives $x\bar{S}^{-1}ySzSw$, which gives xSw because S is an ac-order. Thus, it remains to show that $x \| y$. But we already have $y\bar{S}x$, and xSy cannot hold because it would give $xSySz$, and so xSz by transitivity, contradicting $x\bar{S}z$. We have thus $R = S$ in Case 1. The proof in Case 2 is similar. By the implication (i) $\Rightarrow$ (v) in Theorem 5.2.2, the family $\mathcal{A}$ is thus well-graded if $|\mathcal{X}| = 4$.

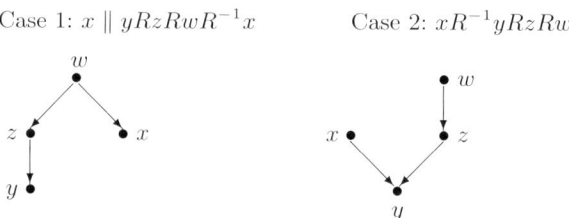

Figure 5.2. The graphs of Case 1 and Case 2 for $|\mathfrak{X}| = 4$ in the proof of Theorem 5.6.5.

Figure 5.3 illustrates the argument establishing that $\mathcal{A}$ is not well-graded when $|\mathfrak{X}| = n \geq 5$. With obvious notation, we have $R_n^{\mathfrak{I}} = \{31, 42\} = S_n^{\mathfrak{I}}$ and $R_n^{\mathfrak{O}} = \{32, 41\} = S_n^{\mathfrak{O}}$, but R_n and S_n are different relations. Condition (v) of Theorem 5.2.2 fails, and since Condition (v) is equivalent to Conditiom (i), $\mathcal{A}$ is not well-graded. □

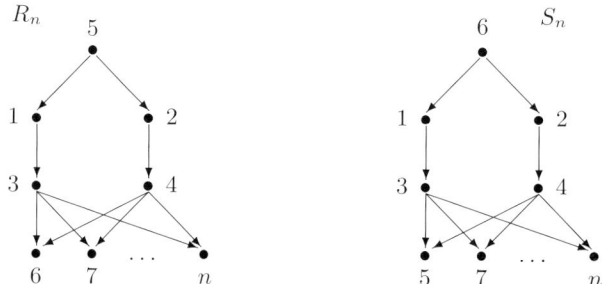

Figure 5.3. The graphs of two distinct ac-orders on an n-element set having their respective inner and outer fringes identical, a contradiction of Condition (v) of Theorem 5.2.2.

Properties closely related, but weaker than well-gradedness do hold for the family $\mathcal{A}$.

5.6.6 Definition. Any set in a family $\mathcal{F}$ is *upgradable* (respectively, *downgradable*) if it has a nonempty outer fringe (respectively, inner fringe). The family $\mathcal{F}$ is called *upgradable* if all its nonmaximal sets are upgradable. It is called *downgradable* if all its nonminimal sets are downgradable. A family $\mathcal{F}$ is *weakly graded* if for any two of its (distinct) sets S and T, there exists a sequence $S_0 = S, S_1, \ldots, S_n = T$ such that $d(S_i, S_{i+1}) = 1$ for $0 \leq i \leq n-1$.

5.6.7 Remarks. (a) Upgradability and downgradability, jointly satisfied, do not imply weak gradedness (Problem 5.19). However, the family $\mathcal{A}$ of all ac-orders on any finite set X is weakly graded because the empty relation is an ac-order.

(b) Evidently, if wellgradedness holds, so do upgradability and downgradability.

(c) The concepts of upgradability and downgradability are relative to a particular family of sets. However, if $\mathcal{F} \subseteq \mathcal{G}$ for two families of sets $\mathcal{F}$ and $\mathcal{G}$, and some set $S \in \mathcal{F}$ is upgradable (respectively, downgradable) in $\mathcal{F}$, then S is also upgradable (respectively, downgradable) in $\mathcal{G}$. For example, any semiorder is upgradable (respectively, downgradable) in $\mathcal{A}$ if is not a linear order (respectively, is not empty).

(d) Any linear order, weak order, or semiorder is an ac-order, which is itself a partial order. Denoting by $\mathcal{L}$, $\mathcal{W}$, $\mathcal{SO}$ and $\mathcal{P}$ for the families of linear orders, weak orders, semiorders, and partial orders, respectively, on a set X, we have in fact

$$\mathcal{L} \subseteq \mathcal{W} \subseteq \mathcal{SO} \subseteq \mathcal{AP}, \tag{5.47}$$

with all inclusions being strict if $|X| \geq 4$.

5.6.8 Theorem. *The family $\mathcal{A}$ of all the ac-orders on a finite set X is both upgradable and downgradable.*

SKETCH OF PROOF. Take any $R \in \mathcal{A}$. Suppose that R is a semiorder which is neither empty nor a linear order. Then, by Theorem 5.5.7 and Remarks 5.6.7 (b), (c) and (d), it is both upgradable and downgradable in $\mathcal{A}$. So, we can suppose that R is not a semiorder. This means that R fails the biorder axiom: there must be x, y, z, and w in X such that xRy and zRw, but neither xRw, nor zRy. In this case, it can be shown that we must have both $xy \in R^J$ and $xw \in R^O$.

The argument proceeds by contradiction. If $R \setminus \{xy\}$ is not an ac-order and is asymmetric (since R is asymmetric) it cannot be 2-connected. This observation leads to consider two cases, bot of which leading to a contradiction. The proof that $xw \in R^O$ is in a similar vein. □

Problems

5.1 Verify that the intersection of a family of partial orders is a partial order.

5.2 Prove that any tight path $S_0 = S, S_1, \ldots, S_k = T$ between two distinct sets S and T of a well-graded family verifies (5.15) for $0 \leq j \leq k-1$.

5.3 Prove by a counterexample that Condition [F] in Remark 5.2.3 does not imply wellgradedness.

5.4 Prove Equation (5.17) in Theorem 5.2.5.

5.5 Prove Lemma 5.3.2.

5.6 Prove that Condition [F] does not imply wellgradedness.

5.7 Suppose that $\mathcal{X}$ is uncountable. Let L be a well-ordering of $\mathcal{X}$, and $\mathcal{L}$ the family of all linear orders on $\mathcal{X}$. Let $\mathfrak{P}(\mathcal{X} \times \mathcal{X})_{\bowtie}$ be the cluster partition of $\mathcal{X}$ according to Definition 3.4.1. The strict linear order L is a partial order. Let $\mathcal{P}$ be the set of all partial orders on $\mathcal{X}$. As a well-ordering, L contains a pair xy such that y cover x. By Lemma 5.3.3, $L \setminus \{xy\}$ is a partial order at finite distance from L. Prove that there are uncountably many partial orders at finite distance from L, and $\langle L \rangle \cap \mathcal{P}$ is thus uncountable.

5.8 Prove that if R is a biorder, then

  (i) both $\bar{R}$ and R^{-1} are biorders;

  (ii) both $R\bar{R}^{-1}$ and $\bar{R}^{-1}R$ are strict weak orders (cf. (1.10));

  (iii) $(\bar{R})^{\mathfrak{I}} = R^{\mathfrak{O}}$ and $(\bar{R})^{\mathfrak{O}} = R^{\mathfrak{I}}$.

5.9 Prove the following result stated in Remark 5.2.3: If a family of sets $\mathcal{F}$ is closed under finite intersection and satisfies $\varnothing \in \mathcal{F}$, then it is well-graded if and only if any set in $\mathcal{F}$ is defined by its inner fringe in the sense of the condition

[IF] $S^{\mathfrak{I}} = T^{\mathfrak{I}} \iff S = T$ $(S, T \in \mathcal{F})$.

(Hint: use Lemma 5.3.1.)

5.10 Prove Theorem 5.4.2.

5.11 Prove Lemma 5.4.3.

5.12 Let $\mathcal{X}$ and $\mathcal{Y}$ be two arbitrary sets. Prove that if L is a linear order on $\mathcal{Z} = \mathcal{X} \cup \mathcal{Y}$, then $R = L \cap (\mathcal{X} \times \mathcal{Y})$ is a biorder from $\mathcal{X}$ to $\mathcal{Y}$. Argue that since $\mathcal{Z}$ can be well-ordered, a biorder always exist between any two sets.

5.13 Prove Theorem 5.4.7.

Problems for Chapter 5

5.14 Prove that if R is a biorder from X to X with a numerical representation

$$xRy \iff f(x) > g(y) \qquad (x, y \in X)$$

with $f : X \to \mathbb{R}$, $g : X \to \mathbb{R}$ and $f(x) - g(x) = f(y) - g(y)$, constant for all $x, y \in X$, then R is a semiorder on X.

5.15 Verify all the equivalences in Lemma 5.5.1.

5.16 Verify that the expressions for the inner fringe and the outer fringe of a semiorder are those given by (5.34) and (5.35).

5.17 Verify that an ac-order R is transitive and irreflexive, and that $R\bar{R}^{-1}$ is also irreflexive.

5.18 Prove that the product of two ac-order is an ac-order.

5.19 Construct an example showing that a family that is both upgradable and downgradable is not necessarily weakly graded. Formulate a (necessary and sufficient) condition for weak gradedness, given upgradability and downgradability (cf. Definition 5.6.6 and Remark 5.6.7 (a)).

5.20 Prove all the inclusions in Eq. (5.47) of Remark 5.6.7(d), including the statement regarding the strict inclusions.

5.21 Fill the gaps and complete the proof of Theorem 5.6.8.

6

Mediatic Graphs

6.1 The Graph of a Medium

This chapter follows closely Falmagne and Ovchinnikov (2007).

6.1.1 Definition. A *graph representation* of a medium $(\mathcal{S}, \mathcal{T})$ is a bijection $\gamma : \mathcal{S} \to V$, where V is a set of vertices of a graph (V, E), such that for distinct states S, T, we have $S\tau = T$ for some token τ if and only if $\{\gamma(S), \gamma(T)\}$ is an edge of the graph; formally,

$$\{\gamma(S), \gamma(T)\} \in E \iff (\exists \tau \in \mathcal{T})(S\tau = T) \quad (S, T \in \mathcal{S}, S \neq T). \quad (6.1)$$

We say then that the graph (V, E), which has no loops, *represents* the medium[1]. A graph (V, E) representing a medium $(\mathcal{S}, \mathcal{T})$ is called the *graph of the medium* $(\mathcal{S}, \mathcal{T})$ (cf. Definition 2.3.2) if $V = \mathcal{S}$, the edges in E are defined as in (6.1), and γ is the identity mapping. Clearly, any medium has its graph. We shall prove in this chapter that the converse also holds, namely: the graph of a medium defines its medium (see Theorem 6.4.5).

We recall that two graphs (V, E) and $(V'E')$ are isomorphic (see Section 1.8.5) if there exists a bijection $\varphi : V \to V'$ such that

$$\{P, Q\} \in E \iff \{\varphi(P), \varphi(Q)\} \in E' \quad (P, Q \in E, P \neq Q). \quad (6.2)$$

6.1.2 Lemma. *Any graph which is isomorphic to a graph representing a medium* $\mathcal{M}$ *also represents* $\mathcal{M}$.

Proof. From (6.1) and (6.2), it follows that if γ is a graph representation of a medium $\mathcal{M}$ by a graph G, and φ is an isomorphism of G onto some graph G', then $\varphi \circ \gamma$ is a graph representation of $\mathcal{M}$ by G'. □

It is intuitively clear that shortest paths in the graph of a medium correspond to concise messages of that medium. Our next lemma formalizes this intuition.

---

[1] An exemplary collection of small mediatic graphs is displayed on Figures A.1, A.2 and A.3 of the Appendix on page 305.

6.1.3 Lemma. *Let $\gamma : \mathcal{S} \to V$ be the representation of a medium $(\mathcal{S}, \mathcal{T})$ by a graph $G = (V, E)$. If $\boldsymbol{m} = \tau_1 \ldots \tau_m$ is a concise message producing a state T from a state S, then the sequence of vertices $(\gamma(S_i))_{0 \leq i \leq m}$, where $S_i = S\tau_0 \tau_1 \ldots \tau_i$, for $0 \leq i \leq m$, forms a shortest path joining $\gamma(S)$ and $\gamma(T)$ in G. Conversely, if a sequence $(\gamma(S_i))_{0 \leq i \leq m}$ is a shortest path connecting $\gamma(S_0) = \gamma(S)$ and $\gamma(S_m) = \gamma(T)$, then $\boldsymbol{m} = \tau_1 \ldots \tau_m$ with $S\tau_0\tau_1 \ldots \tau_i = S_i$, for $0 \leq i \leq m$, is a concise message producing T from S.*

Proof. (Necessity.) Let $\gamma(P_0) = \gamma(S), \gamma(P_1), \ldots, \gamma(P_n) = \gamma(T)$ be a path in G joining $\gamma(S)$ to $\gamma(T)$. Correspondingly, there is a stepwise effective message $\boldsymbol{n} = \rho_1 \cdots \rho_n$ such that $P_i = T\rho_1 \ldots \rho_{n-i}$ for $0 \leq i < n$. The message $\boldsymbol{mn}$ is a return for S. By Axiom [Mb], this message is vacuous. Since $\boldsymbol{m}$ is a concise message for S, we must have

$$\ell(\boldsymbol{m}) = m \leq \ell(\boldsymbol{n}) = n.$$

(Sufficiency.) Let now $\gamma(S_0) = \gamma(S), \gamma(S_1), \ldots, \gamma(S_m) = \gamma(T)$ be a shortest path from $\gamma(S)$ to $\gamma(T)$ in G. Then, there are some tokens τ_i, $1 \leq i \leq m$ such that $S_i \tau_{i+1} = S_{i+1}$ for $0 \leq i < m$. The message $\boldsymbol{m} = \tau_1 \ldots \tau_m$ produces the state T from the state S. An argument akin to that used in the foregoing paragraph shows that $\boldsymbol{m}$ is a concise message for S. □

We now establish a result of the same vein for the regular returns of a medium (cf. Definition 2.6.4).

6.1.4 Definition. We recall that a sequence of vertices $\boldsymbol{s}_m = (v_i)_{0 \leq i \leq m}$ such that $\{v_i, v_{i+1}\}$ are edges in a graph is a circuit if $v_m = v_0$ and all the vertices $v_1, \ldots, v_m$ are different (see Section 1.8.5, page 16). By abuse of language, we say that the edges $\{v_i, v_{i+1}\}$ *belong* to the circuit $\boldsymbol{s}_m$, for $0 \leq i \leq m-1$. The circuit $\boldsymbol{s}_m$ is *even* if it has an even number of edges: $m = 2n$; any two of its edges $\{v_i, v_{i+1}\}$ and $\{v_{i+n}, v_{i+n+1}\}$, $0 \leq i \leq n-1$ are then called *opposite*. A circuit is *minimal* if at least one shortest path between any two of its vertices is a segment of the circuit. A graph is *even* if all its circuits are even.

6.1.5 Lemma. *Let $\gamma : \mathcal{S} \to V$ be the representation of a medium $\mathcal{M} = (\mathcal{S}, \mathcal{T})$ by a graph $G = (V, E)$. If $\boldsymbol{m} = \tau_1 \ldots \tau_{2n}$ is a regular return for some state $S \in \mathcal{S}$, then the sequence of vertices $(\gamma(S_i))_{0 \leq i \leq 2n}$, where $S_i = S\tau_0\tau_1 \cdots \tau_i$, for $0 \leq i \leq 2n$, forms an even, minimal circuit of G. Conversely, if a sequence $(\gamma(S_i))_{0 \leq i \leq 2n}$ is an even minimal circuit of G, then $\boldsymbol{m} = \tau_1 \ldots \tau_m$, with $S\tau_0\tau_1 \cdots \tau_i = S_i$ for $0 \leq i \leq 2n$, is a regular return for S in $\mathcal{M}$.*

Note that $S_0 = S_{2n} = S$.

Proof. In the notation of the lemma, let $\boldsymbol{m}$ be a regular return for some state S. Thus, by definition of a regular return (cf. 2.6.4), $\tau_1 \ldots \tau_n$ and $\tilde{\tau}_{2n} \ldots \tilde{\tau}_{n+1}$ are concise messages for S. By Lemma 6.1.3, the sequence of vertices $(\gamma(S_i))_{0\leq i\leq n}$, where $S_i = S\tau_0\tau_1\cdots\tau_i$, for $0 \leq i \leq n$, forms a shortest path joining $\gamma(S)$ and $\gamma(T)$, with $T = S\tau_1\cdots\tau_n$. Similarly, the sequence $\gamma(S_{2n}), \gamma(S_{2n-1}), \ldots, \gamma(S_{n+1})$ is another shortest path joining $\gamma(S)$ and $\gamma(T)$. Since γ is a 1-1 function, all the vertices $\gamma(S_i)$ are distinct, and so the sequence $(\gamma(S_i))_{0\leq i\leq 2n}$ is an even circuit. This circuit is a minimal one. Indeed, by definition of a regular return, all the messages $\tau_i\tau_{i+1}\cdots\tau_{i+n-1}$ are concise for $S\tau_0\tau_1\cdots\tau_{i-1}$. So, by Lemma 6.1.3, all the sequences $\gamma(S_i), \ldots, \gamma(S_{i+n-1})$ are shortest paths between $\gamma(S_i)$ and $\gamma(S_{i+n-1})$, which implies that at least one shortest path between any two vertices of the circuit $(\gamma(S_i))_{0\leq i\leq 2n}$ is a segment of that circuit. We omit the proof of the converse part of this lemma. The argument is based on the converse part of Lemma 6.1.3 and is similar (see Problem 6.2). □

6.1.6 Remark. A close reading of this proof shows that opposite tokens τ_i, $\tau_{i+n} = \tilde{\tau}_i$ in a regular return correspond to opposite edges $\{\gamma(S_i), \gamma(S_{i+1})\}$, $\{\gamma(S_{i+n}), \gamma(S_{i+1+n})\}$ in the even minimal circuit of the representing graph, with $S_{i+1} = S_i\tau_i$ and $S_{i+n} = S_{i+n+1}\tau_{i+n}$.

6.2 Media Inducing Graphs

Our next task is to characterize the graphs representing media in terms of graph concepts. Some necessary conditions are easily inferred from the axioms of a medium. For example, Axiom [Ma] forces the graph to be connected, and [Mb] demands that it is even. By convention, the graph should not have any loops. However, as shown by the two example below, these three conditions are not sufficient to characterize the graph of a medium.

6.2.1 Two Counterexamples. The graphs corresponding to the digraphs A and B in Figure 6.1 are connected and all their circuits are even. Moreover, they have no loops. Yet, neither A nor B can yield the graph of a medium. We leave to the reader to prove this for Figure 6.1A (Problem 6.3).

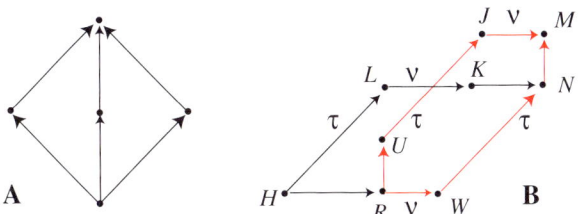

Figure 6.1. Neither of these graphs is that of a medium. The token system corresponding to Digraph **B** contradicts [Ma]. Which of the properties of a medium is contradicted by Digraph **A**?

Here is why in the case of B. The circuit pictured in red is even and minimal. By Lemma 6.1.5, it must represent a regular return in a medium. From Remark 6.1.6, we know that the same token must be matched to opposite edges of the circuit. Accordingly, the same token ν has been assigned to the arcs JM and RW. (To simplify the graph, only one token from each pair of mutually reverse tokens is indicated.) The circuit containing the six vertices L, K, N, W, R and H is also even and minimal. Thus, the arcs LK and RW must be assigned the same token, and since RW has been assigned token ν, that token must also be assigned to TL. The argument governing the placement of the token τ are similar. The consequence, however, is that there is no concise message from L to J: any message producing J from L contains either both ν and $\tilde{\nu}$, or both $\tilde{\tau}$ and τ. Axiom [Ma] is thus violated. This example will be crucial in our understanding of the appropriate axiomatization of a graph capable of representing a medium.

In our failed attempt at representing a medium in Figure 6.1, we have chosen to picture the arcs representing the same token by parallel arcs (forming two sides of an implicit rectangle). The intuition that the opposite arcs of even minimal circuits should be parallel is a sound one, and suggests the construction of an equivalence relation on the set of set of arcs of the digraph. Such a construction is delicate, however, and the two examples of media pictured below by their digraphs must be taken into account.

6.2.2 Examples. Together with the examples of Figure 6.1, Examples A and B in Figure 6.2 will also guide and illustrate our choice of concepts and axioms.

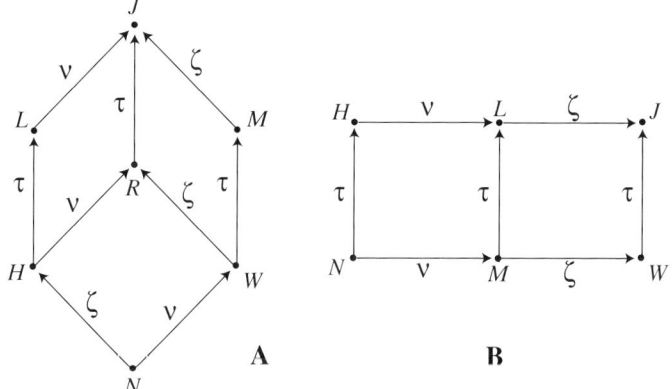

Figure 6.2. Two examples of graphs of media. In Example B, different tokens are assigned to arcs HL and MW, which are are opposite in the circuit N, H, L, J, W, M, N. This circuit is not minimal. Compare with the situation of the arcs LJ and NW in Example A.

6.2.3 Definition. We denote by $\vec{E} = \{ST \mid \{S,T\} \in E\}$ the set of all the arcs of a graph $G = (V, E)$. The *like relation*[2] of the graph G is a relation $\mathfrak{L}$ on $\vec{E}$ defined by

$$ST \mathfrak{L} PQ \iff (\delta(S,P) + 1 = \delta(T,Q) + 1 = \delta(S,Q) = \delta(T,P))$$
$$(\{S,T\}, \{P,Q\} \in E).$$

In Example B in Figure 6.2, we have $NH \mathfrak{L} WJ$ because

$$\delta(H,J) + 1 = \delta(N,W) + 1 = \delta(H,W) = \delta(N,J),$$

but $HL \mathfrak{L} MW$ does not hold because

$$\delta(H,M) = \delta(L,W) = 2 \quad \text{and} \quad 4 = \delta(H,W) \neq \delta(L,M) = 1.$$

The point is that the arcs HL and MW are opposite in the circuit H, L, J, W, M, N, H, but this circuit is not minimal.

The like relation is clearly reflexive and symmetric. Moreover, we have

$$ST \mathfrak{L} PQ \iff TS \mathfrak{L} QP \qquad (\{S,T\}, \{P,Q\} \in E). \qquad (6.3)$$

We now come to the main concept of this chapter. We recall that a graph is bipartite if and only if it is even (König, 1916, and see 1.8.5, page 17).

6.2.4 Definition. Let $G = (V, E)$ be a graph equipped with its like relation $\mathfrak{L}$. The graph G is called *mediatic* if the following three axioms hold.

[G1] G is connected.

[G2] G is bipartite.

[G3] $\mathfrak{L}$ is transitive.

The set of vertices is not assumed to be finite. It is easily verified that any graph isomorphic to a mediatic graph is mediatic (see Problem 6.1).

Axiom [G3] eliminates the counterexample of Figure 6.1B. Indeed, we have

$$LK \mathfrak{L} RW \mathfrak{L} JM \quad \text{but not} \quad LK \mathfrak{L} JM$$

since

$$\delta(L,J) = 4, \quad \delta(K,M) = 2, \quad \delta(L,M) = 3 = \delta(K,J).$$

The following result is immediate.

---

[2] The like relation is analogous to the Djoković-Winkler relation on the edges of a graph, described in Chapter 7; the exact connection between these relations is spelled out in Theorem 7.2.5.

6.2.5 Lemma. *The like relation $\mathfrak{L}$ of a mediatic graph (V, E) is an equivalence relation on $\vec{E}$.*

6.2.6 Definition. We denote by

$$\langle ST \rangle = \{ PQ \in \vec{E} \mid ST \, \mathfrak{L} \, PQ \}$$

the equivalence class containing the arc ST (in the partition of $\vec{E}$ induced by $\mathfrak{L}$).

We will show that a graph representing a medium is mediatic (see Theorem 6.2.9). Our next lemma is the first step.

6.2.7 Lemma. *Let γ be the representation of a medium $\mathcal{M} = (\mathcal{S}, \mathcal{T})$ by a graph $G = (\mathcal{S}, E)$ which is equipped with its like relation $\mathfrak{L}$. Suppose that $\gamma(N)\gamma(S) \, \mathfrak{L} \, \gamma(W)\gamma(Q)$. Then $N\tau = S$ and $W\tau = Q$ for some $\tau \in \mathcal{T}$. In fact, there exists an orderly return $\boldsymbol{q}\tilde{\tau}\tilde{\boldsymbol{w}}\tau$ for S in $\mathcal{M}$, with $S\boldsymbol{q}\tilde{\tau} = S\tilde{\tau}\boldsymbol{w} = W$; thus $\boldsymbol{q}$ and $\boldsymbol{w}$ are concise with $\ell(\boldsymbol{q}) = \ell(\boldsymbol{m})$. Such a circuit is not necessarily regular.*

Proof. We abbreviate our notation and write $S^\gamma = \gamma(S)$ for all $S \in \mathcal{S}$ (for this proof only). By definition, $N^\gamma S^\gamma \, \mathfrak{L} \, W^\gamma Q^\gamma$ implies that $\delta(S^\gamma, Q^\gamma) = \delta(N^\gamma, W^\gamma) = \delta(N^\gamma, Q^\gamma) - 1 = \delta(S^\gamma, W^\gamma) - 1$; so, there are, for some $n \in \mathbb{N}$, two shortest paths

$$S_0^\gamma = S^\gamma, S_1^\gamma, \ldots, S_n^\gamma = Q^\gamma \quad \text{and} \quad N_0^\gamma = N^\gamma, N_1^\gamma, \ldots, N_n^\gamma = W^\gamma$$

between S^γ and Q^γ, and N^γ and W^γ, respectively. Moreover,

$$S_0^\gamma = S^\gamma, S_1^\gamma, \ldots, S_n^\gamma = Q^\gamma, W^\gamma \quad \text{and} \quad N_0^\gamma = N^\gamma, N_1^\gamma, \ldots, N_n^\gamma = W^\gamma, Q^\gamma$$

are also shortest paths. Using Lemma 6.1.3, we can assert the existence of two concise messages $\boldsymbol{q}$ and $\boldsymbol{w}$ such that $S\boldsymbol{q} = Q$ and $N\boldsymbol{w} = W$, with $\ell(\boldsymbol{q}) = \ell(\boldsymbol{w}) = n$. Also, for some tokens τ and μ, we have $N\tau = S$ and $W\mu = Q$ with $\boldsymbol{q}' = \tau\boldsymbol{q}$ and $\boldsymbol{w}' = \tilde{\tau}\boldsymbol{w}$ concise for N and S, respectively, and $\ell(\boldsymbol{q}') = \ell(\boldsymbol{w}') = n + 1$. We are exactly in the situation of Theorem 2.6.2 (see Figure 2.3). Using the implication (iv) $\Rightarrow$ (ii) of this theorem, we obtain $\tau = \mu$. Condition (i) of the same theorem also implies that $\boldsymbol{q}\tilde{\tau}\tilde{\boldsymbol{w}}\tau$ is an orderly return for S, with $S\boldsymbol{q}\tilde{\tau} = S\tilde{\tau}\boldsymbol{w} = W$. The example of Figure 6.3 shows that, with $\boldsymbol{q} = \boldsymbol{w} = \nu\zeta$, such a circuit need not be regular. □

6.2.8 Convention. Any graph representing a medium comes implicitly or explicitly equipped with its like relation $\mathfrak{L}$. When several such graphs are considered, we distinguish their respective like relations by diacritics, such as $\mathfrak{L}'$ or $\mathfrak{L}^*$.

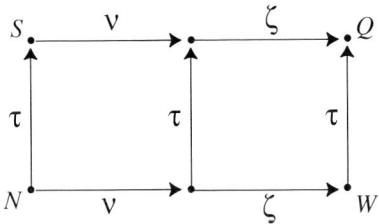

Figure 6.3. Under the hypotheses of Lemma 6.2.7, with $NS\,\mathcal{L}\,WQ$, the orderly return $\boldsymbol{q}\tilde{\tau}\tilde{\boldsymbol{w}}\tau = \nu\zeta\tilde{\tau}\tilde{\nu}\tilde{\zeta}\tau$ for S is not regular. For example, $\zeta\tilde{\tau}\tilde{\zeta}$ is not concise for $S\nu$ (cf. Definition 2.6.4).

6.2.9 Theorem. *Any graph representing a medium is mediatic.*

Proof. Because any graph isomorphic to a mediatic graph is mediatic, we can invoke Lemma 6.1.2 and content ourselves with proving that the graph of a medium is mediatic (which simplifies our notation). Denote the medium by $\mathcal{M} = (\mathcal{S}, \mathcal{T})$, and let $G = (\mathcal{S}, E)$ be its graph. We prove that G satisfies [G1], [G2] and [G3].

[G1] Axiom [Ma] requires that G be connected.

[G2] Axiom [Mb] implies that G must be even. Hence, by König's Theorem, it must be bipartite.

[G3] Suppose that $NS\,\mathcal{L}\,PR\,\mathcal{L}\,WQ$. By Lemma 6.2.7 (applied twice), there must be some tokens τ and μ such that $N\tau = S$, $P\tau = R$, $P\mu = R$ and $W\mu = Q$, so $\tau = \mu$. Let then $\boldsymbol{q}$ and $\boldsymbol{w}'$ be two concise messages from S, and let $\boldsymbol{w}$ and $\boldsymbol{bq}'$ be two concise messages from N, such that

$$S\boldsymbol{q} = Q, \quad S\boldsymbol{w}' = W, \quad N\boldsymbol{w} = W, \quad N\boldsymbol{q}' = Q.$$

The situation is exactly that of Theorem 2.6.2, with the same notation (see Figure 2.3). Because $\tau = \mu$, Condition (ii) of this theorem holds. We conclude that Conditions (iii) and (iv) also hold, which leads to

$$\delta(S, Q) + 1 = \delta(N, W) + 1 = \delta(S, W) = \delta(N, Q).$$

We have thus $NS\,\mathcal{L}\,WQ$; so, $\mathcal{L}$ is transitive. □

We leave the proof of the next lemma as Problem 6.5.

6.2.10 Lemma. *Let $G = (V, E)$ and $G' = (V', E')$ be two mediatic graphs, with their respective like relations $\mathcal{L}$ and $\mathcal{L}'$, and let φ be a bijection of V onto V'. Then φ is an isomorphism of G onto G' if and only if*

$$ST\,\mathcal{L}\,PQ \iff \varphi(S)\varphi(T)\,\mathcal{L}'\,\varphi(P)\varphi(Q) \qquad (S, T, P, Q \in V).$$

6.2.11 Remark. The like relation is the fundamental tool for the study of mediatic graphs. We shall see that any mediatic graph G can be used to construct a medium $\mathcal{M}$ that has G as its graph. Each of the equivalence classes $\langle ST \rangle$ of the partition induced by the like relation contains 'parallel' arcs of the graph, and will turn out to correspond to a particular token τ of the medium under construction, with the class $\langle TS \rangle$ corresponding to the reverse token $\tilde{\tau}$. Before proceeding to such a construction, we establish in Theorem 6.3.1 a useful result which precisely links the isomorphism of media to that of their graphs. We recall (cf. Definition 2.7.1, page 34) that two media $(\mathcal{S}, \mathcal{T})$ and $(\mathcal{S}', \mathcal{T}')$ are isomorphic if there exists a pair (α, β) of bijections $\alpha : \mathcal{S} \to \mathcal{S}'$ and $\beta : \mathcal{T} \to \mathcal{T}'$ such that $S\tau = V$ if and only if $\alpha(S)\beta(\tau) = \alpha(V)$, for all states S and V in $\mathcal{S}$ and tokens τ in $\mathcal{T}$.

6.3 Paired Isomorphisms of Media and Graphs

6.3.1 Theorem. *Suppose that $\mathcal{M} = (\mathcal{S}, \mathcal{T})$ and $\mathcal{M}' = (\mathcal{S}', \mathcal{T}')$ are two media and let $G = (\mathcal{S}, E)$ and $G' = (\mathcal{S}', E')$ be their respective graphs. Then $\mathcal{M}$ and $\mathcal{M}'$ are isomorphic if and only if G and G' are isomorphic; more precisely:*

(i) if (α, β) is an isomorphism of $\mathcal{M}$ onto $\mathcal{M}'$, then $\alpha : \mathcal{S} \to \mathcal{S}'$ is an isomorphism of G onto G' in the sense of (6.2);

(ii) if $\varphi : \mathcal{S} \to \mathcal{S}'$ is an isomorphism of G onto G' in the sense of (6.2), then there exists a bijection $\beta : \mathcal{T} \to \mathcal{T}'$ such that (φ, β) is an isomorphism of $\mathcal{M}$ onto $\mathcal{M}'$.

Proof. (i) Suppose that (α, β) is an isomorphism of $\mathcal{M}$ onto $\mathcal{M}'$. For any two distinct S, T in $\mathcal{S}$, we have successively

$\{S,T\} \in E$
$\iff (\exists \tau \in \mathcal{T})(S\tau = T)$ \qquad (G is the graph of $\mathcal{M}$)
$\iff (\exists \tau \in \mathcal{T})(\alpha(S)\beta(\tau) = \alpha(T))$ \qquad ($\mathcal{M}$ and $\mathcal{M}'$ are isomorphic)
$\iff \{\alpha(S), \alpha(T)\} \in E'$ \qquad (G' is the graph of $\mathcal{M}'$),

and so

$$\{S,T\} \in E \iff \{\alpha(S), \alpha(T)\} \in E' \qquad (S, T \in \mathcal{S},\ S \neq T).$$

We conclude that $\alpha : \mathcal{S} \to \mathcal{S}'$ is an isomorphism of G onto G'.

(ii) Let $\varphi : \mathcal{S} \to \mathcal{S}'$ be an isomorphism of G onto G'. Define a function $\beta : \mathcal{T} \to \mathcal{T}'$ by

$$\beta(\tau) = \tau' \iff (\forall S, T \in \mathcal{S})(S\tau = T \Leftrightarrow \varphi(S)\tau' = \varphi(T)). \qquad (6.4)$$

We first verify that the r.h.s. of the equivalence (6.4) correctly defines β as a bijection of $\mathcal{T}$ onto $\mathcal{T}'$. For any $\tau \in \mathcal{T}$, there exists distinct states S and

T in $\mathcal{S}$ such that $S\tau = T$ and $\{S,T\} \in E$. Fix S and T temporarily. By the isomorphism $\varphi : \mathcal{S} \to \mathcal{S}'$ of G onto G', we have $\{\varphi(S), \varphi(T)\} \in E'$, and because G' is the graph of $\mathcal{M}'$, we necessarily have $\varphi(S)\tau' = \varphi(T)$ for some $\tau' \in \mathcal{T}'$, which is unique by Lemma 2.3.1(iii). The hypothesis that φ is an isomorphism of G onto G' ensures that we must have an equivalence in the r.h.s. of (6.4).

Next, we show that $\beta(\tau)$ does not depend upon the choice of S and T. Let P, Q be another pair of distinct states in $\mathcal{S}$ such that $P\tau = Q$, and let $P = S\boldsymbol{m}$ and $Q = T\boldsymbol{n}$ for some concise messages $\boldsymbol{m} = \tau_1 \ldots \tau_m$ and $\boldsymbol{n} = \mu_1 \ldots \mu_n$. By Lemma 2.2.7, $\tau \boldsymbol{n}$ and $\boldsymbol{m}\tau$ are concise messages. Invoking the implication (ii) $\Rightarrow$ (iii) in Theorem 2.6.2, we get $\ell(\boldsymbol{m}) = \ell(\boldsymbol{n})$ and $\mathcal{C}(\boldsymbol{m}) = \mathcal{C}(\boldsymbol{n})$, and so $m = n$. Denote by $\mathcal{L}$ and $\mathcal{L}'$ the like relations of G and G' respectively. We have thus shown that $ST \mathcal{L} PQ$. By Lemma 6.2.10, we also have

$$\varphi(S)\varphi(T) \mathcal{L}' \varphi(P)\varphi(Q).$$

Because $\varphi(S)\tau' = \varphi(T)$, Lemma 6.2.7 applies, yielding $\varphi(P)\tau' = \varphi(Q)$. Thus, β is a well-defined function.

We still have to prove that β is indeed a bijection. For any $\tau' \in \mathcal{T}'$ there are some $S', T' \in \mathcal{T}'$ such that $S'\tau' = T'$. We have thus $\{S', T'\} \in E'$, and since φ is an isomorphism of G onto G', also $\{\varphi^{-1}(S'), \varphi^{-1}(T')\} \in E$, with $\varphi^{-1}(S')\tau = \varphi^{-1}(T')$ for some $\tau \in \mathcal{T}$. Thus β maps $\mathcal{T}$ onto $\mathcal{T}'$. Suppose now that $\beta(\tau) = \beta(\mu) = \tau' \in \mathcal{T}'$. This implies that for some $S, T, P, Q \in \mathcal{S}$ and $N, M \in \mathcal{S}'$, we must have

$$S\tau = T, \quad P\mu = Q, \text{ and } N\tau' = M, \tag{6.5}$$

together with

$$\varphi(S) = \varphi(P) = N \text{ and } \varphi(T) = \varphi(Q) = M$$

by the definition of β. As φ is a 1-1 function, we obtain $S = P$ and $T = Q$ in (6.5). Using Lemma 2.3.1(ii), we get $\tau = \mu$. Thus, β is a 1-1 function and so a bijection.

The fact that (φ, β) is an isomorphism of $\mathcal{M}$ onto $\mathcal{M}'$ follows from the definition of β by (6.4). We have

$$S\tau = T \iff \varphi(S)\beta(\tau) = \varphi(T) \qquad (S, T \in \mathcal{S})$$

whether or not $\{S, T\} \in E$. $\square$

Having defined the graph of a medium and shown that such a graph is necessarily mediatic, we now go in the opposite direction and construct a medium from an arbitrary mediatic graph.

6.4 From Mediatic Graphs to Media

6.4.1 Definition. Let $G = (\mathcal{S}, E)$ be a mediatic graph equipped with its like relation $\mathcal{L}$. For any $ST \in \vec{E}$, define a function

$$\tau_{ST} : \mathcal{S} \to \mathcal{S} : P \mapsto P\tau_{ST}$$

by the formula

$$P\tau_{ST} = \begin{cases} Q & \text{if } ST \mathcal{L} PQ, \\ P & \text{otherwise.} \end{cases} \qquad (6.6)$$

We denote by $\mathcal{T} = \{\tau_{ST} \mid ST \in \vec{E}\}$ the set containing all those transformations. It is easily verified that the pair $(\mathcal{S}, \mathcal{T})$ is a token system. Such a token system is said to be *induced* by the mediatic graph G. Theorem 6.4.3 establishes that a token system $\mathcal{K}$ induced by a mediatic graph G is in fact a medium. We say that $\mathcal{K}$ is the *medium of the graph* G. Notice that, since $\mathcal{L}$ is an equivalence relation on $\vec{E}$, we have $\tau_{ST} = \tau_{PQ}$ whenever $ST \mathcal{L} PQ$. In such a case, we have in fact $\langle ST \rangle = \langle PQ \rangle$. The choice of a particular pair $ST \in \langle PQ \rangle$ to denote a token τ_{ST} is thus arbitrary. Occasionally, we may in fact vary our notation so as to make an argument more intuitive or immediate. Notice that, as a consequence of this definition, whenever $\{S, T\} \in E$, then also $ST \mathcal{L} ST$, and so $S\tau_{ST} = T$.

This construction is motivated by Theorem 6.4.3. We first prove a lemma.

6.4.2 Lemma. *Let $(\mathcal{S}, \mathcal{T})$ be the token system constructed in Definition 6.4.1. For any $\{S, T\} \in E$ the tokens τ_{ST} and τ_{TS} defined by (6.6) are mutual reverses.*

Proof. For any distinct states S, T, P and Q, we have

$$P\tau_{ST} = Q \iff ST \mathcal{L} PQ \qquad \text{(by definition)}$$
$$\iff TS \mathcal{L} QP \qquad \text{(by (6.3))}$$
$$\iff Q\tau_{TS} = P \qquad \text{(by definition),}$$

and so τ_{TS} is the reverse of τ_{ST}. □

6.4.3 Theorem. *The token system $(\mathcal{S}, \mathcal{T})$ induced, in the sense of 6.4.1, by a mediatic graph $G = (\mathcal{S}, E)$ is a medium.*

Proof. We verify that $(\mathcal{S}, \mathcal{T})$ satisfies Axioms [Ma] and [Mb] of a medium.

[Ma] For any $S, T \in \mathcal{S}$, there is a shortest path $S_0 = S, S_1, \ldots, S_n = T$ between S and T in G. This implies that, for $0 \leq i \leq n-1$, we have $\{S_i, S_{i+1}\} \in E$, which yields $S_i \tau_{S_i S_{i+1}} = S_{i+1}$. It follows that the message $m = \tau_{S_0 S_1} \ldots \tau_{S_{n-1} S_n}$ produces T from S and is stepwise effective. To prove

6.4 From Mediatic Graphs to Media

that m is concise, we must still show that it is consistent and without repetitions. The message m is consistent since otherwise we would have

$$S_h \tau_{MN} = S_{h+1} \quad \text{and} \quad S_k \tau_{NM} = S_{k+1} \tag{6.7}$$

for some indices h and k, with $h < k$, and some $NM \in \vec{E}$. Since τ_{MN} is the reverse of τ_{NM}, the last equality in (6.7) can be rewritten as $S_{k+1}\tau_{MN} = S_k$. Thus, by definition of the tokens in (6.6), the above statement (6.7) leads to $S_h S_{h+1} \mathfrak{L} MN \mathfrak{L} S_{k+1} S_k$ which, by transitivity, gives $S_k S_{k+1} \mathfrak{L} S_{h+1} S_h$. Because $h < k$, we can apply the definition of the like function $\mathfrak{L}$ in 6.2.3 and derive

$$k + 1 - h = \delta(S_{k+1}, S_h) = \delta(S_k, S_{h+1}) = k - 1 - h$$

yielding the absurdity $1 = -1$. Thus, m is consistent. Suppose that m has repeated tokens, say $S_i \tau_{S_i S_{i+1}} = S_{i+1}$ and $S_{i+k} \tau_{S_i S_{i+1}} = S_{i+k+1}$ for some indices $0 \le i < n$ and $0 \le i+k < n$. This would give $S_i S_{i+1} \mathfrak{L} S_{i+k} S_{i+k+1}$, leading to

$$d(S_i, S_{i+k+1}) = k + 1 > k - 1 = d(S_{i+1}, S_{i+k}),$$

while by the definition of $\mathfrak{L}$ we should have $d(S_i, S_{i+k+1}) = d(S_{i+1}, S_{i+k})$, a contradiction. Thus, the message m is concise.

[Mb] Let $m = \tau_{S_0 S_1} \tau_{S_1 S_2} \ldots \tau_{S_{n-1} S_n}$ be a return message for some state S; we have thus $S_0 = S_n = S$. In the terminology of G, we have a closed walk $S = S_0, S_1, \ldots, S_n = S$. We denote this closed walk by $\mathbf{W}$ and we write $\vec{E}_{\mathbf{W}}$ for the set of all its arcs $S_i S_{i+1}$, $0 \le i \le n-1$. By [G2] and König's Theorem, such a closed walk is even; so $n = 2q$ for some $q \in \mathbb{N}$. We prove by induction on q that m is vacuous. The case $q = 1$ (the smallest possible return) is trivial, so we suppose that [Mb] holds for any $1 \le p < q$ and prove that [Mb] also holds for $q = p$. We consider two cases.

Case 1: $\mathbf{W}$ is an isometric subgraph of G. Thus, $\mathbf{W}$ is a minimal circuit of G. Take any token $\tau_{S_i S_{i+1}}$ in m. Since (with the addition modulo k in the indices), we have for $0 \le i < n$

$$\delta(S_{i+1}, S_{i+k}) = \delta(S_i, S_{i+k+1}) = k - 1,$$
$$\delta(S_i, S_{i+k}) = \delta(S_{i+1}, S_{i+k+1}) = k,$$

we obtain $S_i S_{i+1} \mathfrak{L} S_{i+k+1} S_{i+k}$. By the definition of the tokens in (6.6) and the transitivity and symmetry of $\mathfrak{L}$, we get for any $P, Q \in \mathcal{S}$

$$P\tau_{S_i S_{i+1}} = Q \iff S_i S_{i+1} \mathfrak{L} PQ$$
$$\iff S_{i+k+1} S_{i+k} \mathfrak{L} PQ$$
$$\iff P\tau_{S_{i+k+1} S_{i+k}} = Q$$
$$\iff Q\tau_{S_{i+k} S_{i+k+1}} = P.$$

We conclude that $\tau_{S_{i+k} S_{i+k+1}}$ and $\tau_{S_i S_{i+1}}$ are mutual reverses, and so m is vacuous. (Note that the induction hypothesis has not been used here.)

134   6 Mediatic Graphs

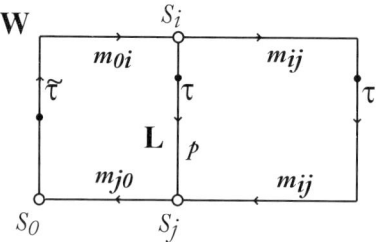

Figure 6.4. Case 2 in the proof of Axiom [Mb] in Theorem 6.4.3: the closed walk **W** is not an isometric subgraph.

Case 2: **W** is not an isometric subgraph of G. Then, there are two vertices S_i and S_j in **W**, with $i < j$, and a shortest path **L** from S_i to S_j in G with $\delta_{ij} = \delta(S_i, S_j) < \min\{j-i, i+n-j\}$ (see Figure 6.4). Thus, $j-i$ and $i+n-j$ are the lengths of the two segments of **W** with endpoints S_i and S_j. For simplicity, we can assume without loss of generality that S_i and S_j are the only vertices of **L** that are also in **W**. Let p the straight message producing S_j from S_i and corresponding to the shortest path **L** in the sense of Lemma 6.1.3. We also split m into the three messages:

$$m_{0i} = \tau_{S_0 S_1} \ldots \tau_{S_{i-1} S_i}$$
$$m_{ij} = \tau_{S_i S_{i+1}} \ldots \tau_{S_{j-1} S_j}$$
$$m_{j0} = \tau_{S_j S_{j+1}} \ldots \tau_{S_{n-1} S_0}.$$

We have thus $m = m_{0i} m_{ij} m_{j0}$. Note that the two messages $m_{0i} p m_{j0}$ and $\widetilde{p} m_{ij}$ have a length strictly smaller that $n = 2q$. By the induction hypothesis, these two messages are vacuous. Accordingly, for any token τ of p, there is an reverse token $\tilde{\tau}$ either in m_{0i} or in m_{j0}. (In Figure 6.4 the token $\tilde{\tau}$ is pictured as being part of m_{0i}.) Considered from the viewpoint of the message $\widetilde{p} m_{ij}$ from S_j, the token $\tilde{\tau}$ is in $\widetilde{p}$ with its reverse τ in m_{ij}. The two reverses of the tokens in p and $\widetilde{p}$, form a pair of mutually reverse tokens $\{\tau, \tilde{\tau}\}$ in m. Such a pair can be obtained for any token τ in p. Augmenting the set of all those pairs by the set of mutually reverse tokens in m_{0i}, m_{ij} and m_{j0}, we obtain a partition of the set $\mathcal{C}(m)$ into pairs of mutually reverse tokens, which establishes that the message m is vacuous.

We have shown that the token system $(\mathcal{S}, \mathcal{T})$ satisfies Axioms [Ma] and [Mb]. The proof is thus complete. □

6.4.4 Remark. In the above proof, the inductive argument used to establish Case 2 of [Mb] may convey the mistaken impression that the situation is always straightforward. The graph pictured in Figure 6.4 is actually glossing over some intricacies. The non-isometric subgraph **W** is pictured in red in Figure 6.5 and is not 'convex.' We can see how the inductive stage splitting the closed walk **W** by the shortest path **L** may lead to form, in each of the two smaller closed walks, pairs $\{\mu, \tilde{\mu}\}$ and $\{\nu, \tilde{\nu}\}$ which correspond in fact to the same pair of tokens in **W**. Since the arcs corresponding to μ and ν are in the like relation $\mathcal{L}$, the mistaken assignment is temporary.

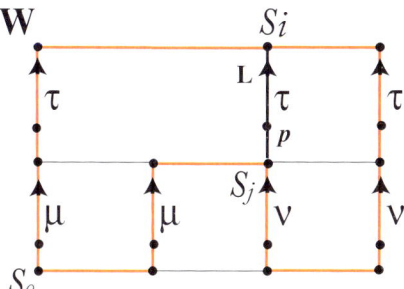

Figure 6.5. The non-isometric subgraph **W** of Case 2 in the proof of [M3] in Theorem 6.4.3 is pictured in red. The inductive stage of the proof leads to form temporarily, in each of the two smaller closed walks delimited by the shortest path **L**, pairs $\{\mu, \tilde{\mu}\}$ and $\{\nu, \tilde{\nu}\}$ corresponding to the same pair of mutually reverse tokens in **W**.

We finally obtain:

6.4.5 Theorem. *Let $\mathcal{S}$ an arbitrary set, with $|\mathcal{S}| \geq 2$. Denote by $\mathfrak{M}$ the set of all media on $\mathcal{S}$, and by $\mathfrak{G}$ the set of all mediatic graphs on $\mathcal{S}$. There exists a bijection $\mathfrak{f} : \mathfrak{M} \to \mathfrak{G} : \mathcal{M} \mapsto \mathfrak{f}(\mathcal{M})$ such that $G = \mathfrak{f}(\mathcal{M})$ is the graph of $\mathcal{M}$ in the sense of Definition 6.1.1 if and only if $\mathcal{M}$ is the medium of the mediatic graph G in the sense of Definition 6.4.1.*

Proof. Because the set $\mathcal{S}$ of states is constant in $\mathfrak{M}$ and confounded with the constant set of vertices in $\mathfrak{G}$, we could reinterpret the function $\mathfrak{f}$ as a mapping of the family $\mathfrak{T}$ of all sets of token $\mathcal{T}$ making $(\mathcal{S}, \mathcal{T})$ a medium, into the family $\mathfrak{E}$ of all sets of edges E making $(\mathcal{S}, E)$ a mediatic graph. However, any set of edges E of a mediatic graph on $\mathcal{S}$ is characterized by its like relation $\mathcal{L}$, or equivalently, by the partition of $\vec{E}$ induced by $\mathcal{L}$. We choose the latter characterization for the purpose of this proof, and denote by $\vec{\mathfrak{E}}_\mathcal{L}$ the set of all the partitions of the sets of arcs $\vec{E}$ induced by the like relations characterizing the sets of edges in the collection $\mathfrak{E}$.

From Lemmas 6.2.9 and 6.4.3, we know that the graph of a medium is mediatic, and that the token system induced by a mediatic graph is a medium. We have to show that the functions

$$\mathfrak{f} : \mathfrak{T} \to \vec{\mathfrak{E}}_\mathcal{L} \quad \text{and} \quad \mathfrak{g} : \vec{\mathfrak{E}}_\mathcal{L} \to \mathfrak{T}$$

implicitly defined by (6.1) and (6.6), respectively, are mutual inverses. Note that, for any $\mathcal{T} \in \mathfrak{T}$, the partition $\mathfrak{f}(\mathcal{T})$ is defined via a function f mapping $\mathcal{T}$

into the the partition $\mathfrak{f}(\mathcal{T})$. Writing as before $\langle ST \rangle$ for the equivalence class containing the arc ST, we have

$$P\tau = Q \iff f(\tau) = \langle PQ \rangle \qquad (\tau \in \mathcal{T};\ P,Q \in \mathcal{S}). \qquad (6.8)$$

Proceeding similarly, but inversely, for the function $\mathfrak{g}$, we notice that it defines, for each $\vec{E}_\mathcal{L}$ in $\vec{\mathcal{E}}_\mathcal{L}$ the set of tokens $\mathfrak{g}(\vec{E}_\mathcal{L})$ via a function g mapping $\vec{E}_\mathcal{L}$ into the set of tokens $\mathfrak{g}(\vec{E}_\mathcal{L})$; we obtain

$$\langle ST \rangle = \langle PQ \rangle \iff Pg(\langle ST \rangle) = Q \qquad (S,T,P,Q \in \mathcal{S}). \qquad (6.9)$$

Combining (6.8) and (6.9) we obtain

$$P\tau = Q \iff f(\tau) = \langle PQ \rangle \iff P(g \circ f)(\tau) = Q \quad (\tau \in \mathcal{T};\ P,Q \in \mathcal{S}).$$

We have thus $g = f^{-1}$ and so $\mathfrak{g} = \mathfrak{f}^{-1}$. Conversely, we have

$$\langle ST \rangle = \langle PQ \rangle \iff Pg(\langle ST \rangle) = Q \iff (f \circ g)(\langle ST \rangle) = \langle PQ \rangle$$
$$(S,T,P,Q \in \mathcal{S}),$$

yielding $f = g^{-1}$ and so $\mathfrak{f} = \mathfrak{g}^{-1}$. □

Problems

6.1 Prove that any graph isomorphic to a mediatic graph is mediatic.

6.2 Prove the converse in Lemma 6.1.5.

6.3 Prove that Digraph **A** in Figure 6.1 cannot give the graph of a medium.

6.4 Prove that the three axioms [G2], [G3] and [G3] defining a mediatic graph are independent.

6.5 Prove Lemma 6.2.10.

6.6 Suppose that $(\mathcal{S}, \mathcal{T})$ is a token system induced by a mediatic graph $(\mathcal{S}, E)$ in the sense of Definition 6.4.1. Prove that any concise message $\tau_1 \ldots \tau_n$ producing a state T from a state S corresponds to a shortest path

$$S = S_0, S_1 = S_0\tau_1, \ldots, S_n = T = S_{n-1}\tau_n$$

from S to T.

6.7 Prove that Definition 6.4.1 and Eq. (6.6) properly define a token system.

6.8 Can any closed walk $S = S_0, S_1, \ldots, S_{2n} = S$ that is not a circuit be decomposed into segments $S_i, S_{i+1}, \ldots, S_{i+2k} = S_i$ that form circuits? If this is not true, give a counterexample and formulate the correct decomposition.

6.9 Is it true that, given a medium and its representing graph, there is a 1-1 correspondence between the orderly returns of the medium and the minimal circuits of the graph? Can you formulate a correct statement in this regard?

6.10 Show how [G3] rules out the Counterexample 6.2.1B.

6.11 Prove the statement made in Remark 6.1.6.

6.12 Derive a contradiction to the axioms of a mediatic graph or of Remark 6.1.6 in the graph pictured in Figure 6.6 on the right. A failed attempt at an assignment of the tokens to the edges should facilitate your analysis of the situation.

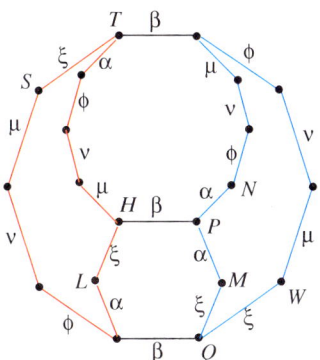

Figure 6.6. Why is this not the graph of a medium?

6.13 Let (G, E) be a graph equipped with its like relation $\mathcal{L}$. Suppose that Axioms [G1] and [G2] of a mediatic graph hold, but not [G3]. Can you obtain a mediatic graph, and thus also a medium, by simply constructing the transitive closure of $\mathcal{L}$? Prove your answer.

7

Media and Partial Cubes

Isometric subgraphs of hypercubes, or 'partial cubes', were first studied by Graham and Pollak (1971) in connection with modeling communication networks. Partial cubes are graphs that are isometrically embeddable into hypercubes (Imrich and Klavžar, 2000). We use partial cubes in our studies of structural properties of wg-families of sets and media. In particular, we characterize mediatic graphs as partial cubes.

7.1 Partial Cubes and Mediatic Graphs

The *n-dimensional cube* (or *n-cube*) is the graph on $\{0,1\}^n$ in which two vertices form an edge if and only if they differ in exactly one position. The n-cube can be equivalently defined as the graph on the power set $\mathfrak{P}(X)$, where X is a set of cardinality n; two sets $P, Q \in \mathfrak{P}(X)$ form an edge if and only if $|P \triangle Q| = 1$. Following R. Rado's suggestion, Djoković (1973) extended the latter definition to arbitrary sets X.

7.1.1 Definition. The *cube* on a set X is the graph

$$\mathcal{H}(X) = (\mathfrak{P}_\text{F}(X), E)$$

where

$$E = \{\{P, Q\} \subseteq \mathfrak{P}_\text{F}(X) \,|\, |P \triangle Q| = 1\}.$$

A graph G is called a *partial cube* if it is isometrically embeddable into the cube $\mathcal{H}(X)$ for some set X. Thus, isometric subgraphs of the cube $\mathcal{H}(X)$ are partial cubes on X.

7.1.2 Example. Figure 7.1 depicts the four-dimensional cube $\mathcal{H}(X)$ on the set $X = \{a, b, c, d\}$. This graph has $2^4 = 16$ vertices and $\frac{1}{2} \cdot 4 \cdot 2^4 = 32$ edges.

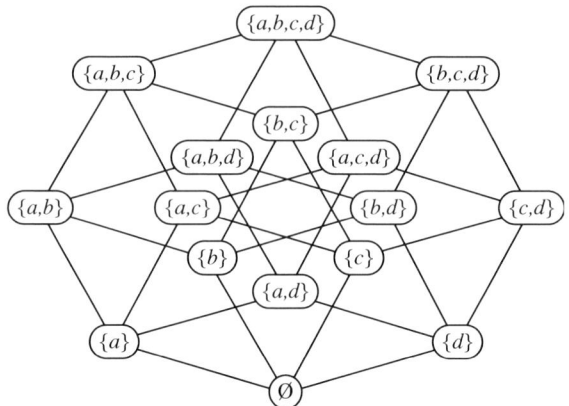

Figure 7.1. The cube $\mathcal{H}(\{a, b, c, d\})$.

7.1.3 Theorem. *The graph distance δ on the cube $\mathcal{H}(\mathcal{X})$ coincides with the distance d between sets in $\mathfrak{P}_F(\mathcal{X})$:*

$$\delta(P, Q) = d(P, Q) = |P \triangle Q|,$$

for all $P, Q \in \mathfrak{P}_F(\mathcal{X})$.

Proof. Take any two distinct sets $P, Q \in \mathfrak{P}_F(\mathcal{X})$, and consider a sequence of sets obtained from P by removing successively all the elements from $P \setminus Q$ and then adding successively all the elements of $Q \setminus P$ until the set Q is obtained. This sequence of sets form a path of length $d(P, Q)$. Therefore, $\delta(P, Q) \leq d(P, Q)$. In any path from P to Q two adjacent vertices differ by one element; the sets P and Q differ by $|P \triangle Q|$ elements. Hence,

$$\delta(P, Q) \geq |P \triangle Q| = d(P, Q).$$

The result follows. □

From this result and the definitions of a partial cube and of a wg-family of sets (cf. Definition 3.1.5), we obtain immediately:

7.1.4 Theorem. *An induced subgraph of a cube $\mathcal{H}(\mathcal{X})$ is a partial cube on $\mathcal{X}$ if and only if the set of its vertices is a wg-family of sets.*

7.1.5 Example. Let $\mathcal{X} = \{x_1, \ldots, x_n\}$ be a set of cardinality n. The family

$$\{\varnothing, \{x_1\}, \{x_1, x_2\}, \ldots, \{x_1, \ldots, x_n\}\}$$

is well-graded and induces a path of length n in $\mathcal{H}(\mathcal{X})$. For instance, the family $\{\varnothing, \{a\}, \{a, b\}, \{a, b, c\}\}$ defines a path of length three in Figure 7.2. Accordingly, any path is a partial cube.

7.1.6 Example. Let $X = \{a, b, c\}$. The family of sets

$$\{\emptyset, \{a\}, \{a, b\}, \{a, b, c\}, \{b, c\}\}$$

is not well-graded and induces a path of length 4 in $\mathcal{H}(X)$ (see Figure 7.2). This path is a partial cube but not a partial cube on the set X.

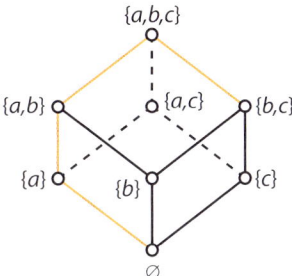

Figure 7.2. A nonisometric path in the cube $\mathcal{H}(\{a, b, c\})$.

By Theorem 3.3.4, any medium is isomorphic to the representing medium of a wg-family of finite sets. It is also clear that the graph of the representing medium of a wg-family $\mathcal{F}$ is the partial cube induced by $\mathcal{F}$. Thus we have the following result:

7.1.7 Theorem. *A graph G represents a medium if and only if G is a partial cube.*

Let G be a partial cube. As such, G can be isometrically embedded into some cube $\mathcal{H}(X)$. This graph admits isometric representations as partial cubes on various sets X. For instance, the graph K_2 (the complete graph on two vertices) can be isometrically embedded in different ways into any cube $\mathcal{H}(X)$ with $|X| > 2$. It is desirable to 'minimize' the class of cubes $\mathcal{H}(X)$ that can be used as target graphs for isometric embeddings of G. We do it by 'retracting the domain' of a wg-family of sets.

7.1.8 Definition. Let $\mathcal{F}'$ be a family of subsets of a set X'. We define the *retraction* of $\mathcal{F}'$ as a family $\mathcal{F}$ of subsets of $X = \cup \mathcal{F}' \setminus \cap \mathcal{F}'$ consisting of the intersections of sets in $\mathcal{F}'$ with X. It is clear that $\mathcal{F}$ satisfies the following two conditions

$$\cap \mathcal{F} = \emptyset \quad \text{and} \quad \cup \mathcal{F} = X. \tag{7.1}$$

From the media theory point of view, the retraction $\mathcal{F}$ of $\mathcal{F}'$ is obtained by eliminating 'inactive' elements from the set X' (cf. Definition 3.2.1).

7.1.9 Theorem. *The retraction $\mathcal{F}$ of a wg-family $\mathcal{F}'$ is a wg-family. The partial cubes induced by $\mathcal{F}'$ and $\mathcal{F}$ are isomorphic and the respective representing media are isomorphic.*

Proof. It suffices to prove that the metric spaces $\mathcal{F}'$ and $\mathcal{F}$ are isometric. We define a mapping $\alpha : \mathcal{F}' \to \mathcal{F}$ by $P \mapsto P \cap \mathcal{X}$. Clearly, α is surjective. We have

$$(P \cap \mathcal{X}) \triangle (Q \cap \mathcal{X}) = (P \triangle Q) \cap \mathcal{X} = (P \triangle Q) \cap (\cup \mathcal{F}' \setminus \cap \mathcal{F}') = P \triangle Q.$$

Thus, $d(\alpha(P), \alpha(Q)) = d(P, Q)$. Consequently, α is an isometry. □

We now show that the classes of mediatic graphs and partial cubes are identical.

7.1.10 Theorem. *A graph is mediatic if and only if it is a partial cube.*

Proof. (Necessity.) By Theorem 6.4.5, a mediatic graph is the graph of a medium. Therefore, by Theorem 7.1.7, it is a partial cube.

(Sufficiency.) Let G be a partial cube. By Theorem 7.1.9, we may assume that G is induced by a wg-family $\mathcal{F}$ of finite subsets of some set $\mathcal{X}$ satisfying conditions (7.1). By Theorems 3.3.1 and 6.2.9, G is a mediatic graph. □

Note that the concepts of mediatic graphs and partial cubes are introduced in different ways: the former one uses intrinsic structures of a graph and their properties, whereas the latter one involves extrinsic objects (cubes). In the next two sections we give intrinsic characterizations of partial cubes.

7.2 Characterizing Partial Cubes

Partial cubes were characterized by Djoković (1973) and Winkler (1984) (see Imrich and Klavžar, 2000, for references to other characterizations of partial cubes). Before presenting these results, we review some notions and facts from graph theory.

Only connected graphs are considered in this section.

7.2.1 Definition. Let $G = (V, E)$ be a graph and δ be its distance function. For any edge $\{a, b\} \in E$ we define a subset W_{ab} of V by

$$W_{ab} = \{w \in V \mid \delta(w, a) < \delta(w, b)\}.$$

Following Eppstein (2005b), we call the sets W_{ab} (and their corresponding induced subgraphs) *semicubes* of the graph G. The semicubes W_{ab} and W_{ba} are called *opposite semicubes*.

Clearly, two opposite semicubes are disjoint. The pairs of opposite semicubes provide a characterization of bipartite graphs.

7.2.2 Theorem. *A graph $G = (V, E)$ is bipartite if and only if, for any edge $\{a, b\}$, the semicubes W_{ab} and W_{ba} form a partition of V.*

Proof. We recall that a connected graph G is bipartite if and only if for every vertex x there is no edge $\{a,b\}$ with $\delta(x,a) = \delta(x,b)$ (see, for instance, Asratian et al., 1998). Now it suffices to note that for any edge $\{a,b\}$ and $x \in V$,
$$\delta(x,a) = \delta(x,b) \quad \Longleftrightarrow \quad x \notin W_{ab} \cup W_{ba}.$$
□

The result of the following lemma is instrumental and will be used often in the chapter.

7.2.3 Lemma. *If $x \in W_{ab}$ for some semicube W_{ab}, then*
$$\delta(x,b) = \delta(x,a) + 1.$$

Proof. For any $x \in W_{ab}$, we have, by the definition of a semicube and the triangle inequality, we have
$$\delta(x,a) < \delta(x,b) \leq \delta(x,a) + \delta(a,b) = \delta(x,a) + 1.$$
The result follows, since δ takes values in $\mathbb{N}$. □

Two binary relations on the set of edges of a graph play a central role in characterizing partial cubes. They are Djoković's relation θ (Djoković, 1973) and Winkler's relation Θ (Winkler, 1984).

7.2.4 Definition. Let $G = (V,E)$ be a graph.
(i) The relation θ on E is defined by
$$\{x,y\}\theta\{u,v\} \Leftrightarrow (\{u,v\} \text{ joins a vertex in } W_{xy} \text{ with a vertex in } W_{yx}) \quad (7.2)$$

(ii) The relation Θ on E is defined by
$$\{x,y\}\Theta\{u,v\} \quad \Leftrightarrow \quad \delta(x,u) + \delta(y,v) \neq \delta(x,v) + \delta(y,u) \quad (7.3)$$

(see Figure 7.3).

The like relation $\vec{\mathfrak{L}}$ (see Definition 6.2.3) on the set $\vec{E}$ of arcs of a graph $G = (V,E)$ and Winkler's relation Θ on the set E of edges are closely related concepts as the following theorem asserts.

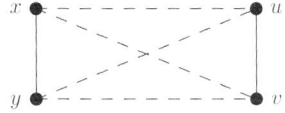

Figure 7.3. Definition of Θ.

7.2.5 Theorem. *Let $\{x,y\}$ and $\{u,v\}$ be two pairs of adjacent vertices of a graph $G = (V,E)$. Then*
$$xy \, \mathfrak{L} \, uv \quad \Rightarrow \quad \{x,y\}\Theta\{u,v\},$$
where $xy, uv \in \vec{E}$ and $\mathfrak{L}$ is the like relation on $\vec{E}$. If G is a bipartite graph, then
$$\{x,y\}\Theta\{u,v\} \quad \Rightarrow \quad (xy \, \mathfrak{L} \, uv \text{ or } xy \, \mathfrak{L} \, vu).$$

Proof. Let $xy \, \mathfrak{L} \, uv$, that is,
$$\delta(x,u) + 1 = \delta(y,v) + 1 = \delta(x,v) = \delta(y,u).$$
By (7.3), we have $\{x,y\}\Theta\{u,v\}$.

Suppose now that G is bipartite and $\{x,y\}\Theta\{u,v\}$. By Theorem 7.2.2, either $x \in W_{uv}$ or $x \in W_{vu}$. If $x \in W_{uv}$, then, by Lemma 7.2.3,
$$\delta(x,v) = \delta(x,u) + 1,$$
and, by (7.3),
$$\delta(y,v) \neq \delta(y,u) + 1,$$
which implies, by Theorem 7.2.2 and Lemma 7.2.3, that
$$\delta(y,u) = \delta(y,v) + 1.$$
By applying the triangle inequality twice, we obtain
$$\delta(x,u) + 1 = \delta(x,v) \leq \delta(y,v) + \delta(x,y) = \delta(y,v) + 1 =$$
$$= \delta(y,u) \leq \delta(x,u) + \delta(x,y) = \delta(x,u) + 1.$$
It follows that
$$\delta(x,u) + 1 = \delta(y,v) + 1 = \delta(x,v) = \delta(y,u),$$
that is, $xy \, \mathfrak{L} \, uv$.

A similar argument shows that $xy \, \mathfrak{L} \, vu$ if $x \in W_{vu}$. □

It is clear that both relations θ and Θ are reflexive and Θ is symmetric; in fact, so is θ:

7.2.6 Lemma. *The relation θ is a symmetric relation on E.*

Proof. Suppose that $\{x,y\}\theta\{u,v\}$ with $u \in W_{xy}$ and $v \in W_{yx}$; the case that $u \in W_{yx}$ and $v \in W_{xy}$ is symmetric. By Lemma 7.2.3 and the triangle inequality, we have
$$\delta(u,x) = \delta(u,y) - 1 \leq \delta(u,v) + \delta(v,y) - 1 = \delta(v,y) =$$
$$= \delta(v,x) - 1 \leq \delta(v,u) + \delta(u,x) - 1 = \delta(u,x).$$
Hence,
$$\delta(u,x) = \delta(v,x) - 1 \quad \text{and} \quad \delta(v,y) = \delta(u,y) - 1.$$
Therefore, $x \in W_{uv}$ and $y \in W_{vu}$. It follows that $\{u,v\}\theta\{x,y\}$. □

7.2.7 Lemma. $\theta \subseteq \Theta$.

Proof. Let $\{x,y\}\theta\{u,v\}$ with $u \in W_{xy}$ and $v \in W_{yx}$. By Lemma 7.2.3,

$$\delta(x,u) + \delta(y,v) = \delta(x,v) - 1 + \delta(y,u) - 1 \neq \delta(x,v) + \delta(y,u).$$

Hence, $\{x,y\}\Theta\{u,v\}$. □

In all but one case, the relations θ and Θ are distinct (see Problem 7.5).

7.2.8 Theorem. *A graph G is bipartite if and only if $\theta = \Theta$.*

Proof. (Necessity.) Suppose that $\neg(\{x,y\}\theta\{u,v\})$. Then, by Theorem 7.2.2, either $u,v \in W_{xy}$ or $u,v \in W_{yx}$. We assume that $u,v \in W_{xy}$ (the other case is treated similarly). By Lemma 7.2.3, we have

$$\delta(x,u) + \delta(y,v) = \delta(y,u) - 1 + \delta(x,v) + 1 = \delta(x,v) + \delta(y,u),$$

so $\neg(\{x,y\}\Theta\{u,v\})$. It follows that $\Theta \subseteq \theta$. Applying Lemma 7.2.7, we get $\theta = \Theta$.

(Sufficiency.) Suppose that G is not bipartite. By Theorem 7.2.2, there is an edge $\{x,y\}$ such that $W_{xy} \cup W_{yx}$ is a proper subset of V. Since G is connected, there is an edge $\{u,v\}$ with $u \notin W_{xy} \cup W_{yx}$ and $v \in W_{xy} \cup W_{yx}$. Clearly, $\neg(\{u,v\}\theta\{x,y\})$. On the other hand,

$$\delta(x,u) + \delta(y,v) \neq \delta(x,v) + \delta(y,u),$$

since $u \notin W_{xy} \cup W_{yx}$ and $v \in W_{xy} \cup W_{yx}$. Thus, $\{x,y\}\Theta\{u,v\}$, a contradiction, since we assumed that $\theta = \Theta$. □

We need one more concept from graph theory.

7.2.9 Definition. Let $G = (V,E)$ be a graph. A subset $W \subseteq V$ is *convex* in G if

$$\delta(a,x) + \delta(x,b) = \delta(a,b) \quad \Rightarrow \quad x \in W,$$

for $a,b \in W$, $x \in V$.

7.2.10 Remark. A semicube of a bipartite graph is not necessarily convex. For instance, some semicubes of the complete bipartite graph $K_{2,3}$ depicted in Figure 6.1A are not convex.

Convexity of semicubes is a central property of graphs in Djoković's characterization of partial cubes (see Theorem 7.2.15(ii) below).

7.2.11 Lemma. *Let $G = (V,E)$ be a bipartite graph such that all its semicubes are convex. If $\{x,y\}\theta\{u,v\}$, then the two pairs of semicubes $\{W_{xy}, W_{yx}\}$ and $\{W_{uv}, W_{vu}\}$ are identical partitions of the set V.*

Proof. We may assume that $u \in W_{xy}$ and $v \in W_{yx}$. Suppose that there is $z \in W_{xy}$ such that $z \notin W_{uv}$. Then, by Theorem 7.2.2, $z \in W_{vu}$ and, by Lemma 7.2.3, $\delta(z,v) = \delta(z,u) - 1$. Thus we have

$$\delta(z,v) + \delta(v,u) = \delta(z,u) - 1 + 1 = \delta(z,u).$$

Since W_{xy} is convex and $z, u \in W_{xy}$, the vertex v must be in W_{xy}, a contradiction. It follows that $W_{xy} \subseteq W_{uv}$. A similar argument shows that $W_{yx} \subseteq W_{vu}$. By Theorem 7.2.2, we have $W_{xy} = W_{uv}$ and $W_{yx} = W_{vu}$. □

The next lemma is an immediate consequence of Lemma 7.2.11 and Theorem 7.2.8.

7.2.12 Lemma. *Let $G = (V, E)$ be a bipartite graph such that all its semicubes are convex. Then the relations θ and Θ are equivalence relations on E.*

The next theorem summarizes basic properties of partial cubes.

7.2.13 Theorem. *If $G = (V, E)$ is a partial cube, then the following hold.*

(i) *G is a bipartite graph.*
(ii) *Each pair of opposite semicubes form a partition of V.*
(iii) *All semicubes are convex subsets of V.*
(iv) *Θ and θ are equivalence relations on E, and*
(v) *There is a 1-1 correspondence between the set of partitions of E defined by opposite semicubes and the quotient-set E/θ, that is, the set of equivalence classes of the relation θ.*

Proof. We may assume that G is a partial cube on some set $\mathcal{X}$, that is, G is an isometric subgraph of the cube $\mathcal{H}(\mathcal{X})$ (cf. Definition 7.1.1).

(i) It suffices to note that if two sets in $\mathcal{H}(\mathcal{X})$ are connected by an edge then they have different parity, that is, one is odd and another is even.

(ii) Follows from (i) and Theorem 7.2.2.

(iii) Note that vertices of G are finite subsets of $\mathcal{X}$. Let W_{AB} be a semicube of G. We may assume that $A = B \cup \{a\}$ for $a \notin B$. By Lemmas 7.2.3 and 3.1.1,

$$C \cap B \subseteq A \subseteq C \cup B$$

for any $C \in W_{AB}$. It follows that $a \in C$ for any $C \in W_{AB}$. In the same vein, $a \notin C$ for any $C \in W_{BA}$. Let $U, V \in W_{AB}$ and C be a vertex of G such that

$$d(U,C) + d(C,V) = d(U,V).$$

Suppose that $C \in W_{BA}$. By Lemma 3.1.1,

$$U \cap V \subseteq C \subseteq U \cup V,$$

a contradiction, since $a \in U$, $a \in V$ but $a \notin C$. Hence, W_{AB} is convex.

(iv) Follows from (iii) and Lemma 7.2.12.
(v) Follows from (iii) and Lemma 7.2.11. □

We need one more property of relations θ and Θ to establish the main result of this section.

7.2.14 Lemma. *Let e and f be two distinct edges of a shortest path P in a graph G. Then neither $e\theta f$ nor $e\Theta f$ hold.*

Proof. Let $P = x_0, x_1, \ldots, x_n$ be a shortest path from x_0 to x_n and $e = \{x_i, x_{i+1}\}$ and $f = \{x_j, x_{j+1}\}$ be two edges in P, where $i < j$. Then $\delta(x_i, x_j) < \delta(x_i, x_{j+1})$ and $\delta(x_{i+1}, x_j) < \delta(x_{i+1}, x_{j+1})$, so $x_i, x_{i+1} \in W_{x_j, x_{j+1}}$. Hence, $\neg(e\theta f)$.

We also have
$$\delta(x_i, x_j) + \delta(x_{i+1}, x_{j+1}) = (\delta(x_{i+1}, x_j) + 1) + (\delta(x_i, x_{j+1}) - 1),$$
which implies that $\neg(e\Theta f)$. □

The following theorem puts forth characterizations of partial cubes due to Djoković (1973) and Winkler (1984) (cf. Theorem 2.10 in Imrich and Klavžar, 2000).

7.2.15 Theorem. *Let $G = (V, E)$ be a connected graph. The following statements are equivalent:*

(i) *G is a partial cube.*
(ii) *G is bipartite and all semicubes of G are convex.*
(iii) *G is bipartite and θ is an equivalence relation.*
(iv) *G is bipartite and Θ is an equivalence relation.*

Proof. (i) ⇒ (ii). Follows from Theorem 7.2.13(i),(iii).
(ii) ⇒ (iii). Follows from Lemma 7.2.12.
(iii) ⇒ (i). We say that two edges e and f are θ-equivalent if $e\theta f$. By Theorem 7.2.2, each pair of opposite semicubes of G form a partition of V. We orient these partitions by calling, in an arbitrary way, one of the two opposite semicubes in each partition a *positive semicube*. Let us assign to each $x \in V$ the set $W^+(x)$ of all the positive semicubes containing x. In the next paragraph we prove that the family $\mathcal{F} = \{W^+(x)\}_{x \in V}$ is well-graded and that the assignment $x \mapsto W^+(x)$ is an isometry between G and $\mathcal{F}$.

Let x and y be two distinct vertices of G. We say that a positive semicube W separates x and y if either $x \in W$, $y \in \overline{W}$ or $x \in \overline{W}$, $y \in W$ ($\overline{W}$ stands for the complement of W). Let P be a shortest path

$$x_0 = x, x_1, \ldots, x_n = y$$

from x to y. By Lemma 7.2.14, no two distinct edges of P stand in the relation θ. Therefore distinct edges of P define distinct positive semicubes that clearly separate x and y. Moreover, any edge defining a positive semicube that separates x and y is θ-equivalent to one of the edges of P. Indeed, otherwise, for this edge, say $\{u,v\}$, we would have $x \in W_{uv}$ and $y \in W_{vu}$ or $y \in W_{uv}$ and $x \in W_{vu}$. In either case, there is an edge in P that joins semicubes W_{uv} and W_{vu}, a contradiction. It follows that any semicube in $W^+(x) \triangle W^+(y)$ is defined by a unique edge in P and any edge in P defines a semicube in $W^+(x) \triangle W^+(y)$. Therefore $\mathcal{F}$ is a wg-family of sets and

$$d(W^+(x), W^+(y)) = \delta(x,y).$$

By Theorem 3.4.12, the family $\mathcal{F}$ is isometric to a wg-family of finite sets. Hence, G is a partial cube.

(iii) $\Leftrightarrow$ (iv). Follows from Theorem 7.2.8. □

The next definition enables another useful characterization of partial cubes.

7.2.16 Definition. Let J be a set of colours. A *J-edge-colouring* of a graph $G = (V, E)$ is a function $E \to J$. We also say that edges of G are *coloured* by elements of J.

7.2.17 Theorem. *A graph $G = (V,E)$ is a partial cube if and only if it is possible to color its edges by elements of some set J in such a way that:*

(i) *the edges of any shortest path of G are of different colours;*
(ii) *in each closed walk of G every colour appears an even number of times.*

Proof. (Necessity.) Without loss of generality, we may assume that $G = (\mathcal{F}, \mathcal{E})$ is an isometric subgraph of a cube $\mathcal{H}(J)$ such that $\cap \mathcal{F} = \emptyset$ and $\cup \mathcal{F} = J$ for a wg-family $\mathcal{F}$. For any edge $\{S,T\}$ of G there is an element $j \in J$ such that $S \triangle T = \{j\}$, so we can colour the edges of G by the elements of J.

(i) Let $S_0 = S, S_1, \ldots, S_n = T$ be a shortest path from S to T in G. For every i, we have $S \cap T \subseteq S_i \subseteq S \cup T$. Therefore,

$$\{j_i\} = S_{i-1} \triangle S_i \subseteq S \triangle T.$$

Since (S_i) is a shortest path, $|S \triangle T| = d(S,T) = n$. It follows that all colours j_i are distinct.

(ii) Let $S_0, S_1, \ldots, S_n = S_0$ be a closed walk W in G and let $E_p = \{S_{p-1}, S_p\}$ be the first edge in W coloured by j, so $S_{p-1} \triangle S_p = \{j\}$. We assume that $j \notin S_{p-1}$ and $j \in S_p$; the other case is treated similarly. Since E_p is the first edge of W coloured by j, we must have $j \notin S_0$. Because the walk W is closed and $j \in S_p$, we must have another occurrence of j in W. Let $E_q = \{S_{q-1}, S_q\}$ be the next edge of W coloured by j. We have $j \in S_{q-1}$ and $j \notin S_q$. By repeating this argument, we partition the occurrences of j in W into pairs, so the total number of these occurrences must be even.

(Sufficiency.) Let S_0 be a fixed vertex of G. For any vertex $S \in V$ and a shortest path p from S_0 to S, we define

$$J_S = \{j \in J \mid j \text{ is a colour of an edge of } p\},$$

and $J_{S_0} = \varnothing$. The set J_S is well-defined. Indeed, let q be another shortest path from S_0 to S and $\tilde{q}$ be its reverse, so $p\tilde{q}$ is a closed walk. By (i) and (ii), J_S does not depend on the choice of p.

The correspondence $\alpha : S \mapsto J_S$ defines an isometric embedding of G into the cube $\mathcal{H}(J)$. Indeed, for $S, T \in V$, let p (resp. q) be a shortest path from S_0 to S (resp. T) and let r be a shortest path from S to T. By (ii) applied to the closed walk $pr\tilde{q}$ and (i), we have

$$j \in J_S \triangle J_T \iff j \text{ is a colour of an edge of } r,$$

so $\delta(S,T) = |J_S \triangle J_T| = d(J_S, J_T)$. □

The next result is a restatement of Theorem 7.1.7, with an alternative proof based on Theorem 7.2.17.

7.2.18 Theorem. *The graph of a medium is a partial cube.*

Proof. Let $(\mathcal{S}, \mathcal{T})$ be a medium and G be its graph. For any edge of G there is a unique pair of tokens $\{\tau, \tilde{\tau}\}$ defining that edge. Let us colour edges of G by elements of the set $J = \{\{\tau, \tilde{\tau}\}\}_{\tau \in \mathcal{T}}$. Since the shortest paths of G correspond to the concise messages of $(\mathcal{S}, \mathcal{T})$ (cf. Lemma 6.1.3), condition (i) of Theorem 7.2.17 is satisfied. A closed walk W in G defines a closed message $\boldsymbol{m}$ for a vertex of W. By Axiom [Mb], the message $\boldsymbol{m}$ is vacuous. Thus every colour appears an even number of times in the walk W. The result follows from Theorem 7.2.17. □

7.3 Semicubes of Media

The concept of a semicube will be very useful in our investigation of media embeddings into lattices later in Chapter 8. In this section we introduce a media counterpart of this concept.

7.3.1 Definition. Let $(\mathcal{S}, \mathcal{T})$ be a medium. For $\tau \in \mathcal{T}$, the subset

$$\mathcal{S}_\tau = \{S \in \mathcal{S} \mid \tau \in \widehat{S}\}$$

of $\mathcal{S}$ is called a *semicube*. The semicubes $\mathcal{S}_\tau$ and $\mathcal{S}_{\tilde{\tau}}$ are called *opposite*.

By Theorem 2.4.3, we have

$$\mathcal{S}_\tau \cap \mathcal{S}_{\tilde{\tau}} = \varnothing \quad \text{and} \quad \mathcal{S}_\tau \cup \mathcal{S}_{\tilde{\tau}} = \mathcal{S} \quad \text{for any } \tau \in \mathcal{T},$$

so pairs of opposite semicubes form partitions of $\mathcal{S}$.

Semicubes are rather special sets of states. In particular, the construction of the semicubes for a pair of tokens τ and $\tilde{\tau}$ can be viewed as a special case of the projection of a medium (cf. Definition 2.11.2) with $\mathcal{U} = \{\tau, \tilde{\tau}\}$. By Theorem 2.11.6(ii) with $\mathcal{U} = \{\tau, \tilde{\tau}\}$, the reduction $(\mathcal{S}_\tau, \mathcal{T}_{\mathcal{S}_\tau})$ of a medium $(\mathcal{S}, \mathcal{T})$ to the semicube $\mathcal{S}_\tau$ is a submedium of $(\mathcal{S}, \mathcal{T})$. Clearly, the graph of the submedium $(\mathcal{S}_\tau, \mathcal{T}_{\mathcal{S}_\tau})$ is a semicube of the graph of the medium. We use this fact to calculate the average length of a finite medium in Theorem 7.3.4.

7.3.2 Definition. Let $\mathcal{M} = (\mathcal{S}, \mathcal{T})$ be a finite medium with n states. The average length $\ell_a(\mathcal{M})$ of the medium $\mathcal{M}$ is defined by

$$\ell_a(\mathcal{M}) = \frac{1}{n(n-1)} \sum_{S \in \mathcal{S}} \sum_{T \in \mathcal{S}} \delta(S, T). \tag{7.4}$$

The average length $\ell_a(G)$ of the partial cube G representing the medium $\mathcal{M}$ is defined by the same formula.

Note that $\ell_a(\mathcal{M})$ is the average length of all the concise messages of the medium $\mathcal{M}$. To establish a useful formula for $\ell_a(\mathcal{M})$, we first prove a general proposition (cf. Imrich and Klavžar, 2000).

Let $Q = \{0,1\}^m$ be the vertex set of the m-dimensional cube endowed with the usual Hamming distance

$$d_H(u,v) = \sum_{i=1}^{m} |u_i - v_i|,$$

where $u = (u_1, \ldots, u_m)$ and $v = (v_1, \ldots, v_m)$ are vertices of the cube.

For $1 \leq i \leq m$, we denote by

$$Q_i' = \{v \in Q \mid v_i = 0\} \quad \text{and} \quad Q_i'' = \{v \in Q \mid v_i = 1\}$$

the opposite facets of Q.

7.3.3 Lemma. *For $X \subseteq Q$, let $X_i' = X \cap Q_i'$ and $X_i'' = X \cap Q_i''$. Then*

$$\frac{1}{2} \sum_{u \in X} \sum_{v \in X} d_H(u,v) = \sum_{i=1}^{m} |X_i'||X_i''|.$$

Proof. We have

$$\sum_{u \in X} \sum_{v \in X} d_H(u,v) = \sum_{u \in X} \sum_{v \in X} \sum_{i=1}^{m} |u_i - v_i| = \sum_{i=1}^{m} \left(\sum_{u \in X} \sum_{v \in X} |u_i - v_i| \right).$$

Note that $|u_i - v_i| = 1$ if and only if $u \in X_i'$, $v \in X_i''$ or $u \in X_i''$, $v \in X_i'$; otherwise, $|u_i - v_i| = 0$. Therefore

$$\sum_{u \in X} \sum_{v \in X} |u_i - v_i| = 2|X_i'||X_i''|.$$

The result follows. □

7.3.4 Theorem. *Let* $\mathcal{M} = (\mathcal{S}, \mathcal{T})$ *be a finite medium with n states and $2m$ tokens. Then*

$$\ell_a(\mathcal{M}) = \frac{2}{n(n-1)} \sum_{\{\tau, \tilde{\tau}\}} |S_\tau||S_{\tilde{\tau}}|,$$

where the sum is taken over the set of all pairs of mutually reverse tokens.

Proof. Let $(Q, \mathcal{T}_Q)$ be the medium with set of states $Q = \{0,1\}^m$ and tokens defined by

$$\gamma_i(u) = \begin{cases} (u_1, \ldots, u_m), & \text{if } u_i = 0, \\ (u_1, \ldots, 1 - u_i, \ldots, u_m), & \text{if } u_i = 1, \end{cases}$$

and

$$\tilde{\gamma}_i(u) = \begin{cases} (u_1, \ldots, 1 - u_i, \ldots, u_m), & \text{if } u_i = 0, \\ (u_1, \ldots, u_m), & \text{if } u_i = 1. \end{cases}$$

It is clear that $(Q, \mathcal{T}_Q)$ is a medium isomorphic to the complete medium $(\mathfrak{P}(Z), \mathcal{W}_{\mathfrak{P}(Z)})$, where $Z = \{1, \ldots, m\}$. By Theorem 3.3.4, there is an isomorphic embedding

$$(\alpha, \beta) : (\mathcal{S}, \mathcal{T}) \to (Q, \mathcal{T}_Q),$$

which defines an isometric embedding $(\mathcal{S}, \delta) \to (Q, d_H)$. It is not difficult to verify that the image $\alpha(S_\tau)$ of a semicube S_τ is a subset of a facet of Q, and that opposite semicubes are mapped into opposite facets of Q. The result follows from Lemma 7.3.3. $\square$

7.4 Projections of Partial Cubes

We begin by extending the usual definition of a projection of a finite dimensional cube onto its face to arbitrary cubes $\mathcal{H}(X)$.

7.4.1 Definition. Let U be a nonempty subset of a set X. A *projection* of the cube $\mathcal{H}(X)$ onto the cube $\mathcal{H}(U)$ is the mapping

$$\varphi : P \mapsto P \cap U \quad \text{for } P \in \mathfrak{P}_F(X).$$

Since

$$(P \triangle Q) \cap U = (P \cap U) \triangle (Q \cap U), \tag{7.5}$$

the projection φ maps an edge $\{P, Q\}$ of $\mathcal{H}(X)$ into either a single vertex of $\mathcal{H}(U)$ or into an edge of $\mathcal{H}(U)$. Thus, for a subgraph $G = (\mathcal{F}, E)$ of $\mathcal{H}(X)$, the projection φ defines a subgraph $\varphi(G)$ of $\mathcal{H}(U)$ that we call the *projection of G into $\mathcal{H}(U)$*.

7.4.2 Theorem. *The projection of a partial cube on X into $\mathcal{H}(U)$ is a partial cube on U.*

Proof. It suffices to prove that the projection of a shortest path in $\mathcal{H}(\mathcal{X})$ is a shortest path in $\mathcal{H}(U)$. Let $R_0 = P, R_1, \ldots, R_n = Q$ be a shortest path in $\mathcal{H}(\mathcal{X})$, so $d(P, Q) = n$. Then $R_i \triangle R_{i+1} = \{x_i\}$ with distinct $x_i \in P \triangle Q$ for $0 \le i < n$. By (7.5),

$$\{x_i\} \cap U = (R_i \triangle R_{i+1}) \cap U = (R_i \cap U) \triangle (R_{i+1} \cap U) = \varphi(R_i) \triangle \varphi(R_{i+1}).$$

Therefore, $\{\varphi(R_i), \varphi(R_{i+1})\}$ is an edge if and only if

$$x_i \in (P \triangle Q) \cap U = \varphi(P) \triangle \varphi(Q).$$

It follows that distinct vertices $\varphi(R_i)$ form a shortest path in $\mathcal{H}(U)$. □

As we proved in this chapter (Theorems 7.1.7 and 7.2.18), the graph of a medium is a partial cube. In what follows we establish a rather transparent interpretation (Theorem 7.4.3) of projections of media (as defined in 2.11.4) in terms of partial cube projections. By Theorem 3.3.4, any medium is isomorphic to the representing medium $(\mathcal{F}, \mathcal{W}_\mathcal{F})$ of some wg-family $\mathcal{F}$ of finite subsets of a set $\mathcal{X}$, so we consider only representing media. Moreover, by Theorem 7.1.9, we may assume that conditions (7.1) are satisfied.

There is a 1-1 correspondence established by $\{\gamma_x, \tilde{\gamma}_x\} \leftrightarrow x$, between symmetric subsets of $\mathcal{W}_\mathcal{F}$ and subsets of $\mathcal{X}$. For a given nonempty, symmetric subset $\mathcal{U}$ of $\mathcal{W}_\mathcal{F}$, we denote the corresponding subset of $\mathcal{X}$ by U.

Let $S \ne T$ be two states in $\mathcal{F}$ and let $\boldsymbol{m} = \tau_1 \ldots \tau_n$ be a concise message producing T from S. Then each τ_i is either γ_{x_i} or $\tilde{\gamma}_{x_i}$ for $x_i \in S \triangle T$. We also have $x_i \ne x_j$ for $i \ne j$ and $\cup_i \{x_i\} = S \triangle T$. From this observation and (7.5), we obtain the equivalences:

$$\mathcal{C}(\boldsymbol{m}) \subseteq (\mathcal{W}_\mathcal{F} \setminus \mathcal{U}) \quad \Leftrightarrow \quad S \triangle T \subseteq \mathcal{X} \setminus U \quad \Leftrightarrow \quad S \cap U = T \cap U.$$

Thus, $S \sim T$ (see Definition 2.11.2) if and only if the projections of S and T into the cube $\mathcal{H}(U)$ are equal. It follows that classes of the equivalence relation $\sim$ are uniquely determined by projections of elements of $\mathcal{F}$ into the cube $\mathcal{H}(U)$. In fact, we have the following result:

7.4.3 Theorem. *The graph of the projection of a medium $(\mathcal{F}, \mathcal{W}_\mathcal{F})$ under $\mathcal{U}$ is the projection of the graph of $(\mathcal{F}, \mathcal{W}_\mathcal{F})$ into the cube $\mathcal{H}(U)$.*

We leave the details of the proof to the reader (Problem 7.17).

The projections φ of partial cubes on a set $\mathcal{X}$ into a cube $\mathcal{H}(U)$ have a particularly simple description in the case when $U = \mathcal{X} \setminus \{a\}$ for some $a \in \mathcal{X}$.

Let G be a partial cube on $\mathcal{X}$ with a vertex set $\mathcal{F}$, where $\mathcal{F}$ is a wg-family of sets satisfying conditions

$$\cap \mathcal{F} = \varnothing \quad \text{and} \quad \cup \mathcal{F} = \mathcal{X}, \tag{7.6}$$

and let $U = \mathcal{X} \setminus \{a\}$. We denote by φ_a the projection of $\mathcal{H}(\mathcal{X})$ onto $\mathcal{H}(U)$. Then

$$\varphi_a(R) = \begin{cases} R, & \text{if } a \notin R, \\ R \setminus \{a\}, & \text{if } a \in R, \end{cases} \quad \text{for } R \in \mathcal{F}.$$

Thus, we have a partition of $\mathcal{F}$ into two families of sets:

$$\{R \in \mathcal{F} \mid \varphi_a(R) = R\} = \{R \in \mathcal{F} \mid a \notin R\}, \tag{7.7}$$

and

$$\{R \in \mathcal{F} \mid \varphi_a(R) \neq R\} = \{R \in \mathcal{F} \mid a \in R\}. \tag{7.8}$$

7.4.4 Theorem. (i) The families defined by (7.7) and (7.8) are opposite semicubes of the partial cube G.

(ii) For any pair $\{W_{PQ}, W_{QP}\}$ of opposite semicubes, there is a unique $a \in X$ such that the pair of sets $\{R \in \mathcal{F} \mid a \notin R\}$ and $\{R \in \mathcal{F} \mid a \in R\}$ form the same partition of $\mathcal{F}$ as W_{PQ} and W_{QP}.

(iii) There is a 1-1 correspondence between the set $X = \cup \mathcal{F}$ and the set of all pairs of opposite semicubes.

Proof. (i) By (7.6), there are $S, T \in \mathcal{F}$ such that $a \notin S$ and $a \in T$. Let $R_0 = S, R_1, \ldots, R_n = T$ be a shortest path in G. There is an index i such that $a \notin R_i$ and $a \in R_{i+1}$. We denote $P = R_i$ and $Q = R_{i+1}$. Clearly, $Q = P + \{a\}$. By the definition of a semicube, Lemma 7.2.3, and (3.2), we have equivalences

$$R \in W_{PQ} \iff d(R, P) < d(R, Q) \iff d(R, P) + d(P, Q) = d(R, Q)$$
$$\iff R \cap Q \subseteq P \subseteq R \cup Q \iff a \notin R,$$

where the last equivalence is easily verified for $Q = P + \{a\}$. It follows that

$$\{R \in \mathcal{F} \mid \varphi_a(R) = R\} = W_{PQ}.$$

A similar argument shows that $\{R \in \mathcal{F} \mid \varphi_a(R) \neq R\} = W_{QP}$.

(ii) Let $\{P, Q\}$ be an edge of G. We may assume that $Q = P + \{a\}$ for some $a \in X$. Then, for any $R \in \mathcal{F}$,

$$R \triangle Q = R \triangle (P + \{a\}) = \begin{cases} (R \triangle P) + \{a\}, & \text{if } a \in R, \\ R \triangle P, & \text{if } a \notin R. \end{cases}$$

Hence $|R \triangle P| < |R \triangle Q|$ if and only if $a \in R$. It follows that

$$W_{PQ} = \{R \in \mathcal{F} \mid a \in R\} \quad \text{and} \quad W_{QP} = \{R \in \mathcal{F} \mid a \notin R\} \tag{7.9}$$

If $\{S, T\}$ is another edge defining the pair $\{W_{PQ}, W_{QP}\}$, then S and T belong to different semicubes in this pair. By (7.9), $S \triangle T = \{a\}$. Thus the element a is uniquely defined by the pair $\{W_{PQ}, W_{QP}\}$.

(iii) Follows immediately from (i) and (ii). □

7.4.5 Remark. Geometrically, the edges $\{P, Q\}$ of the partial cube G of the theorem with $P \triangle Q = \{a\}$ are 'parallel' edges of the cube $\mathcal{H}(\mathcal{X})$ and φ_a projects G along these edges into $\mathcal{H}(U)$, which can be regarded as a 'facet' of $\mathcal{H}(\mathcal{X})$. The resulting graph $\varphi_a(G)$ is isomorphic to the isometric contraction of the partial cube G (Imrich and Klavžar, 2000; Ovchinnikov, 2007).

7.4.6 Remark. Let $U = \mathcal{F} \setminus \{a_1, \ldots, a_n\}$. Then the projection φ of $\mathcal{H}(\mathcal{X})$ is a composition of projections φ_{a_i} in arbitrary order (Problem 7.18). Thus, in the finite case, a projection can be defined as a composition of contractions.

7.5 Uniqueness of Media Representations

Theorem 3.3.4 asserts that any medium $(S, \mathcal{T})$ is isomorphic to the medium $(\mathcal{F}, \mathcal{W}_\mathcal{F})$ of a well-graded family $\mathcal{F}$ of finite subsets of some set X. In this section we show that this representation is unique in some precise sense.

Suppose that $(\mathcal{F}_1, \mathcal{W}_{\mathcal{F}_1})$ and $(\mathcal{F}_2, \mathcal{W}_{\mathcal{F}_2})$ are two representations of $(S, \mathcal{T})$. By Theorem 7.1.9, we may assume that conditions (7.1) are satisfied. The media $(\mathcal{F}_1, \mathcal{W}_{\mathcal{F}_1})$ and $(\mathcal{F}_2, \mathcal{W}_{\mathcal{F}_2})$ are isomorphic, which implies that the sets $\mathcal{X}_1 = \cup \mathcal{F}_1$ and $\mathcal{X}_2 = \cup \mathcal{F}_2$ have the same cardinality. It follows that the cubes $\mathcal{H}(\mathcal{X}_1)$ and $\mathcal{H}(\mathcal{X}_2)$ are isomorphic. Thus we may assume that $\mathcal{X}_1 = \mathcal{X}_2 = \mathcal{X}$ and consider two isomorphic media $(\mathcal{F}_1, \mathcal{W}_{\mathcal{F}_1})$ and $(\mathcal{F}_2, \mathcal{W}_{\mathcal{F}_2})$ representing $(S, \mathcal{T})$ with $\cup \mathcal{F}_1 = \cup \mathcal{F}_2 = \mathcal{X}$ and $\cap \mathcal{F}_1 = \cap \mathcal{F}_2 = \emptyset$. The graphs of these media are isomorphic partial subcubes of the cube $\mathcal{H}(\mathcal{X})$. On the other hand, by Theorems 7.1.10 and 6.4.5, isomorphic partial cubes represent isomorphic media.

The uniqueness problem is formulated geometrically as follows:

7.5.1 Uniqueness Problem. Show that, given two isometric partial subcubes of $\mathcal{H}(\mathcal{X})$, there is an isometry of $\mathcal{H}(\mathcal{X})$ onto itself that maps one of the partial cubes onto the other.

We shall use the following general 'homogeneity' properties of a metric space (cf. Bogatyi, 2002):

7.5.2 Definition. Let Y be a metric space and $\mathcal{K}$ be a nonempty family of subsets of Y. The space Y is said to be $\mathcal{K}$-*homogeneous* if, for every two subsets $A, B \in \mathcal{K}$ and an isometry $A \hookrightarrow B$, this isometry can be extended to an isometry of the entire space Y onto itself. If $\mathcal{K}$ is the family of all singletons in Y, then the space Y is said to be *homogeneous*. If $\mathcal{K}$ is the family of all subsets of Y, then the space Y is said to be *fully homogeneous* (cf. Burago et al., 2001).

We denote by $\mathcal{WG}$ the collection of all wg-families $\mathcal{F}$ satisfying conditions (7.1):

7.5 Uniqueness of Media Representations   155

$$\cap \mathcal{F} = \varnothing \quad \text{and} \quad \cup \mathcal{F} = \mathcal{X}.$$

We shall prove an even stronger statement than the one formulated in 7.5.1, namely, that $\mathcal{H}(\mathcal{X})$ is a $\mathcal{WG}$-homogeneous metric space, that is, any isometry between two partial cubes on $\mathcal{X}$ can be extended to an isometry of the hypercube $\mathcal{H}(\mathcal{X})$ (see Theorem 7.5.11).

7.5.3 Remark. Note, that $\mathcal{H}(\mathcal{X})$ is not a fully homogeneous space, since an isometry between two subsets of $\mathcal{H}(\mathcal{X})$ cannot be extended, in general, to an isometry of the cube $\mathcal{H}(\mathcal{X})$ (cf. Problem 7.10). On the other hand, $\mathcal{H}(\mathcal{X})$ is a homogeneous metric space (cf. Problem 7.11).

A general remark is in order. Let Y be a homogeneous metric space, A and B be two metric subspaces of Y, and α be an isometry from A onto B. Let c be a fixed point in Y. For a given $a \in A$, let $b = \alpha(a) \in B$. Since Y is homogeneous, there are isometries β and γ of Y such that $\beta(a) = c$ and $\gamma(b) = c$. Then $\lambda = \gamma \alpha \beta^{-1}$ is an isometry from $\alpha(A)$ onto $\beta(B)$ such that $\lambda(c) = c$. Clearly, α is extendable to an isometry of Y if and only if δ is extendable. Therefore, in the case of the space $\mathcal{H}(\mathcal{X})$, we may consider only well-graded families of subsets containing the point $\varnothing$ and isometries between these families fixing this point.

To prove the main theorem of this section (Theorem 7.5.11), we first establish some technical results.

7.5.4 Definition. For $\mathcal{F} \in \mathcal{WG}$, with $\varnothing \in \mathcal{F}$ and $|\mathcal{F}| > 1$, we define a function $r_{\mathcal{F}} : \cup \mathcal{F} \to \mathbb{N}$ by
$$r_{\mathcal{F}}(x) = \min\{|A| \,|\, x \in A, A \in \mathcal{F}\}.$$
For $k \in \mathbb{N}$, the sets $\mathcal{X}_k^{\mathcal{F}}$ are defined by
$$\mathcal{X}_k^{\mathcal{F}} = \{x \in \cup \mathcal{F} \,|\, r_{\mathcal{F}}(x) = k\}.$$

We have $\mathcal{X}_i^{\mathcal{F}} \cap \mathcal{X}_j^{\mathcal{F}} = \varnothing$ for $i \neq j$, and $\cup_k \mathcal{X}_k^{\mathcal{F}} = \cup \mathcal{F}$. Some of the sets $\mathcal{X}_k^{\mathcal{F}}$ could be empty for $k > 1$, although $\mathcal{X}_1^{\mathcal{F}}$ is not empty, since, by the well-gradedness property, $\mathcal{F}$ contains at least one singleton (we assumed that $\varnothing \in \mathcal{F}$).

7.5.5 Example. Let $\mathcal{X} = \{a, b, c\}$ and $\mathcal{F}$ be the wg-family of Example 2.1; thus,
$$\mathcal{F} = \{\varnothing, \{a\}, \{b\}, \{a, b\}, \{a, b, c\}\}.$$
We have $r_{\mathcal{F}}(a) = r_{\mathcal{F}}(b) = 1$, $r_{\mathcal{F}}(c) = 3$ and
$$\mathcal{X}_1^{\mathcal{F}} = \{a, b\}, \ \mathcal{X}_2^{\mathcal{F}} = \varnothing, \ \mathcal{X}_3^{\mathcal{F}} = \{c\}.$$

7.5.6 Lemma. *For $A \in \mathcal{F}$ and $x \in A$, we have*
$$r_{\mathcal{F}}(x) = |A| \quad \Rightarrow \quad A \setminus \{x\} \in \mathcal{F}. \tag{7.10}$$

Proof. Let $k = |A|$. Since $\mathcal{F}$ is well-graded, there is a nested sequence $A_0 = \varnothing \subseteq A_1 \subseteq \cdots \subseteq A_k = A$. Since $r_\mathcal{F}(x) = k$, we have $x \notin A_i$ for $i < k$. It follows that $A \setminus \{x\} = A_{k-1} \in \mathcal{F}$. □

Let us recall from Theorem 3.4.5(ii) that

$$A \in [B \cap C, B \cup C] \Leftrightarrow d(B, A) + d(A, C) = d(B, C), \quad (7.11)$$

for all $A, B, C \in \mathfrak{P}(X)$. It follows that

$$A \in [B \cap C, B \cup C] \Leftrightarrow \alpha(A) \in [\alpha(B) \cap \alpha(C), \alpha(B) \cup \alpha(C)] \quad (7.12)$$

for $A, B, C \in \mathcal{F}_1$ and an isometry $\alpha : \mathcal{F}_1 \to \mathcal{F}_2$.

In the sequel, $\mathcal{F}_1$ and $\mathcal{F}_2$ are two well-graded families of finite subsets of X both containing $\varnothing$, and $\alpha : \mathcal{F}_1 \to \mathcal{F}_2$ is an isometry such that $\alpha(\varnothing) = \varnothing$.

7.5.7 Definition. We define a relation σ between $\cup \mathcal{F}_1$ and $\cup \mathcal{F}_2$ (therefore, $\sigma \subseteq \cup \mathcal{F}_1 \times \cup \mathcal{F}_2$) by means of the following construction. By (7.10), for $x \in \cup \mathcal{F}_1$ there is $A \in \mathcal{F}_1$ such that $x \in A$, $r_{\mathcal{F}_1}(x) = |A|$, and $A \setminus \{x\} \in \mathcal{F}_1$. Since $\varnothing \subseteq A \setminus \{x\} \subset A$, we have, by (7.12), $\alpha(A \setminus \{x\}) \subset \alpha(A)$. Since $d(A \setminus \{x\}, A) = 1$, there is $y \in \cup \mathcal{F}_2$ such that $\alpha(A) = \alpha(A \setminus \{x\}) + \{y\}$. In this case we say that $xy \in \sigma$.

7.5.8 Lemma. *If $x \in \mathcal{X}_k^{\mathcal{F}_1}$ and $xy \in \sigma$, then $y \in \mathcal{X}_k^{\mathcal{F}_2}$.*

Proof. Let $A \in \mathcal{F}_1$ be a set of cardinality k defining $r_{\mathcal{F}_1}(x) = k$. Since $|A| = d(\varnothing, A) = d(\varnothing, \alpha(A)) = |\alpha(A)|$ and $y \in \alpha(A)$, we have $r_{\mathcal{F}_2}(y) \leq k$. Suppose that $m = r_{\mathcal{F}_2}(y) < k$. Then, by (7.10), there is $B \in \mathcal{F}_2$ such that $y \in B$, $|B| = m$, and $B \setminus \{y\} \in \mathcal{F}_2$. Clearly,

$$\alpha(A \setminus \{x\}) \cap B \subseteq \alpha(A) \subseteq \alpha(A \setminus \{x\}) \cup B.$$

By (7.12), we have

$$(A \setminus \{x\}) \cap \alpha^{-1}(B) \subseteq A \subseteq (A \setminus \{x\}) \cup \alpha^{-1}(B).$$

It follows that $x \in \alpha^{-1}(B)$, a contradiction, because

$$r_{\mathcal{F}_1}(x) = k \quad \text{and} \quad |\alpha^{-1}(B)| = m < k.$$

Thus $r_{\mathcal{F}_2}(y) = k$, that is, $y \in \mathcal{X}_k^{\mathcal{F}_2}$. □

We proved that, for every $k \geq 1$, the restriction of σ to $\mathcal{X}_k^{\mathcal{F}_1} \times \cup \mathcal{F}_2$ is a relation $\sigma_k \subseteq \mathcal{X}_k^{\mathcal{F}_1} \times \mathcal{X}_k^{\mathcal{F}_2}$.

7.5.9 Lemma. *The relation σ_k is a bijection for every $k \geq 1$.*

Proof. Suppose that there are $z \neq y$ such that $xy \in \sigma_k$ and $xz \in \sigma_k$. Then, by (7.10), there are two distinct sets $A, B \in \mathcal{F}_1$ such that

$$k = r_{\mathcal{F}_1}(x) = |A| = |B|, \quad A \setminus \{x\} \in \mathcal{F}_1, \ B \setminus \{x\} \in \mathcal{F}_1,$$

and

$$\alpha(A) = \alpha(A \setminus \{x\}) + \{y\}, \quad \alpha(B) = \alpha(B \setminus \{x\}) + \{z\}.$$

We have

$$d(\alpha(A), \alpha(B)) = d(A, B) = d(A \setminus \{x\}, B \setminus \{x\})$$
$$= d(\alpha(A) \setminus \{y\}, \alpha(B) \setminus \{z\}).$$

Thus $y, z \in \alpha(A) \cap \alpha(B)$, that is, in particular, that $z \in \alpha(A) \setminus \{y\}$, a contradiction, because $r_{\mathcal{F}_2}(z) = k$ and $|\alpha(A) \setminus \{y\}| = k - 1$.

By applying the above argument to α^{-1}, we prove that σ_k is a bijection. □

It follows from the previous lemma that σ is a bijection from $\cup \mathcal{F}_1 = X$ onto $\cup \mathcal{F}_2 = X$, that is, σ is a permutation on the set X.

7.5.10 Lemma. $\alpha(A) = \sigma(A)$ for any $A \in \mathcal{F}_1$.

Proof. We prove this statement by induction on $k = |A|$. The case $k = 1$ is trivial, since $\alpha(\{x\}) = \{\sigma_1(x)\}$ for $\{x\} \in \mathcal{F}_1$.

Suppose that $\alpha(A) = \sigma(A)$ for all $A \in \mathcal{F}_1$ such that $|A| < k$. Let A be a set in $\mathcal{F}_1$ of cardinality k. By the wellgradedness property, there is a nested sequence $\{A_i\}_{0 \leq i \leq k}$ of distinct sets in $\mathcal{F}_1$ with $A_0 = \emptyset$ and $A_k = A$. Thus, $A = A_{k-1} + \{x\}$ for some $x \in \cup \mathcal{F}_1$). Clearly, $m = r_{\mathcal{F}_1}(x) \leq k$.

If $m = k$, then $\alpha(A) = \alpha(A_{k-1}) \cup \{\sigma(x)\} = \sigma(A)$, by the definition of σ and the induction hypothesis.

Suppose now that $m < k$. There is a set $B \in \mathcal{F}_1$ containing x such that $|B| = m$. By the wellgradedness property, there is a nested sequence $\{B_i\}_{0 \leq i \leq m}$ of distinct sets in $\mathcal{F}_1$ with $B_0 = \emptyset$ and $B_m = B$. We have $x \notin B_i$ for $i < m$, since $m = r_{\mathcal{F}_1}(x)$. Therefore, $B = B_{m-1} + \{x\}$. Clearly,

$$B_{m-1} \cap A \subseteq B \subseteq B_{m-1} \cup A.$$

By (7.12),

$$\alpha(B) \subseteq \alpha(B_{m-1}) \cup \alpha(A).$$

Thus, by the induction hypothesis, we have

$$\sigma(B_{m-1}) + \{\sigma(x)\} = \sigma(B) \subseteq \sigma(B_{m-1}) \cup \alpha(A).$$

Hence, $\sigma(x) \in \alpha(A)$. Since $\alpha(A) = \sigma(A_{k-1}) + \{y\}$ for $y \notin \sigma(A_{k-1})$ and $x \notin A_{k-1}$, we have $y = \sigma(x)$, that is, $\alpha(A) = \sigma(A)$. □

From this lemma, we immediately obtain:

7.5.11 Theorem. *The space $\mathcal{H}(X)$ is $\mathcal{WG}$-homogeneous.*

7.5.12 Remark. In the case of a finite set X, Theorem 7.5.11 is a consequence of Theorem 19.1.2 in Deza and Laurent (1997).

In the proof of Theorem 7.5.11 we used two kinds of isometries of $\mathcal{H}(X)$: isometries that map elements of $\mathcal{H}(X)$ to the empty set, and isometries defined by permutations on X. It is not difficult to show that these isometries generate the isometry group of $\mathcal{H}(X)$:

7.5.13 Theorem. *The isometry group of $\mathcal{H}(X)$ is generated by permutations on the set X and functions*

$$\varphi_A : S \mapsto S \triangle A, \qquad (S \in \mathcal{H}(X)).$$

We omit the proof (see Problem 7.19).

7.6 The Isometric Dimension of a Partial Cube

7.6.1 Definition. (Djoković, 1973) Let X be a set. The *dimension* of the hypercube $\mathcal{H}(X)$ is the cardinality of the set X. The *isometric dimension* $\dim_I(G)$ of a partial cube G is the minimum possible dimension of a hypercube $\mathcal{H}(X)$ in which G is isometrically embeddable.

Clearly, isomorphic partial cubes have the same isometric dimension. Thus, by Theorem 7.1.9, we may consider only partial cubes that are induced by wg-families of finite sets satisfying conditions (7.1). We recall that, by the implication (i) $\Rightarrow$ (iii) of Theorem 7.2.15, the Djoković's relation θ of a graph $G = (E, V)$ is an equivalence relation on E if G is a partial cube. The notation E/θ refers to the partition on E induced by the equivalence θ (cf. 1.8.4). By Theorems 7.2.13(v) and 7.4.4, we have the following result.

7.6.2 Theorem. *Let $G = (V, E)$ be a partial cube induced by a wg-family $\mathcal{F}$ of finite subsets of a set X such that $\cap \mathcal{F} = \varnothing$ and $\cup \mathcal{F} = X$. Then*

$$\dim_I(G) = |E/\theta| = |X|,$$

where θ is Djoković's equivalence relation on E.

7.6.3 Corollary. *Let $\mathcal{M} = (\mathcal{S}, \mathcal{T})$ be a medium and G be its representing graph. Then*

$$\dim_I(G) = |\mathcal{T}|/2.$$

Proof. By Theorem 3.3.4, we may assume that $\mathcal{M} = (\mathcal{F}, \mathcal{W}_\mathcal{F})$ for a wg-family $\mathcal{F}$ of finite subsets of a set X satisfying conditions (7.1). Each element $x \in X$ defines a pair of tokens $\{\tau_x, \tilde{\tau}_x\}$. The result follows from Theorem 7.6.2. □

Problems

7.1 Let M is a medium with n states and $2m$ tokens.

(i) Show that $m+1 \leq n \leq 2^m$.
(ii) Show that $m+1$ and 2^m are exact bounds for n.

7.2 Let T be a tree with n vertices. Show that T is a partial cube and find $\dim_I(T)$.

7.3 Let C_{2n} be a cycle of even length. Show that C_{2n} is a partial cube and find $\dim_I(C_{2n})$. What are equivalence classes of relations θ and Θ? What are semicubes of C_{2n}?

7.4 Let C_{2n+1} be a cycle of odd length. Show that the relation θ is the identity relation on the set of edges of C_{2n+1} and that the relation Θ is not transitive.

7.5 Let $G = K_{2,3}$ be a complete bipartite graph with $2+3=5$ vertices. Show that the relation Θ is not transitive on the set of edges of G. Describe semicubes of the graph G.

7.6 Give an example of a graph G for which the second part of Theorem 7.2.5 does not hold.

Figure 7.4. The graph in Problem 7.7.

7.7 Show that the graph depicted in Figure 7.4 is not a partial cube.

7.8 Show that a connected, bipartite graph in which every edge is contained in at most one cycle is a partial cube.

7.9 Let $X = \{a, b, c\}$. Show that the 3-cube $\mathcal{H}(X)$ is fully homogeneous.

7.10 Let $X = \{a, b, c, d\}$. Show that the 4-cube $\mathcal{H}(X)$ is not fully homogeneous.

7.11 Prove that any cube $\mathcal{H}(X)$ is a homogeneous metric space.

7.12 Let P be a path of length n and M be the medium represented by P. Find the average length $\ell_a(M)$ of M.

7.13 Show that the average length of the cycle C_{2n} is $\frac{n^2}{2n-1}$.

Figure 7.5. The graph in Problem 7.15.

7.14 Show that the average length of the complete medium on n states is $\frac{n2^{n-1}}{2^n-1}$.

7.15 Let G be the graph shown in Figure 7.5.

(i) Show that G is a partial cube.
(ii) Find the isometric dimension $\dim_I(G)$.
(iii) Find the average length of the corresponding medium.

7.16 Let P and Q be two vertices of the cube $\mathcal{H}(X)$. Show that the subgraph induced by the interval $[P \cap Q, P \cup Q]$ is isomorphic to the cube $\mathcal{H}(U)$ where $U = (P \cup Q) \setminus (P \cap Q)$.

7.17 Restore the details of the proof of Theorem 7.4.3.

7.18 Let φ_a be the projection of $\mathcal{H}(X)$ onto $\mathcal{H}(U)$, where $U = \mathcal{F} \setminus \{a\}$, and let φ be the projection of $\mathcal{H}(X)$ onto $\mathcal{H}(U)$, where $U = \mathcal{F} \setminus \{a_1, \ldots, a_n\}$.

(i) Show that $\varphi_a \varphi_b = \varphi_b \varphi_a$ for all $a, b \in X$.
(ii) Show that $\varphi = \varphi_{a_1} \cdots \varphi_{a_n}$.

7.19 Prove Theorem 7.5.13.

8
Media and Integer Lattices

Because any finite medium has an isomorphic representation as a finite dimensional partial cube, it is also representable as a finite isometric subgraph of some integer lattice. Obviously, the dimension of a lattice representation may be much lower than the dimension of a partial cube representing the medium, making this representation an invaluable tool in visualizing large media. We already used lattice representations for this purpose in previous chapters.

In this chapter (following Eppstein, 2005b) we establish the minimum dimension of a lattice representation of a given partial cube. We extend these results to handle infinite as well as finite partial cubes and to handle both oriented and unoriented media.

8.1 Integer Lattices

In order to define the dimension of infinite partial cubes, we need a notion of infinite dimensional integer lattices. As the Cartesian product of infinitely many copies of $\mathbb{Z}$ does not form a connected graph, we cannot use it directly in our definitions. Instead we base our definition on a single connected component of this Cartesian product. Essentially the same notion is called a *weak Cartesian Product* by Imrich and Klavžar (2000).

8.1.1 Definition. The *support* of a real valued function f on a set $\mathcal{X}$ is the subset of $\mathcal{X}$ on which f takes a nonzero value. We denote by $\mathbb{Z}_{\mathcal{X}}$ the collection of all functions $f : \mathcal{X} \to \mathbb{Z}$ with finite support. The *integer lattice on* $\mathcal{X}$ is the graph $\mathbb{Z}(\mathcal{X})$ having $\mathbb{Z}_{\mathcal{X}}$ as its set of vertices and whose set of edges contains all $\{f, g\}$ such that

$$|f(x) - g(x)| = 1 \quad \text{for some } x \in \mathcal{X}, \text{ and}$$
$$f(y) = g(y) \quad \text{for all } y \in \mathcal{X} \setminus \{x\}.$$

The *dimension* of the integer lattice $\mathbb{Z}(\mathcal{X})$ is the cardinality of $\mathcal{X}$. If $\mathcal{X}$ has finite cardinality d, we write $\mathbb{Z}(\mathcal{X}) = \mathbb{Z}^d$.

8 Media and Integer Lattices

8.1.2 Theorem. *The integer lattice $\mathbb{Z}(\mathcal{X})$ is a connected graph with the graph distance given by*

$$\delta_\mathbb{Z}(f,g) = \sum_{x \in \mathcal{X}} |f(x) - g(x)|. \tag{8.1}$$

Proof. The set of arguments where the values of f and g differ,

$$A = \{x \in \mathcal{X} \mid f(x) \neq g(x)\},$$

is a subset of the union of the supports of f and g, and therefore finite; thus $\delta_\mathbb{Z}(f,g)$ is a finite integer for every f and g.

We first prove by induction on δ_Z that the value $\delta_\mathbb{Z}(f,g)$ defined as a sum in the lemma is greater than or equal to to the graph distance. That is, we assume as an induction hypothesis that every f' and g' with $\delta_Z(f',g') < D$ can be connected by a path of length at most $\delta_Z(f',g')$, and show that this hypothesis implies the existence of a path of length $\delta_Z(f,g) = D$ connecting f to g. As a base case, if $\delta_Z(f,g) = 0$, then f and g are clearly the same vertex and form the endpoints of a path of length zero. Otherwise, let f and g be two distinct vertices in $\mathbb{Z}(\mathcal{X})$ with $\delta_Z(f,g) = D$. Let a be any element of A, as defined above, and swap f and g if necessary so that that $f(a) > g(a)$. Define the function $h \in \mathbb{Z}_\mathcal{X}$ by the formula

$$h(x) = \begin{cases} f(x) - 1, & \text{if } x = a, \\ f(x), & \text{otherwise.} \end{cases}$$

Then $\delta_\mathbb{Z}(h,g) = \delta_\mathbb{Z}(f,g) - 1$. By the induction hypothesis there exists a path in $\mathbb{Z}(\mathcal{X})$ of length $\delta_\mathbb{Z}(f,g) - 1$ between h and g. Concatenating this path with the edge in $\mathbb{Z}(\mathcal{X})$ between f and h yields a path of length $\delta_\mathbb{Z}(f,g)$ between f and g.

To finish the proof, we prove in the other direction that, for every f and g the value $\delta_\mathbb{Z}(f,g)$ is less than or equal to the graph distance. Let $h_0 = f, h_1, \ldots, h_n = g$ be a path from f to g in $\mathbb{Z}(\mathcal{X})$. We must show that this path has length at least $\delta_\mathbb{Z}(f,g)$. But since h_1 can differ from h_0 in only a single function value, $\delta_\mathbb{Z}(h_1,g) \geq \delta_\mathbb{Z}(f,g) - 1$ and the result follows by induction on n.

Since $\delta_\mathbb{Z}(f,g)$ is bounded above and below by the graph distance, it is equal to that distance. □

8.1.3 Remark. In the case of a finite-dimensional lattice $\mathbb{Z}^d$ the distance defined by (8.1) is the usual ℓ_1-distance on $\mathbb{Z}^d$.

8.2 Defining Lattice Dimension

As the following theorem shows, isometric embedding into integer lattices is closely related to isometric embedding into hypercubes.

8.2 Defining Lattice Dimension

8.2.1 Definition. Let a given set $\mathcal{X}$ be given, and let A be a subset of $\mathcal{X}$. Then the *indicator function* $\chi_A : \mathcal{X} \mapsto \{0,1\}$ is defined by

$$\chi_A(x) = \begin{cases} 1, & \text{if } x \in A, \\ 0, & \text{otherwise.} \end{cases}$$

Observe that the support of χ_A is A itself.

8.2.2 Theorem. *For any set $\mathcal{X}$, the hypercube $\mathcal{H}(\mathcal{X})$ can be embedded isometrically into $\mathbb{Z}(\mathcal{X})$, and the integer lattice $\mathbb{Z}(\mathcal{X})$ can be embedded isometrically into $\mathcal{H}(\mathcal{X} \times \mathbb{Z})$.*

Proof. The function $\alpha : A \mapsto \chi_A$, where χ_A is the indicator function of $A \subseteq \mathcal{X}$, is clearly an isometric embedding of the cube $\mathcal{H}(\mathcal{X})$ into the integer lattice $\mathbb{Z}(\mathcal{X})$.

In the other direction, for $f \in \mathbb{Z}(\mathcal{X})$ define the set

$$B_f = \{(x,k) \in \mathcal{X} \times \mathbb{Z} \mid k \leq f(x)\}.$$

Then

$$B_f \triangle B_g = \{(x,k) \in \mathcal{X} \times \mathbb{Z} \mid g(x) < k \leq f(x) \text{ or } f(x) < k \leq g(x)\}.$$

Let $A_f = B_f \triangle B_0$, where 0 stands for the zero function on $\mathcal{X}$. Clearly, A_f is a finite set, $A_f \triangle A_g = B_f \triangle B_g$, and

$$d(A_f, A_g) = |A_f \triangle A_g| = |B_f \triangle B_g| = \sum_{x \in \mathcal{X}} |f(x) - g(x)| = \delta_{\mathbb{Z}}(f,g),$$

so $\beta : f \mapsto A_f$ is an isometric embedding of the integer lattice $\mathbb{Z}(\mathcal{X})$ into the cube $\mathcal{H}(\mathcal{X} \times \mathbb{Z})$. □

8.2.3 Corollary. *The graph $\mathbb{Z}(\mathcal{X})$ is a partial cube.*

8.2.4 Corollary. *A graph is a partial cube if and only if it is isometrically embeddable into an integer lattice.*

Proof. If a graph is a partial cube, it can be isometrically embedded into some $\mathcal{H}(\mathcal{X})$, and (by Theorem 8.2.2 and transitivity of isometric embedding) also into the integer lattice $\mathbb{Z}(\mathcal{X})$. On the other hand, if a graph embeds into an integer lattice $\mathbb{Z}(\mathcal{X})$, then by Theorem 8.2.2 it also embeds into $\mathcal{H}(\mathcal{X} \times \mathbb{Z})$ and is therefore a partial cube. □

As every partial cube can be embedded into an integer lattice, it makes sense to ask what the minimum dimension of such an embedding can be. That is, we use these embeddings to define a notion of *lattice dimension* for partial cubes.

164   8 Media and Integer Lattices

8.2.5 Definition. The *lattice dimension* $\dim_{\mathbb{Z}}(G)$ of a partial cube G is the minimum cardinality of a set $\mathcal{X}$ such that G is isometrically embeddable into $\mathbb{Z}(\mathcal{X})$.

We can use the connection between hypercubes and integer lattices to bound the lattice dimension in terms of the isometric dimension:

8.2.6 Corollary. *For any partial cube G*

$$\dim_{\mathbb{Z}}(G) \leq \dim_I(G).$$

Proof. By Theorem 7.6.2, G can be isometrically embedded into $\mathcal{H}(\mathcal{X})$, where $\mathcal{X}$ is a set of cardinality $\dim_I(G)$. The result follows from the isometric embedding of $\mathcal{H}(\mathcal{X})$ into $\mathbb{Z}(\mathcal{X})$ given by Theorem 8.2.2. □

Theorem 8.3.5, later in this chapter, strengthens Corollary 8.2.6 for finite partial cubes by quantifying the difference between $\dim_{\mathbb{Z}}(G)$ and $\dim_I(G)$.

8.2.7 Example. To illustrate the definitions above, we consider a simple case when G is a finite tree $T = (V, E)$. It is not difficult to see (cf. Deza and Laurent, 1997) that a finite tree T is indeed a partial cube and $\dim_I(T) = |E|$.

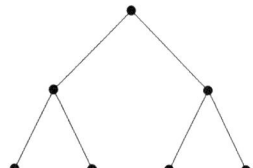

Figure 8.1. Tree T.

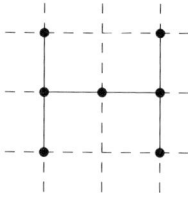

Figure 8.2. A nonisometric embedding of T into $\mathbb{Z}^2$.

Let T be the tree shown in Figure 8.1. A possible embedding of T into $\mathbb{Z}^2$ is shown in Figure 8.2. This embedding is not isometric. An isometric embedding of T into $\mathbb{Z}^2$ is shown in Figure 8.3. Clearly, $\dim_{\mathbb{Z}}(T) = 2$, whereas $\dim_I(T) = 6$. Note that 2 is half the number of leaves in T.

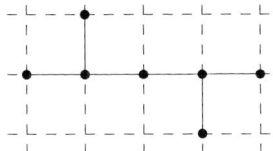

Figure 8.3. An isometric embedding of T into $\mathbb{Z}^2$.

The lattice dimension of any tree may be easily calculated:

8.2.8 Theorem. *Let T be a finite tree with m leaves. Then*
$$\dim_{\mathbb{Z}}(T) = \lceil m/2 \rceil.$$

A lower bound of $\lceil m/2 \rceil$ on $\dim_{\mathbb{Z}}(T)$ is easily established, as each leaf must be the unique positive or negative extreme point in some dimension, so completing the proof merely requires embedding any tree in a lattice of the stated dimension. For a detailed proof see Hadlock and Hoffman (1978), Deza and Laurent (1997) (Proposition 11.1.4), and Ovchinnikov (2004).

The following lemma can be useful in providing a lower bound on the dimension of more general media and partial cubes. We omit the proof (see Problem 8.9).

8.2.9 Lemma. *Let G' be an isometric subgraph of a partial cube G. Then $\dim_{\mathbb{Z}}(G') \leq \dim_{\mathbb{Z}}(G)$.*

8.2.10 Corollary. *Suppose a partial cube G has a vertex v of degree d. Then $\dim_{\mathbb{Z}}(G) \geq \lceil d/2 \rceil$.*

Proof. The tree T formed by v and its neighbors has d leaves; no two of these leaves can be adjacent in G since G is bipartite, so T forms an isometric subgraph of G. The result follows from Theorem 8.2.8 and Lemma 8.2.9. □

8.2.11 Example. We now describe a more complicated number-theoretic example of a partial cube, involving the partitions of an integer, a well-studied concept in number theory. As we show, these partitions can be represented as a partial cube; lattice embedding turns out to be more convenient than set labeling as a way of proving that the graph we form is a partial cube, and naive methods suffice to estimate the lattice dimension of this partial cube.

8.2.12 Definition. A *partition* of a positive integer k is a sequence $p_0 p_1 p_2 \ldots$ of positive integers, with $p_i \geq p_{i+1}$, summing to k. For instance, there are seven partitions of the integer 5: 5, 41, 32, 311, 221, 2111, and 11111. For any k, we may form a graph $\mathcal{P}_k$ having the partitions of k as its vertices; two partitions $p_0 p_1 p_2 \ldots$ and $p'_0 p'_1 p'_2 \ldots$ are adjacent in this graph if there exist $\delta \in \{1, -1\}$ and $i > 0$ such that $p_0 = p'_0 + \delta$, $p_i = p'_i - \delta$, and $p_j = p'_j$ for $j \notin \{0, i\}$. Figure 8.4 depicts one such graph, $\mathcal{P}_7$.

166   8 Media and Integer Lattices

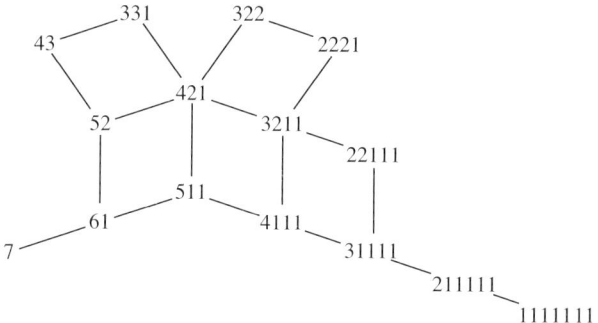

Figure 8.4. The graph $\mathcal{P}_7$ of partitions of the integer 7.

In the result below, we use standard "big theta" notation: $f(x) = \Theta(g(x))$ if and only if both $f(x) = O(g(x))$ and $g(x) = O(f(x))$.

8.2.13 Theorem. $\mathcal{P}_k$ *is a partial cube with* $\dim_{\mathbb{Z}}(\mathcal{P}_k) = \Theta(\sqrt{k})$.

Proof. To show an upper bound on the dimension, we exhibit an embedding of $\mathcal{P}_k$ into a lattice of dimension $2\lfloor\sqrt{k}\rfloor$ and prove that this embedding is isometric. To do so, define

$$q_j = |\{x_i \mid i > \sqrt{k} \text{ and } p_i \geq j\}|.$$

We embed $\mathcal{P}_k$ into a lattice by placing a partition $p_0 p_1 p_2 \ldots$ at the lattice coordinates $p_1 p_2 \ldots p_{\lfloor\sqrt{k}\rfloor} q_1 q_2 q_3 \ldots q_{\lfloor\sqrt{k}\rfloor}$. We note that, for $j > \sqrt{k}$, $q_j = 0$, as any partition of k can have at most $\sqrt{k}$ values that are each larger than $\sqrt{k}$, so all such large values occur too early in the sorted sequence of p_i to be included in the definition of q_j.

If P and P' are partitions adjacent in $\mathcal{P}_k$, via an adjacency defined by the pair (δ, i), then P and Q differ only by a single unit in the single coordinate p_i (if $i \leq \sqrt{k}$) or q_x where $x = \max(p_i, p'_i)$ (otherwise). Therefore, the endpoints of any path of length λ in $\mathcal{P}_k$ can be embedded at distance at most λ in the lattice.

In the other direction, suppose P and P' are partitions at distance λ in the lattice; we prove by induction on λ that P and Q can be connected by a path of length λ in $\mathcal{P}_k$. As a base case, when $\lambda = 0$, P and P' must be the same partition: for, p_i for $i > \sqrt{k}$ can be recovered by the formula $p_i = \max\{j \mid i < q_j + \lfloor\sqrt{k}\rfloor\}$, and p_0 can be recovered by the formula $p_0 = k - \sum_{i>0} p_i$.

To complete the induction for $\lambda > 0$, let $i = \max\{j \mid p_j \neq p'_j\}$, and suppose (without loss of generality) that $p_j > p'_j$. Then we find a new state Q adjacent to P in $\mathcal{P}_k$ defined by a pair $(\delta = -1, i)$. This adjacent Q has the lattice coordinate p_i (if $i \leq \sqrt{k}$) or q_{p_i} (otherwise) one step closer to the coordinates of P'. By induction, there exists a path of length $\lambda - 1$ from Q to P', which together with the edge from P to Q forms the desired path.

Thus, we have proven that $\mathcal{P}_k$ is an isometric subgraph of $\mathbb{Z}^{2\lfloor\sqrt{k}\rfloor}$, so it is a partial cube of at most this lattice dimension. To complete the proof, we must show a matching lower bound on the dimension. Choose n such that $n(n+1)/2 \le k$, $(n+1)(n+2)/2 > k$; then $n = \sqrt{2n} + O(1)$. The triangular partition
$$1 + 2 + 3 + \cdots + (n-1) + (k - (n(n-1)/2)) = k$$
has $2n - 1$ neighbors in $\mathcal{P}_7$, so a lower bound $\dim_{\mathbb{Z}}(\mathcal{P}_k) \ge n = \Omega(\sqrt{k})$ follows from Corollary 8.2.10. □

8.3 Lattice Dimension of Finite Partial Cubes

In this section we establish a more precise relation between the isometric and lattice dimensions of a finite partial cube G. Specifically, we show that the difference between these two dimensions can be counted by the number of edges in a matching of an auxiliary graph, which we define below.

8.3.1 Definition. Let $G = (V, E)$ be a partial cube. Denote by $\mathrm{Sc}(G)$ the *semicube graph*, a graph the vertices of which are the set of semicubes of G (see Definition 7.2.1). Two vertices W_{ab} and W_{cd} are connected by an edge in $\mathrm{Sc}(G)$ if
$$W_{ab} \cup W_{cd} = V \quad \text{and} \quad W_{ab} \cap W_{cd} \ne \varnothing. \tag{8.2}$$

An example of a semicube graph is depicted in Figure 8.5.

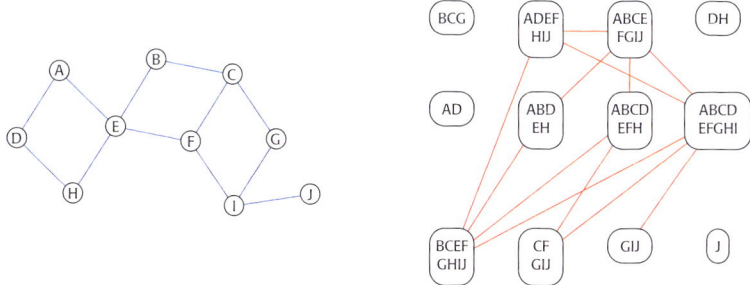

Figure 8.5. A partial cube (left) and its semicube graph (right). The label of each vertex in the semicube graph is formed by the letters representing the corresponding set of partial cube vertices; for instance, the vertex on the lower left of the semicube graph, labeled BCEFGHIJ, corresponds to the semicube $\{B, C, E, F, G, H, I, J\}$.

It is easy to verify that condition (8.2) is equivalent to each of the following two conditions:
$$W_{ba} \subset W_{cd} \quad \text{or} \quad W_{dc} \subset W_{ab}, \tag{8.3}$$

where $\subset$ stands for the proper inclusion.

Let G be a finite isometric subgraph of $\mathbb{Z}^d$, where $d = \dim_\mathbb{Z}(G)$. The projection of G into the k'th coordinate axis is a path P_k in $\mathbb{Z}$. Since d is the lattice dimension of G, the length l_k of P_k is a positive number. Thus G is actually embedded into the complete grid graph

$$P = P_1 \times P_2 \times \cdots \times P_d, \tag{8.4}$$

where $\times$ stands for the Cartesian product. It is easy to see that the equivalence classes of Djoković's relation θ for the graph G are in one-to-one correspondence with the edges in the paths P_k. By Theorem 7.6.2,

$$\dim_I(G) = \sum_{k=1}^{d} l_k.$$

Without loss of generality, we may assume that each P_k is in the form

$$P_k = (0, 1, \ldots, l_k).$$

8.3.2 Definition. A *matching* in an undirected graph is a set of edges, no two of which share any endpoint.

8.3.3 Lemma. *Let G be a finite partial cube which is isometrically embedded into $\mathbb{Z}^d$ for some d. Then there exists a (possibly empty) matching M in the semicube graph $\mathrm{Sc}(G)$, such that*

$$d = \dim_I(G) - |M|.$$

Proof. It can be easily seen that all semicubes in G are intersections of G with complete grid graphs $P'_{k,j}$ and $P''_{k,j}$ where

$$P'_{k,j} = P_1 \times \cdots \times Q_k^j \times \cdots \times P_d, \quad P''_{k,j} = P_1 \times \cdots \times R_k^j \times \cdots \times P_d,$$
$$Q_k^j = (0, 1, \ldots, j),\ 0 \leq j < l_k, \text{ and } R_k^j = (j, \ldots, l_k),\ 0 < j \leq l_k.$$

We define $L_{k,j} = P'_{k,j} \cap G$ and $U_{k,j} = P''_{k,j} \cap G$. Clearly, $L_{k,j} \cup U_{k,j} = V$ and $L_{k,j} \cap U_{k,j} \neq \emptyset$.

Let M be the set of edges in $\mathrm{Sc}(G)$ connecting the semicubes $L_{k,j}$ and $U_{k,j}$. Clearly, the set M is a matching in $\mathrm{Sc}(G)$, that is, it is a set of independent edges. Therefore,

$$|M| = \sum_{k=1}^{d} (l_k - 1) = \dim_I(G) - d.$$

$\square$

8.3 Lattice Dimension of Finite Partial Cubes   169

8.3.4 Lemma. *If M is a matching in the semicube graph $\mathrm{Sc}(G)$ of a finite partial cube G, then there exists an isometric embedding of G into $\mathbb{Z}^d$, where*

$$d = \dim_I(G) - |M|.$$

Proof. Let H be the graph obtained from M by adding an edge between each pair of opposite semicubes W_{ab} and W_{ba}. Clearly, any vertex of H has a degree one or two. Suppose that the semicubes W_{ab} and W_{cd} are connected in M. The diagram below shows a typical fragment of the graph H.

$$W_{ba} \longrightarrow W_{ab} \longrightarrow W_{cd} \qquad (8.5)$$

By (8.3), $W_{ba} \subset W_{cd}$, that is, W_{cd} is a proper superset of W_{ba} (the orientation in (8.5) is consistent with this set inclusion). This is also true for alternating vertices in any path in H, starting at W_{ba}. Hence, H has no cycles and therefore is a union of disjoint paths Q_k, $1 \le k \le d$, for some d. There are $2\dim_I(G)$ semicubes of G, of which $2|M|$ are matched in M. There are two endpoints per path Q_k, and the set of these endpoints consists of $2\dim_I(G) - 2|M|$ unmatched vertices in $\mathrm{Sc}(G)$. Thus, $d = \dim_I(G) - |M|$.

Note that each path Q_k starts and ends with an edge connecting two opposite semicubes. Thus, the length l_k of each path is an odd number.

Choosing, for each path, an orientation defined by one of the set inclusions in (8.3) (cf. (8.5)), we number the vertices of $\mathrm{Sc}(G)$ so that $S_{k,i}$ denotes the ith vertex of path Q_k ($0 \le i \le l_k$). It is convenient to define $S_{k,-1} = S_{k,l_k+1} = V$. By (8.3), we have the two chains of proper inclusions for vertices in a given path Q_k:

$$S_{k,0} \subset S_{k,2} \subset \cdots \subset S_{k,2x} \subset \cdots \subset S_{k,l_k-1} \subset S_{k,l_k+1} = V, \qquad (8.6)$$

and

$$V = S_{k,-1} \supset S_{k,1} \supset \cdots \supset S_{k,2x-1} \supset \cdots \supset S_{k,l_k-2} \supset S_{k,l_k} \qquad (8.7)$$

For a given vertex $v \in V$, let $S_{k,2x}$ be the first set in (8.6) containing v. Let us define $\lambda_k(v) = x$. Then

$$\lambda(v) = (\lambda_1(v), \ldots, \lambda_d(v))$$

maps V into $\mathbb{Z}^d$. It remains to verify that λ defines an isometric embedding of G into $\mathbb{Z}^d$.

Note that if $v \in S_{k,2x}$, where $x = \lambda_k(v)$, then $v \in S_{k,2x-1}$. Indeed, otherwise, we would have $v \in S_{k,2x-2}$, since $S_{k,2x-2} \cup S_{k,2x-1} = V$. This contradicts the definition of $\lambda_k(v)$. A similar argument shows that $v \notin S_{k,2x+1}$. Thus $\lambda_k(v)$ can be equivalently defined as a unique number x such that $v \in S_{k,2x} \cap S_{k,2x-1}$.

Let $u, v \in V$. Any semicube S that separates u and v in G (that is, either $u \in S$, $v \notin S$ or $u \notin S$, $v \in S$) belongs to a single path, say, Q_k. The number of semicubes in the form $S_{k,2x}$ that separate u and v is $|\lambda_k(u) - \lambda_k(v)|$. It is the same number for semicubes in the form $S_{k,2x-1}$ (cf. (8.7)). Therefore, the total number of semicubes in G separating u and v is $2\sum_k |\lambda_k(u) - \lambda_k(v)|$, which is twice the distance between $\lambda(u)$ and $\lambda(v)$ in $\mathbb{Z}^d$. On the other hand, if an edge $\{a,b\}$ belongs to a shortest path connecting vertices u and v, then semicubes W_{ab} and W_{ba} separate u and v and any semicube separating u and v is in one of these forms. Hence, the number of semicubes separating u and v is twice the distance between vertices u and v in G. We proved that

$$\delta_{\mathbb{Z}^d}(\lambda(u), \lambda(v)) = \delta_G(u,v).$$

Thus, λ is an isometric embedding of G into $\mathbb{Z}^d$. □

Lemmas 8.3.3 and 8.3.4 together yield the following theorem:

8.3.5 Theorem. *For any finite partial cube G,*

$$\dim_{\mathbb{Z}}(G) = \dim_I(G) - |M|,$$

where M is a maximum matching in the semicube graph $\mathrm{Sc}(G)$.

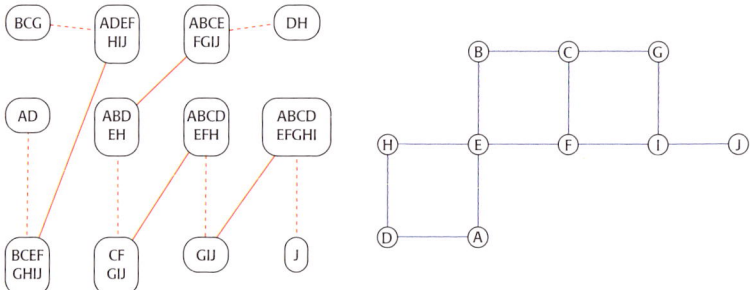

Figure 8.6. A matching in the semicube graph of Figure 8.5, and the corresponding embedding of a partial cube into $\mathbb{Z}^2$, from Eppstein (2005b).

Figure 8.6 shows a maximum matching in the semicube graph for the example from Figure 8.5, and the corresponding minimum dimension lattice embedding of the partial cube from that figure.

8.3.6 Remark. Figure 8.3 shows an isometric embedding of the tree T into a (4×2)-grid. The same tree is isometrically embedded into a (3×3)-grid in Figure 8.7. Although the grid P in (8.4) is not uniquely defined by a partial cube G, as these two figures show, the set $L = \{l_1, \cdots, l_d\}$, where $d = \dim_{\mathbb{Z}}(G)$, determines both the lattice and isometric dimensions of G. Namely,

$$\dim_{\mathbb{Z}}(G) = |L| \quad \text{and} \quad \dim_I(G) = l_1 + \cdots + l_d.$$

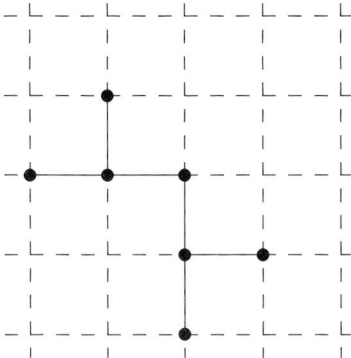

Figure 8.7. Another isometric embedding of T into $\mathbb{Z}^2$.

8.4 Lattice Dimension of Infinite Partial Cubes

We may characterize the lattice dimension of infinite partial cubes by their finite isometric subgraphs. First, we establish some auxiliary results.

Let $G = (V, E)$ be a connected graph. For $v_0 \in V$ and any finite integer k we set
$$N_k(v_0) = \{v \in V \mid \delta(v, v_0) \leq k\}$$
and call this set the *ball of radius k and center v_0*. It is clear that $N_0(v_0) = \{v_0\}$ and $N_i(v_0) \subseteq N_{i+1}(v_0)$.

8.4.1 Lemma. *If the ball $N_k(v_0)$ is infinite, then there is a vertex in this ball of infinite degree.*

Proof. Since $N_0(v_0)$ is finite and $N_k(v_0)$ is infinite, there is i such that $N_i(v_0)$ is finite and $N_{i+1}(v_0)$ is infinite. It suffices to note that any vertex in $N_{i+1}(v_0)$ is adjacent to a vertex in $N_i(v_0)$. □

The *convex hull* of a set $X \subseteq V$ is the intersection of all convex subsets of V containing X. The following lemma recasts a result from Kuzmin and Ovchinnikov (1975) in the terms of this chapter.

8.4.2 Lemma. *Let G be a partial cube. The convex hull of any finite set of vertices of G is a finite set.*

Proof. We may assume that $G = (\mathcal{F}, \mathcal{W}_\mathcal{F})$ for a wg-family of finite subsets of some set X. Let $\mathcal{P} \subset \mathcal{F}$ be finite. By Theorem 3.4.5(ii), the set
$$\{R \in \mathcal{F} \mid \cap \mathcal{P} \subseteq R \subseteq \cup \mathcal{P}\}$$
is convex. Clearly, it is finite and contains $\mathcal{P}$. The result follows. □

Note that, in general, the result of Lemma 8.4.2 does not hold for an infinite bipartite graph which is not a partial cube (cf. Problem 8.3).

8.4.3 Theorem. *For a partial cube G and any integer k, $\dim_{\mathbb{Z}}(G) \le k$ if and only if $\dim_{\mathbb{Z}}(G') \le k$ for all finite isometric subgraphs G' of G. If $\dim_{\mathbb{Z}}(G)$ is infinite, then $\dim_{\mathbb{Z}}(G) = \dim_I(G)$.*

Proof. For the first claim of the theorem, it suffices to prove that, if all finite isometric subgraphs G' can be embedded into $\mathbb{Z}^d$, then so can G itself. Let v_0 be a fixed vertex of G. Then if the ball $N_i(v_0)$ were infinite, Lemma 8.4.1 would allow us to find isometric subtrees of G consisting of a single vertex with $2d + 2$ neighbors. By Theorem 8.2.8 these would have dimension $d + 1$. But this contradicts the assumption that all isometric subgraphs of G have dimension at most d; this contradiction proves the finiteness of the ball $N_i(v_0)$. As all balls $N_i(v_0)$ are finite, Lemma 8.4.2 implies that the convex hulls G_i of these balls are also finite. These convex hulls form a nested sequence

$$G_0 = \{v_0\} \subseteq G_1 \subseteq \cdots \subseteq G_i \subseteq \cdots$$

of finite isometric subgraphs such that $\cup_i G_i = G$.

Form a tree L, the vertices of which are embeddings of G_i into $\mathbb{Z}^d$ such that v_0 is placed at the origin. The parent of a vertex corresponding to an embedding $f : G_i \mapsto \mathbb{Z}^d$ is the embedding of G_{i-1} formed by restricting f to the vertices of G_{i-1}. L is infinite if G is infinite, but each vertex has finitely many children, so by König's Infinity Lemma (Diestel, 2000) it has an infinite path. Any two embeddings of this path that both place the same vertex v of G must place v at the same lattice point as each other, and every vertex of G is placed by all but finitely many embeddings of the path, so we can derive from this path a consistent embedding of all vertices of G into $\mathbb{Z}^d$. This embedding is isometric, as any path in G can be covered by some subgraph G_i for sufficiently large i.

It remains to prove the case of the theorem in which $\dim \mathbb{Z}(G)$ is infinite. By Corollary 8.2.6, $\dim_I(G) \ge \dim_{\mathbb{Z}}(G)$. Consider an isometric embedding of G into $\mathbb{Z}(\mathcal{X})$, where $|\mathcal{X}| = \dim_{\mathbb{Z}}(G)$. Composing this embedding with the embedding of Theorem 8.2.2 from $\mathbb{Z}(\mathcal{X})$ into $\mathcal{H}(\mathcal{X} \times \mathbb{Z})$,

$$\dim_I(G) \le |\mathcal{X} \times \mathbb{Z}| = |\mathcal{X}| = \dim_{\mathbb{Z}}(G)$$

and the result follows. □

8.5 Oriented Media

In the case of oriented finite media, it is natural to desire an integer lattice embedding that respects the orientation.

8.5.1 Definition. Let $\mathcal{M}$ be a finite oriented medium endowed with the orientation $\{\mathcal{T}^+, \mathcal{T}^-\}$. Then, an *oriented embedding* of $\mathcal{M}$ is an isometric embedding of its underlying unoriented medium into $\mathbb{Z}^d$ for some d, with the property that, for each state S and positive token $\tau \in \mathcal{T}^+$, $S\tau$ is placed at a point with coordinate values greater than or equal to those of S. The *oriented lattice dimension* $\dim_{\mathbb{Z}^+}(\mathcal{M})$ is the minimum dimension of a lattice $\mathbb{Z}^d$ in which $\mathcal{M}$ is orientally embeddable.

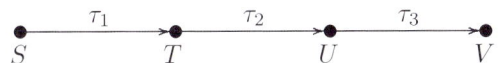

Figure 8.8. A medium on four states embedded in $\mathbb{Z}$.

Consider, for instance, the medium in Figure 8.8 with orientation defined by $\mathcal{T}^+ = \{\tau_1, \tilde{\tau}_2, \tau_3\}$. It is clear that with this orientation the medium in Figure 8.8 is not embeddable into $\mathbb{Z}$; however it has an oriented embedding into $\mathbb{Z}^2$. Thus, its oriented lattice dimension is 2.

As it turns out, the same idea of finding matchings in an associated semicube graph can be used to compute the oriented lattice dimension.

8.5.2 Definition. We define *oriented semicube graph* $\mathrm{Sc}^+(\mathcal{M})$ of an oriented medium $\mathcal{M}$ with orientation $\{\mathcal{T}^+, \mathcal{T}^-\}$ the be a graph having all semicubes of $\mathcal{M}$ as the set of its vertices. Two vertices $\mathcal{S}_\tau$ and $\mathcal{S}_{\tau'}$ are connected in $\mathrm{Sc}^+(\mathcal{M})$, where $\tau \in \mathcal{T}^+$ and $\tau' \in \mathcal{T}^-$, if $\mathcal{S}_\tau \cup \mathcal{S}_{\tau'} = \mathcal{S}$ and $\mathcal{S}_\tau \cap \mathcal{S}_{\tau'} \neq \varnothing$.

The oriented semicube graph can equivalently be seen as the subgraph of the semicube graph $\mathrm{Sc}(\mathcal{M})$ in which we keep only edges between semicubes $\mathcal{S}_\tau$ and $\mathcal{S}_{\tau'}$ for pairs τ and τ' that are oppositely oriented (see Figure 8.9).

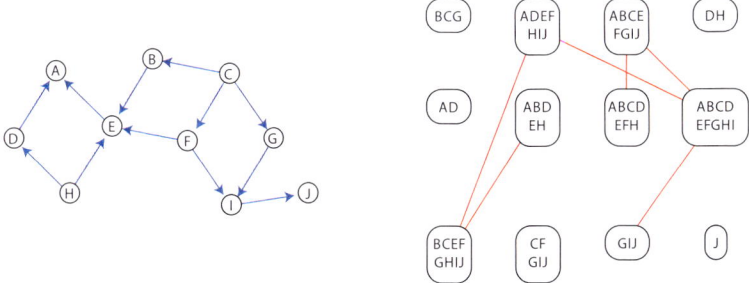

Figure 8.9. The oriented semicube graph of an oriented partial cube (cf. Figure 8.5).

8.5.3 Lemma. *Let $\mathcal{M}$ be a finite oriented medium which is orientally embedded into $\mathbb{Z}^d$ for some d. Then there exists a (possibly empty) matching M in the oriented semicube graph $\mathrm{Sc}^+(\mathcal{M})$, such that*

$$d = \dim_{\mathbb{Z}^+}(\mathcal{M}) - |M|.$$

Proof. As in the proof of Lemma 8.3.3, we observe that all semicubes of $\mathcal{M}$ can be represented geometrically as intersections with complete grid graphs $L_{k,j}$ and $U_{k,j}$, and let M be the set of edges in $\mathrm{Sc}^+(\mathcal{M})$ connecting semicubes $L_{k,j}$ and $U_{k,j}$. It follows from the orientation of the embedding of $\mathcal{M}$ that each such edge connects semicubes S_τ and $S_{\tau'}$ for pairs τ and τ' that are oppositely oriented. The equality of d with $\dim_{\mathbb{Z}^+}(\mathcal{M}) - |M|$ follows in the same manner as in the proof of Lemma 8.3.3. □

8.5.4 Lemma. *If M is a matching in the oriented semicube graph $\mathrm{Sc}^+(\mathcal{M})$ of a finite oriented medium $\mathcal{M}$, then there exists an oriented embedding of $\mathcal{M}$ into $\mathbb{Z}^d$, where*
$$d = \dim_I(G) - |M|.$$

Proof. Since $\mathrm{Sc}^+(\mathcal{M})$ is a subgraph of $\mathrm{Sc}(\mathcal{M})$, M is also a matching in $\mathrm{Sc}(\mathcal{M})$, and we can use Lemma 8.3.4 to construct an isometric embedding of $\mathcal{M}$ from it. It follows from the definition of $\mathrm{Sc}^+(\mathcal{M})$ and the construction of the embedding in the proof of that lemma that the sequence of tokens along each coordinate dimension of the resulting embedding have a consistent orientation. By negating the coordinates in which this orientation is negative, we construct an oriented embedding of $\mathcal{M}$. □

Lemmas 8.5.3 and 8.5.4 together yield the following theorem.

8.5.5 Theorem.
$$\dim_{\mathbb{Z}^+}(\mathcal{M}) = \dim_I(\mathcal{M}) - |M|,$$
for any finite oriented medium $\mathcal{M}$, where M is a maximum matching in the oriented semicube graph $\mathrm{Sc}^+(\mathcal{M})$.

As in the unoriented case, we may extend the definition of oriented lattice dimension to infinite oriented media by taking the maximum dimension of any finite submedium of $\mathcal{M}$; we omit the details.

Problems

8.1 Let G be the graph shown below. Find $\dim_I(G)$, $\dim_\mathbb{Z}(G)$, and $\ell_a(G)$.

8.2 Let C_n be a cycle of length n.

(i) Prove that C_n is a partial cube if and only if n is even (cf. Problems 7.3 and 7.4 in Chapter 7).

(ii) Prove that $\dim_I(C_{2n}) = \dim_{\mathbb{Z}}(C_{2n}) = n$.

(iii) If a cycle C_n is an isometric subgraph of a partial cube G, show that $n \leq 2 \dim_{\mathbb{Z}}(G)$.

8.3 Let A and B be two disjoint sets. We denote $K_{A,B}$ the complete bipartite graph on the set of vertices $A + B$. Show that the result of Lemma 8.4.2 does not hold for an infinite bipartite graph $K_{A,B}$.

8.4 Prove that the three conditions in Equations (8.2) and (8.3) are equivalent.

8.5 Let G be a partial cube and $\{W_{ab}, W_{xy}\}$, $\{W_{ba}, W_{uv}\}$ be two edges in the semicube graph $\mathrm{Sc}(G)$. Prove that $\{W_{xy}, W_{uv}\}$ is an edge in $\mathrm{Sc}(G)$.

8.6 Describe the graph $\mathrm{Sc}(C_n)$ where C_n is the cycle of length n.

8.7 What is the semicube graph of the cube $\mathcal{H}(X)$?

8.8 Prove Theorem 8.2.8. (*Hint:* Use Theorem 8.3.5.)

8.9 Prove Lemma 8.2.9.

8.10 Let P be a path of length d and $G = P^n$ be the complete grid $\underbrace{P \times \cdots \times P}_{n \text{ factors}}$. Show that $\lim_{n \to \infty} \dfrac{\ell_a(G)}{n} = \dfrac{d(d+2)}{3(d+1)}$ (cf. Problem 7.14 in Chapter 7).

8.11 Let T be a rooted tree with ℓ leaves, viewed as an oriented medium. Show that $\dim_{\mathbb{Z}^+}(T) = \ell$.

8.12 Find the oriented lattice dimension of the graph shown in Figure 1.5.

8.13 Let G be the graph shown in Figure 2.1.

(i) Find the lattice dimension $\dim_{\mathbb{Z}}(G)$.

(ii) Find the oriented lattice dimension $\dim_{\mathbb{Z}^+}(G)$ for the orientation given by $\mathcal{F}^+ = \{\tau_1, \tau_3, \tau_6\}$.

8.14 Prove that, for any oriented medium $\mathcal{M}$,

$$\dim_{\mathbb{Z}}(\mathcal{M}) \leq \dim_{\mathbb{Z}^+}(\mathcal{M}) \leq 2 \dim_{\mathbb{Z}}(\mathcal{M}).$$

9

Hyperplane arrangements and their media

Hyperplane arrangements in a finite dimensional vector space, as illustrated by Example 1.2.1, are a rich and useful source of media examples. We prove that the 'regions' of a hyperplane arrangements define a medium, give examples of such media, and compute their isometric and lattice dimensions. Then we apply geometric techniques to show that families of 'labeled interval orders' and weak orders can be cast as media. This part of the chapter supplements the results presented in Chapter 5.

9.1 Hyperplane Arrangements and Their Media

We begin with general definitions and facts about hyperplane arrangements. For details and proofs, the reader is referred to Björner et al. (1999); Bourbaki (2002); Grünbaum (1972); Orlik and Terano (1992); Stanley (1996); Zaslavsky (1975); Ziegler (1995). In our presentation we follow Ovchinnikov (2005, 2006).

9.1.1 Definition. An *arrangement* $\mathcal{A}$ of hyperplanes is a locally finite family $\mathcal{A} = \{H_i\}_{i \in J}$ of affine hyperplanes in $\mathbb{R}^n$, that is, any open ball in $\mathbb{R}^n$ intersects only a finite number of hyperplanes in $\mathcal{A}$. We say that an arrangement is *finite* if it has only finitely many hyperplanes. A *central arrangement* is an arrangement of hyperplanes all of which contain the origin.

9.1.2 Remark. A central arrangement is necessarily finite. There are at most countably many hyperplanes in any arrangement $\mathcal{A}$, so the index set J is either finite or countable. Every hyperplane can be represented (though not in a unique way) by an affine linear function $l_i(\boldsymbol{x}) = \sum_{j=1}^{n} a_{ij} x_j + b_i$, that is,

$$H_i = \{\boldsymbol{x} \in \mathbb{R}^n \,|\, l_i(\boldsymbol{x}) = 0\}.$$

We assume a fixed choice of representation for each hyperplane and refer to

178   9 Hyperplane arrangements and their media

$$H_i^+ = \{x \in \mathbb{R}^n \,|\, l_i(x) > 0\} \quad \text{and} \quad H_i^- = \{x \in \mathbb{R}^n \,|\, l_i(x) < 0\},$$

as the open *half-spaces* separated by H_i.

The hyperplane arrangement $\mathcal{A}$ partitions $\mathbb{R}^n$ into subsets called 'cells'. It is convenient to define cells in terms of their sign vectors.

9.1.3 Definition. Given an arrangement $\mathcal{A}$ in $\mathbb{R}^n$, we associate to every point $x \in \mathbb{R}^n$ its *sign vector (function)* $X(x)$ defined by

$$X(x): J \to \{+, -, 0\}^J \;:\; i \mapsto X_i(x) = \begin{cases} + & \text{if } x \in H_i^+, \\ - & \text{if } x \in H_i^-, \\ 0 & \text{if } x \in H_i. \end{cases}$$

The set of all points $x \in \mathbb{R}^n$ having the same sign vector is called a *cell* in the partition of $\mathbb{R}^n$ defined by $\mathcal{A}$. Each cell is convex and open relative to the affine subspace spanned by the cell. The *dimension* of a cell is the number of nonzeros in its sign vector. The n-dimensional cells are called *regions* of $\mathcal{A}$.

9.1.4 Definition. A hyperplane $H_i \in \mathcal{A}$ *separates* regions P and Q if $P \subseteq H_i^+$ and $Q \subseteq H_i^-$, or $P \subseteq H_i^-$ and $Q \subseteq H_i^+$. Equivalently, H_i separates regions P and Q if the sign vectors for P and Q differ only in their ith coordinate. Two regions P and Q are *adjacent* if there is a unique hyperplane H_i in $\mathcal{A}$ separating them.

9.1.5 Definition. The *region graph* G of an arrangement $\mathcal{A}$ has the set $\mathcal{P}$ of regions as the set of vertices. The set of edges of G consists of all the pairs $\{P, Q\}$ of adjacent regions in $\mathcal{P}$.

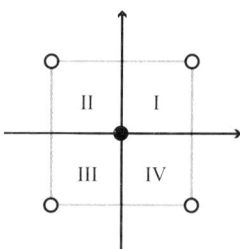

Figure 9.1. The Boolean arrangement $\mathbb{B}_2$ and its region graph.

9.1.6 Example. Let $\mathcal{A}$ be the arrangement of two coordinate lines in the plane shown in Figure 9.1. The regions (2-dimensional cells) of $\mathcal{A}$ are the quadrants I–IV. The four 1-dimensional cells are the positive and negative parts of the coordinate lines. The only 0-dimensional cell is the point $(0,0)$. The graph of this arrangement is the cycle C_4 (cf. Definition 9.1.12 below).

9.1 Hyperplane Arrangements and Their Media

We now construct a token system $(\mathcal{P}, \mathcal{T})$ representing the regions and adjacencies of an arrangement (cf. Example 1.2.1).

9.1.7 Definition. Let $\mathcal{A} = \{H_i\}_{i \in J}$ be a hyperplane arrangement. We define the token system $(\mathcal{P}, \mathcal{T})$ associated with $\mathcal{A}$. The set $\mathcal{P}$ of states is the collection of regions of $\mathcal{A}$. We define the set $\mathcal{T}$ of tokens to be the union $\cup_{i \in J} \{\tau_i^-, \tau_i^+\}$, with, for every $i \in J$, the two functions τ_i^- and τ_i^+ defined by

$$\tau_i^+ : \mathcal{P} \to \mathcal{P} : P \mapsto P\tau_i^+ = \begin{cases} Q & \text{if } P \subset H_i^-,\ Q \subset H_i^+, \text{ with } P, Q \text{ adjacent,} \\ P & \text{otherwise;} \end{cases}$$

$$\tau_i^- : \mathcal{P} \to \mathcal{P} : P \mapsto P\tau_i^- = \begin{cases} Q & \text{if } P \subset H_i^+,\ Q \subset H_i^-, \text{ with } P, Q \text{ adjacent,} \\ P & \text{otherwise.} \end{cases}$$

It is easily verified that τ_i^+ and τ_i^- are genuine tokens (well-defined functions, distinct from the identity on $\mathcal{P}$), and mutual reverses (Problem 9.1).

The following theorem is the main result of this section.

9.1.8 Theorem. *Let $\mathcal{A} = \{H_i\}_{i \in J}$ be a hyperplane arrangement in $\mathbb{R}^n$. The token system $(\mathcal{P}, \mathcal{T})$ of Definition 9.1.7 is a medium.*

Proof. We show that axioms [Ma] and [Mb] defining a medium are satisfied.

[Ma]. To find a concise message transforming a state S into another state V, let s and v be points chosen within regions S and V, so that line segment sv does not pass through any cell of the arrangement of dimension lower than $n-1$. This is always possible, for if we fix s to be any point in S, then the union of all line segments connecting s with points in V and passing through cells of dimension less than $n-1$ has dimension less than n. Let the sequence of arrangement hyperplanes intersected by line segment sv be $H_1, \ldots H_k$, with each H_i in this sequence separating points s and v. This sequence is finite, since $\mathcal{A}$ is locally finite; clearly, all the hyperplanes in the sequence are distinct. For $1 \leq i \leq k$, we define

$$\tau_i = \begin{cases} \tau_i^- & \text{if } s \in H_i^-,\ v \in H_i^+, \\ \tau_i^+ & \text{if } s \in H_i^+,\ v \in H_i^-. \end{cases}$$

Then $\tau_1 \ldots \tau_k$ is a concise message transforming S into V. Therefore, Axiom [Ma] is satisfied.

[Mb]. Suppose that a message $\boldsymbol{m} = \tau_1 \ldots \tau_k$ is stepwise effective for some state S and ineffective for that state, so $S = S\boldsymbol{m}$. We need to show that $\boldsymbol{m}$ is vacuous. Let s_i be a point in $S_i = S\tau_0 \ldots \tau_i$ for $0 \leq i < k$ and $s_k = s_0$. We connect successive points s_i's by directed line segments to obtain a closed oriented piecewise linear curve C in $\mathbb{R}^n$. Each directed line segment in C intersects a single hyperplane $H \in \mathcal{A}$ and corresponds to a unique token τ in $\mathcal{C}(\boldsymbol{m})$. The next occurrence of a line segment in C intersecting H corresponds to the reverse token $\tilde{\tau} \in \mathcal{C}(\boldsymbol{m})$. Since C is closed, there are even number of line segments intersecting H, equally many in each direction. Thus, we can partition the tokens in $\boldsymbol{m}$ into pairs of mutually reverse tokens, so [Mb] is satisfied.

As the two axioms hold, $(\mathcal{P}, \mathcal{T})$ is a medium. □

The region graph G of an arrangement $\mathcal{A}$ represents the medium $(\mathcal{P}, \mathcal{T})$ via Definitions 6.1.1 and 9.1.7. In view of Theorem 7.1.7, Theorem 9.1.8 can be reformulated as follows.

9.1.9 Theorem. *The region graph of a hyperplane arrangement is a partial cube.*

9.1.10 Remark. The same proof goes through, essentially unchanged, to show that hyperplane arrangements restricted to a convex subset of $\mathbb{R}^n$ similarly form media and partial cube region graphs (Ovchinnikov, 2005). For example, Figure 9.2 shows an arrangement of three lines in the plane, restricted to the convex subset bounded by an ellipse, that provides a geometric realization of the medium shown earlier in Figure 2.1.

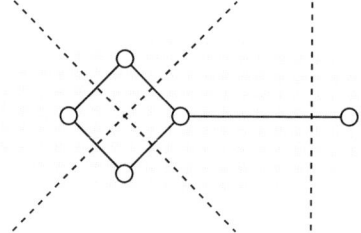

Figure 9.2. An arrangement of three lines (shown dashed), restricted to the shaded ellipse, produces a medium isomorphic to the one in Figure 2.1.

9.1.11 Remark. The region graph G of a finite arrangement $\mathcal{A} = \{H_i\}_{i \in J}$ is the *tope graph* of the oriented matroid associated with the arrangement $\mathcal{A}$ (see Handa, 1990; Björner et al., 1999). It is known that these graphs form isometric subgraphs of hypercubes; for instance, see Proposition 4.2.3 in Björner et al. (1999) or Theorem 28 of Gärtner and Welzl (1994). Theorem 9.1.9 extends this result to infinite arrangements.

Particularly important classes of finite hyperplane arrangements in $\mathbb{R}^n$ are introduced below. We give examples of infinite arrangements later in this section.

9.1 Hyperplane Arrangements and Their Media

9.1.12 Definition. The *Boolean arrangement* $\mathbb{B}_n$ is the central arrangement of the coordinate hyperplanes in $\mathbb{R}^n$ defined by the linear functions $l_i(\boldsymbol{x}) = x_i$ (cf. Example 4.3.4).

The central arrangement of all the hyperplanes in $\mathbb{R}^n$ that are defined by the linear functions $l_{ij}(\boldsymbol{x}) = x_i - x_j$ for $i < j$ is called the *braid arrangement* $\mathcal{B}_n$ (also known as the *Coxeter arrangement*).

For $d = (d_1, \ldots, d_n)$ with $d_i > 0$, the *deformation* $\mathcal{A}_{n,d}$ of the braid arrangement $\mathcal{B}_n$ is defined by the affine linear functions

$$l_{ij} = x_i - x_j - d_j, \quad \text{for } i \neq j. \tag{9.1}$$

The next three examples illustrate these definitions.

9.1.13 Example. The region graph of the Boolean arrangement $\mathbb{B}_2$ is shown in Figure 9.1. In general, the region graph of $\mathbb{B}_n$ is the cube on the set $\{1, \ldots, n\}$.

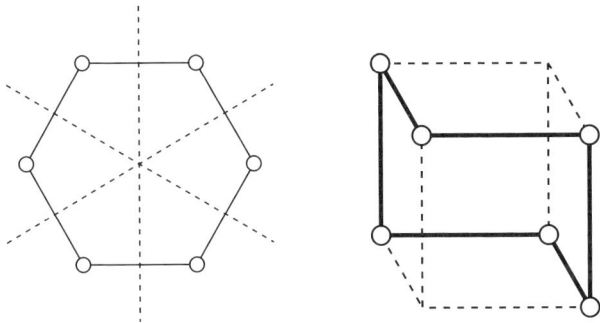

Figure 9.3. Left: The braid arrangement $\mathcal{B}_3$ and its region graph, a cycle C_6. Right: an isometric embedding of C_6 into the cube.

9.1.14 Example. The region graph of the braid arrangement $\mathcal{B}_3$ is shown in Figure 9.3 (left). The three dashed lines are intersections of the planes in $\mathbb{R}^3$ defining the braid arrangement $\mathcal{B}_3$ with the plane $x_1 + x_2 + x_3 = 0$. The region graph of $\mathcal{B}_3$ is the cycle C_6 which is the graph of the permutohedron Π_2 (cf. Figure 3.1). Its isometric embedding into the cube is shown in Figure 9.3 (right).

9.1.15 Example. Let $d = (1, 1, 1)$. The region graph of the deformation $\mathcal{A}_{3,d}$ of the braid arrangement $\mathcal{B}_3$ is shown in Figure 9.4. Again, the dashed lines are intersections of the planes in $\mathbb{R}^3$ defining the arrangement $\mathcal{A}_{3,d}$ with the plane $x_1 + x_2 + x_3 = 0$. By Theorem 9.1.9, this graph is a partial cube (cf. Figure 1.5). Thus, this geometric method can provide an alternative proof of Theorem 5.5.7 (see Corollary 9.3.4).

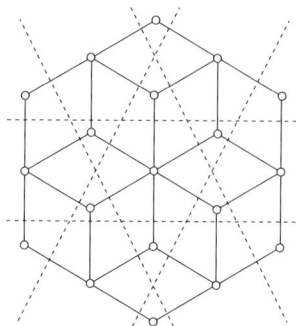

Figure 9.4. The region graph of $\mathcal{A}_{3,d}$.

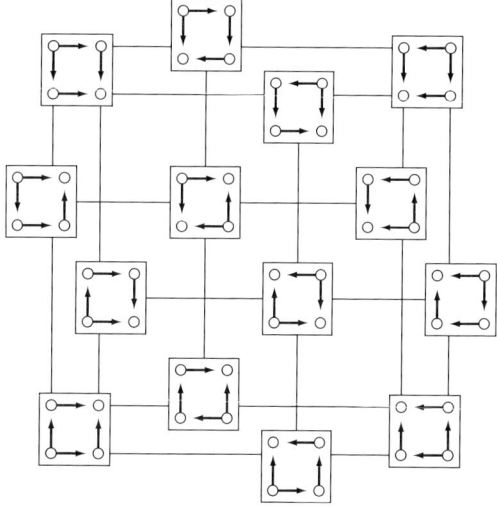

Figure 9.5. The acyclic orientations of the graph C_4; two orientations are connected by an edge when they differ by a single edge flip.

9.1.16 Example. (cf. Orlik and Terano, 1992, p.57.) We now discuss a graph-theoretic token system which can be easily proven to be a medium by representing it as an arrangement. Consider any directed acyclic graph D. Then, if one reverses any edge of the *transitive reduction* of D (the Hasse diagram of the partial order represented by D), the result is again a directed acyclic graph. However, reversing an edge that is not part of the transitive reduction forms a cycle, leading to a directed graph that is not acyclic. This suggests a token system in which the states are directed acyclic graphs and the transformations are acyclicity-preserving edge reversals. For instance, Figure 9.5 shows fourteen directed acyclic graphs, the directed acyclic orientations of a square, with two graphs connected by an edge in the drawing whenever they differ in the orientation of a single edge.

9.1.17 Definition. Let G be an undirected graph. Then an *acyclic orientation* of G is a directed acyclic graph formed by assigning a direction to each edge of G. We construct from G a token system $(\mathcal{P}, \mathcal{T})$, where $\mathcal{P}$ consists of the set of all acyclic orientations of G. $\mathcal{T}$ contains two tokens for each edge of G, one token for each assignment of a direction to each edge. If τ is the token corresponding to the direction (a, b) on edge $\{a, b\}$, then the action of τ on an acyclic orientation S is to reverse edge (b, a), if (b, a) is a directed edge of the state S and if the result of reversing that edge is another acyclic orientation; otherwise τ leaves S unchanged.

9.1.18 Theorem. *The token system $(\mathcal{P}, \mathcal{T})$ defined from the acyclic orientations of a finite graph G with n vertices is isomorphic to a token system defined from a subarrangement of the braid arrangement $\mathcal{B}_n$.*

Proof. Associate each vertex of G with the index of one of the coordinates in a Cartesian coordinate system for $\mathbb{R}^n$. Let $\mathcal{A}$ be the arrangement of hyperplanes $x_i = x_j$, for each pair $\{i, j\}$ forming an edge of G; that is, $\mathcal{A}$ is a subarrangement of the braid arrangement $\mathcal{B}_n$. If x is a point in a region of $\mathcal{A}$, we may find an orientation of G from x by orienting an edge $\{i, j\}$ from x_i to x_j whenever $x_i > x_j$, and from x_j to x_i otherwise; note that, since x is in one of the regions, x_i cannot equal x_j. It is easy to see that this orientation does not depend on the choice of x within its region, so in this way we can associate regions of $\mathcal{A}$ with acyclic orientations of G, in such a way that adjacent regions correspond to orientations that differ in the orientation of a single edge.

We leave it to the reader to show that this association indeed defines an isomorphism between the two token systems (see Problem 9.2). □

Thus, for example, the token system shown in Figure 9.5 can be represented by an arrangement in which four hyperplanes, corresponding to the four edges of the square, lie in a space of four dimensions corresponding to the four vertices of the square. Stanley (1996) calls the arrangement $\mathcal{A}$ formed from G in the proof above a *graphical arrangement*.

9.1.19 Corollary. *The token system of acyclic orientations of any finite undirected graph forms a medium.*

9.1.20 Remark. The family of strict linear orderings of a finite set is isomorphic to the family of acyclic orientations of a complete graph. Thus, Theorem 9.1.18 can be viewed as a geometric generalization of the result that the strict linear orders of a finite set form a medium, a result that we generalized in a different way, to linear orders of infinite sets, in Theorem 3.5.6.

9.1.21 Remark. Not all partial cubes are the region graphs of hyperplane arrangements. For instance, the graph of the medium shown in Figure 2.1 is not the region graph of any hyperplane arrangement in $\mathbb{R}^n$ for any n (cf. Problem 9.4).

184   9 Hyperplane arrangements and their media

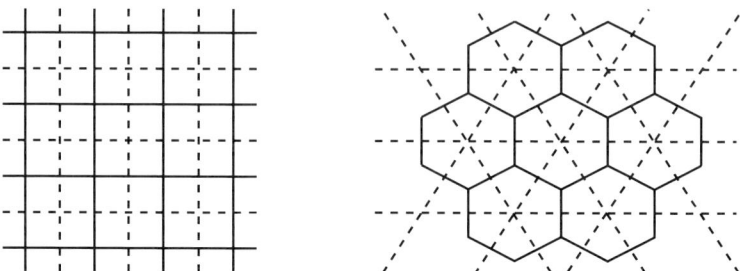

Figure 9.6. Fragments of two infinite line arrangements and their region graphs.

We conclude this section with two examples of infinite arrangements and their graphs.

9.1.22 Examples. In each of the drawings of Figure 9.6, an infinite arrangement of parallel lines is depicted by dashed lines. The corresponding region graphs are 1-skeletons of tilings of the plane by squares (left) and hexagons (right).

9.2 The Lattice Dimension of an Arrangement

Media derived from arrangements have many attractive combinatorial and geometric properties which make some tasks simpler than in the general case of media. To exemplify this statement, we compute in this section the lattice dimension (cf. Definition 8.2.5) of the region graph of an arrangement $\mathcal{A}$.

Let G be the region graph of arrangement $\mathcal{A}$. By Corollary 7.6.3, the isometric dimension (cf. Definition 7.6.1) of the graph G is equal to the cardinality of the set $\mathcal{A}$, that is, $\dim_I(G) = |\mathcal{A}|$. Therefore, by Theorem 8.3.5, in order to evaluate $\dim_\mathbb{Z}(G)$, we need to find the cardinality of a maximum matching M in the semicube graph $\mathrm{Sc}(G)$.

9.2.1 Lemma. *The semicubes in the region graph G of an arrangement $\mathcal{A}$ are the sets of all regions that lie on one side of hyperplanes in $\mathcal{A}$.*

Proof. Let P and Q be two adjacent regions and H be the unique hyperplane in $\mathcal{A}$ that separates these regions. Let R be a region that lies on the same side of H as P. Any hyperplane that separates R and P is different from H. Therefore, it separates regions R and Q, since H is the only hyperplane in $\mathcal{A}$ separating P and Q. The distance between two regions is equal to the number of hyperplanes in $\mathcal{A}$ separating these two regions (see Problem 9.5). It follows that $\delta(R,P) < \delta(R,Q)$. Similarly, $\delta(R,Q) < \delta(R,P)$, if the regions R and Q lie on the same side of H. Hence the semicube W_{PQ} consists of all regions lying on the same side of H as P. □

Thus, the semicubes in G can be identified with half-spaces defined by the hyperplanes in $\mathcal{A}$.

9.2.2 Lemma. *Let $\mathcal{A}$ be a finite arrangement of parallel hyperplanes in $\mathbb{R}^n$, G be its region graph, and M be a maximum matching in $\mathrm{Sc}(G)$. Then*
$$|M| = |\mathcal{A}| - 1.$$

Proof. As we indicated before, $\dim_I G = |\mathcal{A}|$. Clearly, G is a path. Hence, $\dim_{\mathbb{Z}}(G) = 1$. We have
$$|M| = \dim_I(G) - \dim_{\mathbb{Z}}(G) = |\mathcal{A}| - 1,$$
by Theorem 8.3.5. □

9.2.3 Theorem. *Let $\mathcal{A} = \mathcal{A}_1 \cup \cdots \cup \mathcal{A}_m$ be a union of hyperplane arrangements in $\mathbb{R}^n$ such that hyperplanes in each $\mathcal{A}_i$ are parallel and hyperplanes in distinct $\mathcal{A}_i$'s are not parallel. Let G be the region graph of $\mathcal{A}$. Then $\dim_{\mathbb{Z}}(G) = m$.*

Proof. By Theorem 8.4.3, we need only consider finite isometric subgraphs of G; any such subgraph can be viewed as part of the arrangement of a finite subset of $\mathcal{A}$, so we need only consider the case in which each $\mathcal{A}_i$ is finite. By Definition 8.3.1, two semicubes are connected by an edge in $\mathrm{Sc}(G)$ if they have a nonempty intersection and their union is the set of all regions of $\mathcal{A}$. By Lemma 9.2.1, if two hyperplanes in $\mathcal{A}$ are not parallel, the four semicubes defined by these hyperplanes are not mutually adjacent in $\mathrm{Sc}(G)$. Therefore, any matching in $\mathrm{Sc}(G)$ consists of edges connecting semicubes defined by parallel hyperplanes in $\mathcal{A}$. Let M be a maximum matching in $\mathrm{Sc}(G)$. By Lemma 9.2.2,
$$|M| = \sum_{i=1}^{m}(|\mathcal{A}_i| - 1) = |\mathcal{A}| - m.$$
By Theorem 8.3.5,
$$\dim_{\mathbb{Z}}(G) = \dim_I(G) - |M| = |\mathcal{A}| - (|\mathcal{A}| - m) = m.$$
□

As hyperplanes in a central arrangement cannot be parallel, the result can be simplified for that case.

9.2.4 Corollary. *If $\mathcal{A}$ is a central arrangement and G is its region graph, then*
$$\dim_I(G) = \dim_{\mathbb{Z}}(G) = |\mathcal{A}|.$$

9.2.5 Example. Let G be the graph of the arrangement shown in Figure 9.7. This arrangement consists of three pairs of parallel planes. Thus, $\dim_I(G) = 6$ and $\dim_{\mathbb{Z}}(G) = 3$. The embedding of G into $\mathbb{Z}^3$ is shown in Figure 9.8. Clearly, G cannot be isometrically embedded into the 2-dimensional integer lattice.

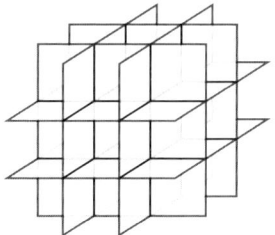

Figure 9.7. A hyperplane arrangement consisting of three pairs of parallel planes.

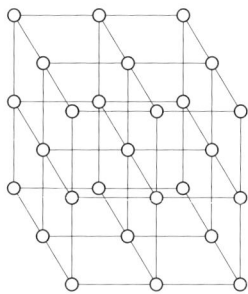

Figure 9.8. The region graph of the arrangement from Figure 9.7.

9.3 Labeled Interval Orders

In this section we show that the family of all labeled interval orders on a finite set is well-graded. Our presentation follows Ovchinnikov (2005), where a similar result is established for a wide class of families of partial orders.

Stanley (1996) introduced labeled interval orders as a generalization of semiorders. We define these relations as follows.

9.3.1 Definition. Let ρ be a positive function on a finite set $\mathcal{X}$. (We may refer to ρ as a *length function* or *threshold*.) A binary relation R on $\mathcal{X}$ is a ρ-*labeled interval order* if there exists a real-valued function f on $\mathcal{X}$ such that

$$aRb \quad \Leftrightarrow \quad f(a) > f(b) + \rho(b), \qquad (9.2)$$

for all $a, b \in \mathcal{X}$. Any function f defining the relation R by (9.2) is said to be a *representing* function for R.

We denote by $\mathcal{IO}_\rho$ the family of ρ-labeled interval orders on $\mathcal{X}$. If ρ is a constant function, then (9.2) defines a semiorder on $\mathcal{X}$ (cf. (5.29)). The family of all semiorders on $\mathcal{X}$ is denoted $\mathcal{SO}$.

Let $V = \mathbb{R}^{\mathcal{X}}$ be the n-dimensional vector space of all real-valued functions on $\mathcal{X}$. Any function f in V defines a labeled interval order R by (9.2). Let $J = (\mathcal{X} \times \mathcal{X}) \setminus \{(x,x)\}_{x \in \mathcal{X}}$. For $ab \in J$, we define affine linear functions l_{ab} on V by

$$l_{ab}(f) = f(a) - f(b) - \rho(b),$$

and hyperplanes H_{ab} by

$$H_{ab} = \{f \in V \mid l_{ab}(f) = 0\}.$$

Note that all hyperplanes H_{ab} are distinct. They form a hyperplane arrangement $\mathcal{A}$ which is the deformation $\mathcal{A}_{n,\rho}$ (see Definition 9.1.12) of the braid arrangement $\mathcal{B}_n$ in V. Corresponding half-spaces, H_{ab}^+ and H_{ab}^-, are defined by

$$H_{ab}^+ = \{f \in V \mid f(a) > f(b) + \rho(b)\}$$

and

$$H_{ab}^- = \{f \in V \mid f(a) < f(b) + \rho(b)\},$$

respectively.

For a given region P of $\mathcal{A}$, we define

$$J_P = \{ab \in J \mid P \subseteq H_{ab}^+\}.$$

Since P is a region, $P \subseteq H_{ab}^+$ or $P \subseteq H_{ab}^-$ for any $ab \in J$. It follows that all functions in P represent the same labeled interval order R as defined by (9.2). Clearly, $R = J_P$, that is, $\mathcal{J} = \{J_P \mid P \in \mathcal{P}\} \subseteq \mathcal{JO}_\rho$. In other words, $\mathcal{J}$ is a family of labeled interval orders which are representable by functions from regions of the arrangement $\mathcal{A}$.

To prove that $\mathcal{J} = \mathcal{JO}_\rho$, we need to show that for any function representing a given labeled interval order there is a function in a region of $\mathcal{A}$ representing the same order.

9.3.2 Lemma. *Let R be a labeled interval order and f be its representing function. There is a function g representing R such that $g \in V \setminus \bigcup_{ab \in J} H_{ab}$.*

Proof. Suppose first $R = \varnothing$. Then, for any function f representing R, we have, by (9.2), $f(x) \le f(y) + \rho(y)$ for all $x, y \in \mathcal{X}$. Let us define

$$g(x) = \lambda = \max\{f(y) \mid y \in \mathcal{X}\}, \quad \text{for all } x \in \mathcal{X}.$$

Clearly, g represents the empty relation and does not belong to any of hyperplanes in the form H_{ab}.

Let $R \in \mathcal{JO}_\rho$, $R \ne \varnothing$, and let f be a function representing R. Let δ be a number satisfying inequalities

$$\max_{xy \in R} \left\{ \frac{\rho(y)}{f(x) - f(y)} \right\} < \delta < 1, \tag{9.3}$$

and let $g = \delta f$. We show first that g represents R.

Suppose that $ab \in R$. then

$$g(a) - g(b) - \rho(b) = \delta[f(a) - f(b)] - \rho(b) > 0,$$

by the first inequality in (9.3). On the other hand, if
$$g(a) > g(b) + \rho(b)$$
for some $a, b \in X$, then $f(a) > f(b)$ and
$$0 < g(a) - g(b) - \rho(b) = \delta[f(a) - f(b)] - \rho(b)$$
$$< f(a) - f(b) - \rho(b),$$
by the second inequality in (9.3). Hence, $ab \in R$. This shows that g represents the relation R.

Suppose $g \in H_{ab}$ for some $ab \in J$, that is, that $g(a) = g(b) + \rho(b)$, or, equivalently, $\delta[f(a) - f(b)] = \rho(b)$. Since $\delta < 1$, we have
$$f(a) - f(b) > \rho(b),$$
that is, $ab \in R$. Since g represents R, we have $g(a) > g(b) + \rho(b)$, a contradiction. Therefore, $g \in V \setminus \cup_{ab \in J} H_{ab}$. □

We proved that $\mathcal{IO}_\rho = \mathcal{J}$. The next result follows immediately from Theorem 9.1.8.

9.3.3 Theorem. *The family $\mathcal{IO}_\rho$ of labeled interval orders on X is well-graded. The representing token system $(\mathcal{IO}_\rho, \mathcal{W}_{\mathcal{IO}_\rho})$ is a medium.*

As a corollary we have a result of Doignon and Falmagne (1997) (see also Theorem 5.5.7):

9.3.4 Corollary. *The family $\mathcal{SO}$ of semiorders on X is well-graded.*

Finally, note, that there are $n(n-1)$ hyperplanes in the arrangement $\mathcal{A} = \mathcal{A}_{n,\rho}$ and $\frac{n(n-1)}{2}$ pairs of parallel hyperplanes in this arrangement. By Theorem 9.2.3,
$$\dim_I(\mathcal{IO}_\rho) = n(n-1) \quad \text{and} \quad \dim_{\mathbb{Z}}(\mathcal{IO}_\rho) = \frac{n(n-1)}{2}.$$

9.4 Weak Orders and Cubical Complexes

A (strict) weak order on a set X is a 1-connected, asymmetric binary relation on X (cf. (1.10) and the equivalent Definition 5.1.1). The following theorem is a well-known result from representational measurement theory (see Theorem 3.1 in Roberts, 1979).

9.4.1 Theorem. *A binary relation R on a finite set X is a weak order if and only if there exists a function $f : X \to \mathbb{R}$ such that*
$$aRb \quad \Leftrightarrow \quad f(a) > f(b), \tag{9.4}$$
for all $a, b \in X$.

Any function f defining the relation R by (9.4) is said to be a *representing function* for R. The family of weak orders on a given finite set X is denoted by $\mathcal{WO}$. This family is ordered by the inclusion relation and we often treat it as a partially ordered set.

9.4.2 Example. Let $X = \{a, b, c\}$. There are 13 weak orders on X; they are represented by vertices of the diagram shown in Figure 9.9 (cf. Figure 2 in Kemeny and Snell, 1972). In this diagram, for instance, $a \sim b \prec c$ denotes the weak order $R = \{ca, cb\}$, which is representable by any function f such that

$$f(a) = f(b) < f(c).$$

The empty weak order $\varnothing$ is denoted by $a \sim b \sim c$ and representable by constant functions. The diagram in Figure 9.9 is the Hasse diagram of the partial order $\mathcal{WO}$.

The family $\mathcal{WO}$ shown in Figure 9.9 is not well-graded. Indeed, the distance between the weak order $R = \{ca, cb\}$ and the empty weak order $\varnothing$ is 2, but there is no weak order different from R and $\varnothing$ that lies between them. Nevertheless, it is possible to model in a natural way the family of weak orders on a finite set X by a wg-family of sets. In fact, the diagram in Figure 9.9 gives a hint, as it has an obvious cubical structure.

To realize this approach we introduce two useful descriptions of weak orders.

Let R be a weak order on X and let f be its representing function. We write $x \prec y$ if $f(x) < f(y)$ and $x \sim y$ if $f(x) = f(y)$ (cf. Figure 9.9). It is clear that $\sim$ is an equivalence relation on the set X and $\prec$ defines a linear order on the quotient set $X/\sim$ of equivalence classes of the relation $\sim$. Let

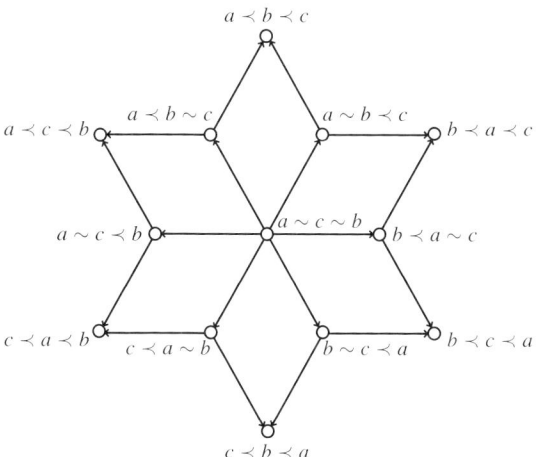

Figure 9.9. The diagram of weak orders on $X = \{a, b, c\}$.

$X_1, \ldots, X_k$ be the equivalence classes of the relation $\sim$ enumerated according to the relation $\prec$. Then

$$xRy \quad \Leftrightarrow \quad x \in X_i,\ y \in X_j, \quad \text{for some } i < j. \tag{9.5}$$

In this case, we write $R = \langle X_1, \ldots, X_k \rangle$ and say that R is a *weak k-order* and that X_i's are *indifference classes* of R. In particular, weak n-orders are linear orders, and the only weak 1-order is the empty weak order. The set of all weak k-orders on $\mathcal{X}$ is denoted by $\mathcal{WO}(k)$.

Another useful representation of $\mathcal{WO}$ is given by the cells of the braid arrangement $\mathcal{B}_n$. We begin with a simple example.

9.4.3 Example. The cells of the braid arrangement $\mathcal{B}_3$ are shown in Figure 9.10 (cf. Figure 9.3). The 3-cells (regions) are labeled $R_1, \ldots, R_6$, 2-cells (facets of regions) are shown by the rays, and the only 1-cell is represented by the dot at the center of the drawing. Note that this figure is the intersection of $\mathcal{B}_3$ with the plane $x_1 + x_2 + x_3 = 0$ (cf. Example 9.1.14).

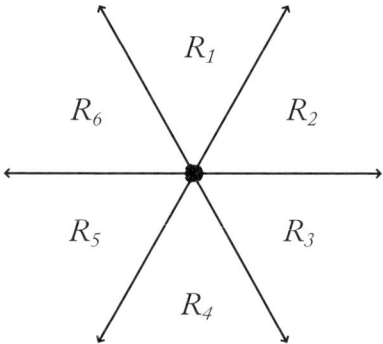

Figure 9.10. Cells of the braid arrangement $\mathcal{B}_3$

As in the previous section, $V = \mathbb{R}^{\mathcal{X}}$ is the n-dimensional vector space of all real-valued functions on $\mathcal{X}$. Let R_0 be a fixed asymmetric and strongly connected binary relation on $\mathcal{X}$. For $ab \in R_0$ we define hyperplanes H_{ab} in V by

$$H_{ab} = \{ f \subset V \mid f(a) = f(b) \}.$$

Clearly, these hyperplanes are distinct and form the braid arrangement $\mathcal{B}_n$. Corresponding half-spaces are defined as follows:

$$H_{ab}^+ = \{ f \in V \mid f(a) > f(b) \} \quad \text{and} \quad H_{ab}^- = \{ f \in V \mid f(a) < f(b) \},$$

for $ab \in R_0$.

The sign vectors

$$X_{ab}(f) = \begin{cases} +, & \text{if } f \in H_{ab}^+, \\ -, & \text{if } f \in H_{ab}^-, \\ 0, & \text{if } f \in H_{ab}, \end{cases} \quad ab \in R_0,$$

define cells of $\mathcal{B}_n$ (cf. Definition 9.1.3). The proof of the next lemma is left as Problem 9.9.

9.4.4 Lemma. *Two sign vectors $X_{ab}(f)$ and $X_{ab}(g)$ are equal if and only if functions f and g represent the same weak order on X.*

9.4.5 Definition. Let $\mathcal{A}$ be an arrangement. The *face poset* $\mathcal{F}(\mathcal{A})$ is the set of all cells of $\mathcal{A}$ ordered by inclusion of their topological closures.

By Lemma 9.4.4, there is one-to-one correspondence between cells of $\mathcal{B}_n$ and weak orders on $\mathcal{X}$. This correspondence has a clear geometric flavor as illustrated by drawings in Figures 9.3, 9.9, and 9.10. In the case of the braid arrangement $\mathcal{B}_n$, the posets $\mathcal{WO}$ and $\mathcal{F}(\mathcal{B}_n)$ are isomorphic (see Problem 9.11).

We obtain a representation of the family of weak orders on $\mathcal{X}$ by a wg-family of sets, by showing that the Hasse diagram of (isomorphic) posets $\mathcal{F}(\mathcal{B}_n)$ and $\mathcal{WO}$ is a partial cube. For this we use representations of weak orders as ordered partitions of the set $\mathcal{X}$ (cf. Janowitz, 1984; Ovchinnikov, 2006). This representation will provide us with a tool for constructing a medium on weak orders.

The following theorem is the statement of Problem 19 on p.115 in Mirkin (1979). The proof is straightforward and omitted (see Problem 9.14).

9.4.6 Theorem. *A weak order $R = \langle X_1, \ldots, X_k \rangle$ contains a weak order R' if and only if*

$$R' = \left\langle \bigcup_{j=1}^{i_1} X_j, \bigcup_{j=i_1+1}^{i_2} X_j, \ldots, \bigcup_{j=i_m}^{k} X_j \right\rangle$$

for some sequence of indices $1 \leq i_1 < i_2 \cdots < i_m \leq k$.

One can say (see Mirkin, 1979, Chapter 2) that $R' \subset R$ if and only if the indifference classes of R' are "enlargements of the adjacent indifference classes" of R.

9.4.7 Corollary. *A weak order $W = \langle X_1, \ldots, X_k \rangle$ covers a weak order W' in $\mathcal{WO}$ if and only if*

$$W' = \langle X_1, \ldots, X_i \cup X_{i+1}, \ldots, X_k \rangle$$

for some $1 \leq i < k$.

The proof is left as Problem 9.15.

Let R be a weak order. We denote by J_R the set of all weak 2-orders that are contained in R.

9.4.8 Theorem. *A weak order admits a unique representation as a union of weak 2-orders, that is, for any $R \in \mathcal{WO}$ there is a uniquely defined set $J \subseteq \mathcal{WO}(2)$ such that*
$$R = \bigcup_{U \in J} U. \tag{9.6}$$

Proof. Clearly, the empty weak order has a unique representation in the form (9.6) with $J = \varnothing$.

Let $R = \langle X_1, \ldots, X_k \rangle$ be a weak order with more than one indifference class. By Theorem 9.4.6, each weak order in J_R is in the form
$$R_i = \langle \cup_1^i X_j, \cup_{i+1}^k X_j \rangle, \quad 1 \leq i < k.$$

By (9.5), we have
$$xRy \quad \Leftrightarrow \quad (\exists p)\, xR_p y \quad \Leftrightarrow \quad x(\cup_{i=1}^k R_i)y.$$

Thus $R = \cup_{i=1}^k R_i$. This proves (9.6) with $J = J_R$.

Let $R = \langle X_1, \ldots, X_k \rangle$ be a weak order in the form (9.6). It is clear that $J \subseteq J_R$. Suppose that $R_s \notin J$, for some s. Let $x \in X_s$ and $y \in X_{s+1}$. Then $xy \in R$ and $xy \notin R_i$, for $i \neq s$, a contradiction. It follows that $J = J_R$, which proves uniqueness of representation (9.6). □

Let $\mathcal{J}$ be the family of subsets of the set $\mathcal{WO}(2)$ in the form J_R, ordered by inclusion. The following theorem is an immediate consequence of Theorem 9.4.8.

9.4.9 Theorem. *The correspondence $R \mapsto J_R$ is an isomorphism of posets $\mathcal{WO}$ and $\mathcal{J}$.*

Clearly, the empty weak order on X corresponds to the empty subset of $\mathcal{WO}(2)$ and the set $\mathcal{LO}$ of all linear orders on X is in 1-1 correspondence with maximal elements in $\mathcal{J}$.

9.4.10 Theorem. *The family $\mathcal{J}$ is an independence system.*

Proof. Let J' be a subset of J_R for some $R \in \mathcal{WO}$. Then $R = \bigcup_{U \in J_R} U$. Let $R' = \bigcup_{U \in J'} U$. As a union of negatively transitive relations, the relation R' itself is negatively transitive. It is asymmetric, since $R' \subseteq R$. Thus, R' is a weak order. By Theorem 9.4.8, $J' = J_{R'} \in \mathcal{J}$. It follows that the family $\mathcal{J}$ is closed under taking subsets, the defining property of an independence system (cf. Definition 4.1.1). □

9.4.11 Theorem. *Let $(\mathcal{J}, \mathcal{W}_\mathcal{J})$ be the representing medium (cf. Definition 3.3.2) of the family $\mathcal{J}$. Then $(\mathcal{J}, \mathcal{W}_\mathcal{J})$ is an i-closed medium.*

Proof. This is an immediate consequence of Theorems 9.4.10 and 4.1.18. □

Let $\alpha : \mathcal{WO} \to \mathcal{J}$ be the isomorphism from Theorem 9.4.9. We obtain a token system $(\mathcal{WO}, \mathcal{T})$ by 'pulling back' tokens from $\mathcal{W}_\mathcal{J}$ to $\mathcal{WO}$, as it was done in 3.5.10 for linear media. Namely, for $\gamma \in \mathcal{W}_\mathcal{J}$, we define τ_γ by

$$\tau_\gamma = \alpha \circ \gamma \circ \alpha^{-1},$$

that is, $R\tau_\gamma = \alpha^{-1}(\alpha(R)\gamma)$ for $R \in \mathcal{WO}$. Let $\mathcal{T} = \{\tau_\gamma\}_{\gamma \in \mathcal{W}_\mathcal{J}}$. We define $\beta(\tau_\gamma) = \gamma$. Clearly, (α, β) is an isomorphism from the token system $(\mathcal{WO}, \mathcal{T})$ to the medium $(\mathcal{J}, \mathcal{W}_\mathcal{J})$. Therefore, $(\mathcal{WO}, \mathcal{T})$ is an i-closed medium itself. Tokens in $\mathcal{T}$ are defined by the weak 2-orders on the set $\mathcal{X}$.

In what follows, we describe effective actions of tokens in $\mathcal{T}$ in terms of weak orders.

Let $V = \langle A, B \rangle$ be a weak 2-order on $\mathcal{X}$. We regard sets A and B as sets of 'dominated' and 'dominating' elements, respectively, (relative, of course, to the weak 2-order V). The weak 2-order V defines two mutually reverse tokens τ_V and $\tilde{\tau}_V$ in $\mathcal{T}$ by means of the bijection β^{-1}. We consider effective actions of τ_V and $\tilde{\tau}_V$ on the states in $\mathcal{WO}$ separately.

(i) The case of τ_V. Let $R = \langle X_1, \ldots, X_k \rangle$ be a weak order on which τ_V is effective, that is, $R\tau_V \neq R$. Then $V \notin J_R$ and $J_R \cup \{V\} \in \mathcal{J}$, according to the definition of token actions in $(\mathcal{J}, \mathcal{W}_\mathcal{J})$. Thus, by Theorem 9.4.8, $R \cap V$ is a weak order and R covers $R \cap V$ in $\mathcal{WO}$. The indifference classes of $R \cap V$ are in the forms $X_i \cap A$ and $X_i \cap B$. Since $R \neq R \cap V$, there is p such that $X'_p = X_p \cap A$ and $X''_p = X_p \cap B$ form a partition of X_p. By Corollary 9.4.7,

$$R\tau_V = \langle X_1, \ldots, X'_p, X''_p, \ldots, X_k \rangle,$$

where $X_i \subset A$ for $i < p$, $X_i \subset B$ for $i > p$. In other words, the action of τ_V partitions the indifference class X_p of R into subsets of 'dominated' and 'dominating' elements according to V. The remaining indifference classes of R consist entirely of either 'dominated' or 'dominating' alternatives with respect to V (cf. Figure 9.11).

(ii) The case of $\tilde{\tau}_V$. Again, let $R = \langle X_1, \ldots, X_k \rangle$ be a weak order on which $\tilde{\tau}_V$ is effective. Then $V \in J_R$ which implies $R \subset V$. By Theorem 9.4.6, there is $1 \leq p < k$ such that $V = \langle \cup_{i=1}^p X_i, \cup_{i=p+1}^k X_i \rangle$. By Theorem 9.4.8, the weak order $R\tilde{\tau}_V$ is the intersection of weak 2-orders in the form $R_q = \langle \cup_{i=1}^q X_i, \cup_{i=q+1}^k X_i \rangle$ with $q \neq p$. The indifference classes of $R\tilde{\tau}_V$ are intersections of the indifference classes of R_q's with $q \neq p$. It follows that

$$R\tilde{\tau}_V = \langle X_1, \ldots, X_p \cup X_{p+1}, \ldots, X_k \rangle.$$

Since $A = \cup_{i=1}^p X_i$ and $B = \cup_{i=p+1}^k X_i$, the action of $\tilde{\tau}_V$ joins the indifference classes X_p and X_{p+1}. These classes consist of maximal (with respect

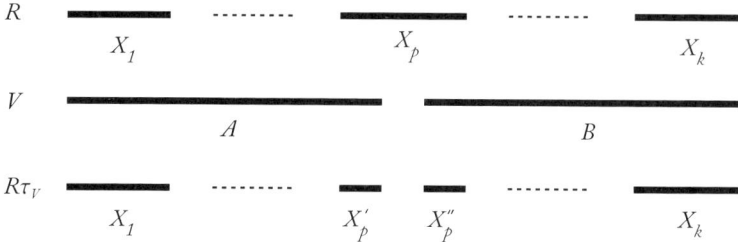

Figure 9.11. An effective action of the token τ_V

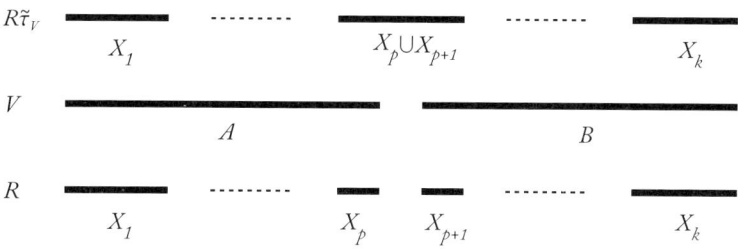

Figure 9.12. An effective action of the token $\tilde{\tau}_V$

to R) 'dominated' elements in A and minimal 'dominating' elements in B, respectively, (cf. Figure 9.12).

The braid arrangement $\mathcal{B}_n$ has a rich geometric structure that provides for various representations of the poset $\mathcal{WO}$. First, by Theorem 9.4.9, the poset $\mathcal{WO}$ can be modeled by a set of vertices of the (oriented) unit cube of the dimension $d = |\mathcal{WO}(2)| = 2^n - 2$ (cf. Problem 9.13) by using characteristic functions $\chi(J_R)$ of the sets J_R. Let $L \in \mathcal{WO}(n)$ be a linear order on $\mathcal{X}$. Then J_L is a maximal element in $\mathcal{J}$ and, by the proof of Theorem 9.4.11, the convex hull of $\{\chi(J_R) \mid R \in \mathcal{WO}\ R \subseteq L\}$ is a subcube C_L of $[0,1]^d$. The dimension of C_L is $n-1$. The collection of all cubes C_L with $L \in \mathcal{WO}(n)$ and all their subcubes form a cubical complex $\mathcal{C}(\mathcal{WO})$ which is a subcomplex of $[0,1]^d$. Clearly, $\mathcal{C}(\mathcal{WO})$ is a complex of the dimension $n-1$ and the (oriented) 1-skeleton of this complex is isomorphic to the Hasse diagram of $\mathcal{WO}$.

The dimension $\dim \mathcal{C}(\mathcal{WO}) = n-1$ is much smaller than the dimension $d = 2^n - 2$ of the space $\mathbb{R}^d$ in which $\mathcal{C}(\mathcal{WO})$ was realized. Simple examples indicate that $\mathcal{C}(\mathcal{WO})$ can be realized in a space of a smaller dimension. For instance, a weak order R on $\mathcal{X}$ is a subset of the set $(\mathcal{X} \times \mathcal{X}) \setminus \{(x,x)\}_{x \in \mathcal{X}}$ and therefore can be represented by a 0/1-vector in $\mathbb{R}^{n(n-1)}$. Note that the convex hull of 0/1-vectors representing weak orders is a *weak order polytope* (Fiorini and Fishburn, 2004). Clearly, we obtained a realization of the complex $\mathcal{C}(\mathcal{WO})$ in the $n(n-1)$-dimensional vector space.

We obtain a realization of $\mathcal{C}(\mathcal{WO})$ even in a smaller space by 'unfolding' the complex in the previous representation. Figure 9.9 is instructive. Consider

9.4 Weak Orders and Cubical Complexes

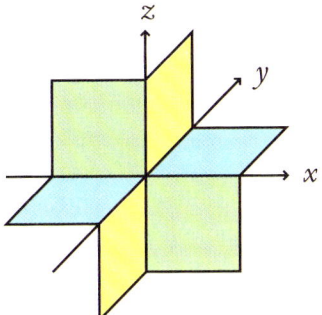

Figure 9.13. "Monkey Saddle".

it as a drawing of three-dimensional diagram with weak orders

$$b \prec a \sim c, \quad a \sim b \prec c, \quad a \prec b \sim c$$

represented by the three coordinate unit vectors. Then we have a realization of the complex $\mathcal{C}(\mathcal{WO})$ for $\mathcal{X} = \{a, b, c\}$, which is shown in Figure 9.13. This is a 'flat' analog of the popular smooth surface $z = x^3 - 3xy^2$ know under the name "monkey saddle". This example suggests that the complex $\mathcal{C}(\mathcal{WO})$ can be realized in $\frac{n(n-1)}{2}$-dimensional space (cf. Problem 9.12).

It is clear that it is impossible to place six squares on the plane in such a way that they represent the complex $\mathcal{C}(\mathcal{WO})$ (here, $n = 3$). Thus, in general, there is no geometric realization of the $(n-1)$-dimensional complex $\mathcal{C}(\mathcal{WO})$ in $\mathbb{R}^{n-1}$. On the other hand, there is a complex in $\mathbb{R}^{n-1}$ which is combinatorially equivalent to $\mathcal{C}(\mathcal{WO})$.

According to Ziegler (1995, p.18), "k-faces (of the permutohedron Π_{n-1}) correspond to ordered partitions of (the set $\mathcal{X}$) into $n-k$ nonempty parts" (see also Barbut and Monjardet, 1970, p.54). In other words, each face of the permutohedron Π_{n-1} represents a weak order on $\mathcal{X}$. Linear orders on $\mathcal{X}$ are represented by the vertices of Π_{n-1} and the trivial weak order on $\mathcal{X}$ is represented by Π_{n-1} itself. Weak 2-orders are in one-to-one correspondence with the facets of Π_{n-1}. Let L be a vertex of Π_{n-1}, and consider the set of barycenters of all faces of Π_{n-1} containing L. A direct computation (cf. Problem 9.19) shows that the convex hull of these points is a (combinatorial) cube of dimension $n - 1$. This also follows from a more general statement of Corollary 7.18 in Ziegler (1995).

The complex depicted in Figure 9.14 (cf. Figure 9.9) is a combinatorial equivalent of the complex $\mathcal{C}(\mathcal{WO})$ for $n = 3$. The colored quadrilaterals represent maximal cubes in $\mathcal{C}(\mathcal{WO})$ (cf. Figure 9.13). The union of these quadrilaterals is the permutohedron Π_2.

9 Hyperplane arrangements and their media

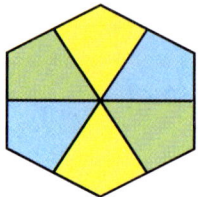

Figure 9.14. The complex $\mathcal{C}(\mathcal{WO})$ for $n = 3$

It is not difficult to calculate the isometric and lattice dimensions of the partial cube $\mathcal{WO}$. We have (see Problems 9.13, 9.12, and 9.16):

$$\dim_I(\mathcal{WO}) = 2^n - 2 \quad \text{and} \quad \dim_{\mathbb{Z}}(\mathcal{WO}) = \frac{n(n-1)}{2} \qquad (9.7)$$

Problems

9.1 Prove that the functions τ_i^+ and τ_i^- introduced in Definition 9.1.7 are genuine tokens (that is, distinct from the identity on $\mathcal{P}$), and mutual reverses.

9.2 Complete the proof of Theorem 9.1.18.

9.3 Let P be a non connected partial order on a finite set X and let $\mathcal{P}$ be the set of all linear extensions of P. (There are at least two such extensions; see Trotter, 1992, p.9). We define a token system $(\mathcal{P}, \mathcal{T})$ in which the set of tokens

$$\mathcal{T} = \{\tau_{ab} \mid ab \in X \times X, a \neq b, \{ab, ba\} \cap P = \varnothing\}$$

is defined, for all $\tau_{ab} \in \mathcal{T}$, by

$$L\tau_{ab} = Q \iff \begin{array}{l} (L \setminus \{ba\}) \cup \{ab\} = Q \\ (L, Q \in \mathcal{P}, \ L \setminus Q = \{ba\}, Q \setminus L = \{ab\}). \end{array}$$

Prove that $(\mathcal{P}, \mathcal{T})$ is a medium.

9.4 Prove that the region graph of a hyperplane arrangement contains a vertex incident to a single edge, if and only if the hyperplanes of the arrangement are all parallel to each other. Show from this that the medium of Figure 2.1 does not come from a hyperplane arrangement in $\mathbb{R}^n$.

9.5 Let G be the region graph of an arrangement $\mathcal{A}$. Show that the graph distance between two vertices of G equals the number of hyperplanes in $\mathcal{A}$ separating the corresponding regions of $\mathcal{A}$.

9.6 Show that the medium formed from an n-dimensional braid arrangement is isomorphic to the medium of linear orders on n items defined in Chapter 3.

9.7 Let $\mathcal{A}$ be the arrangement of lines (cf. Example 9.1.22):
$$y = \sqrt{3}x + i, \quad y = -\sqrt{3} + j, \quad y = k, \quad i, j, k \in \mathbb{Z},$$
and let $\mathcal{A}^{(n)}$ be a subarrangement $\mathcal{A}$ defined by the conditions $|i| < n$, $|j| < n$, and $|k| < n$. The region graph H_n of $\mathcal{A}^{(n)}$ is the nth benzenoid graph from the coronene/circumcoronene series. Show that H_n has $6n^2$ vertices (Imrich and Klavžar, 2000).

9.8 Let C be a cycle in the graph of $\mathbb{Z}^2$ and let G_C be a subgraph formed by the vertices and edges lying on and in the interior of C. Show that G_C is a partial cube.

9.9 Prove Lemma 9.4.4.

9.10 Let $\mathcal{P}$ be the set of all partial orders on a given finite set X with more than two elements, and let ρ be a positive function on X. Prove the following statements:

(i) $\mathcal{WO}$ is a proper subset of $\mathcal{JO}_\rho$.

(ii) If $|X| = 3$, then $\mathcal{SO} = \mathcal{JO}_\rho = \mathcal{P}$.

(iii) If $|X| \geq 4$, then $\mathcal{JO}_\rho$ is a proper subset of $\mathcal{P}$.

(iv) If $|X| = 4$, then $\mathcal{SO} \subseteq \mathcal{JO}_\rho$.

(v) If $|X| = 4$ and ρ is not a constant function, then $\mathcal{SO}$ is a proper subset of $\mathcal{JO}_\rho$.

(vi) If $|X| \geq 5$, then there exist ρ such that
$$\mathcal{SO} \setminus \mathcal{JO}_\rho \neq \emptyset \quad \text{and} \quad \mathcal{JO}_\rho \setminus \mathcal{SO} \neq \emptyset.$$

9.11 Show that the partially ordered set $\mathcal{WO}$ and the face poset $\mathcal{F}(\mathcal{B}_n)$ of the braid arrangement are isomorphic.

9.12 Let $|X| = n$. Show that the cubical complex $\mathcal{C}(\mathcal{WO})$ can be realized in $\mathbb{Z}^d$, where $d = \frac{n(n-1)}{2}$ (cf. Figure 9.13).

9.13 Let $|X|$ be a finite set of cardinality n. Prove that

(i) $|\mathcal{WO}(2)| = 2^n - 2$.

198   9 Hyperplane arrangements and their media

(ii) $|\mathcal{WO}(n-1)| = \frac{n!(n-1)}{2}$.

(iii) $|\mathcal{WO}| = \sum_{k=1}^{n} S(n,k) k!$, where $S(n,k)$ is a Stirling number of the second kind (cf. Stanley (1986)).

9.14 Prove Theorem 9.4.6.

9.15 Prove Corollary 9.4.7.

9.16 Prove (9.7).

9.17 Let $n = |\mathcal{X}|$. Show that

$$l_a(\mathcal{LO}) = \frac{n! n(n-1)}{4(n!-1)} \sim 0.25(n^2 - 1),$$

where l_a is the average length function (cf. Definition 7.3.2) and $\sim$ stands for "asymptotically equivalent".

9.18 Let $\mathcal{X}$ be a finite set, $n = |\mathcal{X}|$, and $f(n) = |\mathcal{WO}|$ (cf. Problem 9.13(iii)).

(i) Show that, for $n = 3$, $l_a(\mathcal{WO}) = \frac{30}{13}$, where l_a is the average length function (cf. Definition 7.3.2).

(ii) Show that

$$l_a(\mathcal{WO}) = \frac{2 \sum_{k=1}^{n-1} \binom{n}{k} f(k) f(n-k) [f(n) - f(k) f(n-k)]}{f(n)[f(n)-1]}.$$

(iii) Use a computer algebra program to plot the first 100 values of $l_a(\mathcal{WO})$. What is your conjecture about the asymptotic behavior of $l_a(\mathcal{WO})$?

9.19 Let L be a vertex of the permutohedron Π_n. Let C_L be the convex hull of the baricenters of all faces of Π_n containing L. Show that the face poset of the polyhedron C_L is isomorphic to the face poset of the n-dimensional cube.

10
Algorithms

In this chapter, we investigate efficient algorithms for solving media-theoretic problems. We follow Eppstein and Falmagne (2002) and concentrate on a number of fundamental algorithms for representing media and for converting between different representations of a medium. We also study basic media-theoretic concepts such as concise messages and closedness. In the next chapter we will apply these techniques in the algorithmic visualization of media.

In general, in the study of algorithms, we are interested in worst-case efficiency: how can we design algorithms that are guaranteed to solve the problem of interest correctly, using a number of computational steps that is guaranteed to be bounded by a slowly-growing function of the input size? Since, as we have seen in Chapter 6, media can be represented as graphs, all the standard graph-theoretic algorithms can be applied (such as breadth first search for finding shortest paths to each state from a single starting state). We are not interested here in duplicating material on these algorithms, as it can be found in any standard algorithms text (e.g., Cormen et al., 2003). Instead, we focus on problems that are specific to media (e.g., closedness and lattice dimension) or problems where the knowledge that the input is a medium can be used to gain some computational efficiency with respect to general graph-theoretic algorithms (e.g., finding concise messages between every pair of states).

We assume implicitly throughout this chapter that all media under consideration are finite.

10.1 Comparison of Size Parameters

We have spoken of analyzing the running time of algorithms in terms of the size of the input. However, , when the input is a medium, several parameters can be taken as defining its size. We consider here three different size parameters of media and prove relations between these parameters that allow us to compare running times that involve more than one of these parameters.

10.1.1 Definition. We write that $f(x) = O(g(x))$ whenever there exist constants x_0 and c_0 such that, for all $x > x_0$, $f(x) \leq c_0 g(x)$.

10.1.2 Remark. As is standard in the study of algorithms, all time bounds will be specified using the O-notation defined above. This convention allows our analysis to be performed independently of the details of the computer system on which an algorithm is implemented, and to avoid the need for a more precise definition of what constitutes a single operation.

10.1.3 Definition. We will assume that an algorithm of interest is operating on a medium $\mathcal{M} = (\mathcal{S}, \mathcal{T})$. As we have seen, such a medium may be represented alternatively as a partial cube graph $G = (V, E)$. In the graph algorithms literature, the numbers of vertices and edges in a graph are conventionally denoted n and m, respectively, but we eschew that notation as these letters are overloaded with other uses (for instance, the dimension of a Euclidean space). Instead, we will represent these quantities in our analysis as $|V|$ and $|E|$ respectively. Similarly, we represent the numbers of states and tokens of medium $\mathcal{M}$ by $|\mathcal{S}|$ and $|\mathcal{T}|$ respectively. Apparently, this gives us four size parameters with which to analyze our algorithms, but only three of these parameters are independent, as $|V| = |\mathcal{S}|$. We will use $|V|$ and $|\mathcal{S}|$ interchangeably in our analysis, preferring one or the other according to whether the context suggests more of a graph-theoretic or media-theoretic interpretation.

While three size parameters amount to a great simplification relative to the multiplicity of different media we may wish to analyze, they nevertheless complicate the comparison of algorithm run times. If we may solve the same problem in two ways, one via an algorithm analyzed in terms of $|V|$ and $|E|$, and another via an algorithm analzed in terms of $|\mathcal{S}|$ and $|\mathcal{T}|$, how may we compare the two algorithms to determine which one is better? To this end, we show the following relations between these size parameters. All logarithms should be assumed to have base 2, although the base is irrelevant when appearing within O-notation.

We begin with a technical lemma. The bound it states appears to be well known (compare Lemma 3 of Matoušek, 2006) but for completeness we prove it here.

10.1.4 Lemma. *In any family $\mathcal{F}$ of n sets, the number of unordered pairs of sets (P, Q) with $|P \triangle Q| = 1$ is at most $\frac{1}{2} n \log_2 n$.*

Proof. Let $M(n)$ denote the maximum possible number of unordered pairs of sets (P, Q) with $|P \triangle Q| = 1$ for a family of n sets. We use induction on n to prove that $M(n) \leq \frac{1}{2} n \log_2 n$. As a base case, for $n = 1$, a family of one set has no such pairs, so $M(1) = 0$.

If $n > 1$, let $\mathcal{F}$ be a family of sets realizing the maximum value of $M(n)$, choose $x \in \cup \mathcal{F}$, belonging to some but not all sets in $\mathcal{F}$, and divide $\mathcal{F}$ into two subfamilies $\mathcal{F}_x$ and $\mathcal{F}_{\bar{x}}$ where $\mathcal{F}_x$ consists of the members of $\mathcal{F}$ that contain x and $\mathcal{F}_{\bar{x}}$ consists of the remaining members of $\mathcal{F}$. Let $|\mathcal{F}_x| = a$ and $|\mathcal{F}_{\bar{x}}| = b$. Then each set P in $\mathcal{F}$ may be used to form at most one pair $(P, P\Delta\{x\})$ such that one set in the pair belongs to $\mathcal{F}_x$ and the other set in the pair belongs to $\mathcal{F}_{\bar{x}}$, so the number of pairs of this type is at most $\min(a,b)$.

The number of pairs of sets, differing by a single element, that do not have one set in each subfamily is at most $M(a) + M(b)$. Therefore,

$$\begin{aligned} M(n) &\le \max_{a+b=n} M(a) + M(b) + \min(a,b) \\ &= \max_{1\le c\le \frac{n}{2}} M(c) + M(n-c) + c \\ &\le \max_{1\le c\le \frac{n}{2}} \frac{1}{2}c\log_2 c + \frac{1}{2}(n-c)\log_2(n-c) + c \\ &\le \max_{1\le c\le \frac{n}{2}} \frac{1}{2}c(\log_2 n - 1) + \frac{1}{2}(n-c)\log_2 n + c \\ &= \frac{1}{2}n\log_2 n. \end{aligned}$$

Here $c = \min\{a,b\}$, the replacement of $M(c)$ by $\frac{1}{2}c\log_2 c$ and of $M(n-c)$ by $\frac{1}{2}(n-c)\log_2(n-c)$ uses the induction hypothesis, and the replacement of $\log_2 c$ by $\log_2 n - 1$ uses the fact that $c \le n/2$. □

10.1.5 Theorem. *The following relations hold between the size parameters $|V|$, $|E|$, $|\mathcal{S}|$, and $|\mathcal{T}|$ of a medium $\mathcal{M} = (\mathcal{S}, \mathcal{T})$ represented by a partial cube graph $G = (V, E)$:*

$$\log|\mathcal{S}| = O(|\mathcal{T}|), \tag{10.1}$$

$$|\mathcal{T}| = O(|\mathcal{S}|), \tag{10.2}$$

$$|\mathcal{S}| = |V| = O(|E|), \tag{10.3}$$

and

$$|E| = O(|V|\log|V|). \tag{10.4}$$

Proof.

(10.1): It follows from Theorem 3.3.3 that every state of the medium can be uniquely associated with a subset of the $|\mathcal{T}|/2$ positive tokens of some orientation of the medium. Therefore, $|\mathcal{S}| \le 2^{|\mathcal{T}|/2}$, or equivalently, $\log|\mathcal{S}| \le |\mathcal{T}|/2$.

(10.2): Choose an initial state S_0. For each other state S_i let τ_i be a token that acts on S_i in a concise message transforming S_i to S_0; then the set of pairs $S_i, S_i\tau_i$ forms the set of edges in a tree T; each path from S_0 to S_i in T is a shortest path in the graph representing the medium.

For each token τ, let S be a state for which $S\tau \neq S$. Then the path from S to $S\tau$ formed by concatenating the paths in T from S to S_0 and from S_0 to $S\tau$ corresponds to a message in the medium that as the single-token message τ. Therefore, somewhere in tree T there is a state S_i for which $\tau_i = \tau$. But tree T has $|\mathcal{S}| - 1$ edges, so there can be at most $|\mathcal{S}| - 1$ token-reverse pairs in the medium.

(10.3): The fact that $|V| = O(|E|)$ can be seen from the same tree T of shortest paths described above: it contains $|V| - 1$ edges, each of which is an edge of G and G may have other edges not in T. Therefore, $|V| \leq |E| + 1 = O(|E|)$.

(10.4): Let $\mathcal{F}$ be a well-graded set family representing the given medium. Then each edge in E corresponds to a pair of sets in $\mathcal{F}$ that differ in a single element, so the result follows from Lemma 10.1.4. □

Further relations may be derived from these, as is suggested in Problem 10.2. In what follows we assume the standard RAM model of computation, in which we may store medium states and tokens in single machine words, and perform machine word operations in constant time per operation. In cases where our bounds depend on the number of bits that may be stored in a single machine word, we use W to denote this number.

10.2 Input Representation

Before we can define and analyze algorithms on media, we need to know something about how those algorithms represent the media they work with, and what basic operations are available to them. Whatever input representation we choose should be space-efficient, should allow our algorithms to quickly list the elements of the sets $\mathcal{S}$ and $\mathcal{T}$, and quickly find the result $S\tau$ of applying token τ to state S.

10.2.1 Definition. The following two representations of a medium are defined directly in terms of the action of tokens on states. The first representation is natural and simpler, while the second applies well-known data structures for greater space efficiency at minimal cost in time per operation.

Transition table

In this representation, we store a two-dimensional matrix M, with rows indexed by states and columns indexed by tokens. Each entry $M[S,t]$ stores the state St reached by the action of that token on that state. Thus, the result of any such action may be found by table lookup in constant time. The amount of storage space necessary to store this table is $O(|\mathcal{S}| \cdot |\mathcal{T}|)$. In order to quickly list all states and tokens of the medium, we may wish to augment this matrix with separate lists of the medium's states and tokens. In many implementation languages, matrices may be most efficiently indexed by integers rather than whatever objects we are using to represent the states and tokens; in this case we may wish to store auxiliary translation tables mapping the states and tokens into integers in the ranges $[0, |\mathcal{S}| - 1]$ and $[0, |\mathcal{T}| - 1]$ respectively. These translation tables do not change by more than a constant factor the overall storage requirements of this data structure.

Hashed transition table

Although the transition table's matrix M allows fast lookups, it uses a large amount of space for media that have many ineffective transitions. Instead, we may replace it by a *hash table*, a standard data structure for storing sparse tables of (key, value) pairs. Such a structure consists of an array H and a *hash function* h that maps keys to table indices; the (key, value) pair corresponding to key k is generally stored at $H[h(k)]$. The hash function must be carefully chosen so that few pairs k_i, k_j have a *collision* $h(k_i) = h(k_j)$, and various strategies have been developed in the hash table literature for choosing h and for handling such collisions when they arise; see a standard algorithms text such as Cormen et al. (2003) for details. In our application of hashing to media, the keys will be the pairs S, τ such that $S\tau \neq S$, and the values stored for those keys will be $S\tau$. Thus, in $H[h(S, \tau)]$ we store the triple $(S, \tau, S\tau)$. To look up the result of applying a token τ to a state S, we examine the table entry $H[h(S, \tau)]$, and determine whether it contains a triple matching S and τ. If it does, we can determine $S\tau$ from the third entry of the triple. If $H[h(S, \tau)]$ does not match, we apply the hash table's collision handling procedure to either determine an alternate location for key (S, τ) or determine that (S, τ) was not one of the keys used to index the hash table. If (S, τ) is a valid key involved in a collision, this collision handling procedure will find its associated value $S\tau$; if, instead, (S, τ) is not one of the keys, we can determine that $S\tau = S$.

It is conventional in computer science to analyze hash tables as taking a constant amount of time per operation (albeit slower than a table lookup as each operation involves computing the hash function) using space proportional to the number of (key, value) pairs. This analysis can be made rigorous by assuming that the hash function's values are independent random variables and analyzing the expected time per operation of an appropriate collision handling procedure; generally the level of randomness required to make this analysis rigorous is less than what is incorporated into the hash tables implemented in most programming libraries, but nevertheless this style of analysis matches well with practical experience of hash tables. Applying this analysis to hashed transition tables, we get constant expected time to look up the effect of a token on a state. The space is $O(|E|)$, as we need store exactly $2|E|$ (key, value) pairs, one per effective state-token transition, where E denotes the set of edges of a partial cube representation of the medium.

Table 10.1. Transition table for the medium of Figure 2.1.

	τ_1	τ_2	τ_3	τ_4	τ_5	τ_6
S:	S	S	V	S	X	S
T:	T	W	T	T	T	T
V:	V	V	V	S	W	V
W:	T	W	W	X	W	V
X:	X	X	W	X	X	S

10.2.2 Example. Table 10.1 depicts a transition table for the medium depicted in Figure 2.1. The table has one row for each of the five states of the medium, and one column for each of the medium's six tokens. Table 10.2 shows the (key, value) pairs of a hashed transition table for the same medium.

Table 10.2. The (key, value) pairs of a hashed transition table for the medium of Figure 2.1.

key	(S, τ_3)	(S, τ_5)	(T, τ_2)	(V, τ_4)	(V, τ_5)	(W, τ_1)	(W, τ_4)	(W, τ_6)	(X, τ_3)	(X, τ_6)
value	V	X	W	S	W	T	X	V	W	S

10.2.3 Definition. We have seen in Chapter 7 that media may be represented mathematically as partial cube graphs. This graphical representation may also be useful in computer algorithms. While there are many ways of representing graphs in computers (Cormen et al., 2003), we concentrate here on variations of one of the most common and most useful graph representations: the adjacency list.

Adjacency list

We represent the graph $G = (V, E)$ of the medium by a collection of vertex and edge objects. Each edge object includes pointers to the two incident vertex objects, and each vertex object has a pointer to a list object, each item of which points to one of the edge objects incident to that vertex. This representation requires space $O(|V| + |E|) = O(|E|)$ and is therefore as space-efficient as the hashed transition table. However, without further elaboration it is somewhat lacking in usefulness for media-theoretic algorithms, because it is not straightforward to look up the effect of a token on a state: the states are represented as graph vertices but the tokens are not represented at all.

Vertex-labeled adjacency list

In definition 7.1.1, a partial cube was defined in terms of an isometric embedding of a graph G into a hypercube $\mathcal{H}(\mathcal{X})$, which can equivalently be thought of as a labeling of the vertices of G by binary numbers (representing the hypercube vertices) in such a way that the graph distance between vertices equals the Hamming distance between the vertices' labels. It is natural to incorporate this definition into our input representation. We can do this by storing an adjacency list together with a binary number label associated with each vertex object. These numbers will be $|\mathcal{T}|/2$ bits long, as long as there exist no positions at which the bits of all vertices' labels agree. Thus, if each bit of each label is stored in a separate memory cell, this representation requires $O(|E| + |V| \cdot |\mathcal{T}|) = O(|\mathcal{S}| \cdot |\mathcal{T}|)$ storage cells total. However, it may be somewhat more efficient than the transition table representation in models of computation in which we may store some number W of bits per memory word; in this case the total number of words needed for this representation is $O(|E| + |\mathcal{S}| \cdot |\mathcal{T}|/W)$. We can define, for each position i of these numbers and for each bit $j \in \{0, 1\}$, a token $\tau_{i,j}$ that transitions from a vertex with label x to a vertex in which the ith bit of the label of x has been replaced by j, if such a vertex exists. In order to look up the effect of a transition, we may wish to augment this data structure with a hash table mapping vertex labels to the associated vertices, so that we can determine quickly whether a vertex with a modified label exists in the graph. Compared to the transition table, this data structure has the advantage that we can list all states reachable by effective transitions from a given state, in time proportional to the number of effective transitions, without having to take the time to examine ineffective tokens. However the space is higher relative to the hashed transition table, motivating our final representation.

Edge-labeled adjacency list

We augment an adjacency list representation by storing, associated with each edge object, the two tokens whose transitions connect states represented by the edge's endpoints, and the direction of each token's transition. As with the adjacency list, the space is $O(|E|)$. This data structure enables us to find, not just the states reachable by effective transitions from a given state, but the tokens that lead to those transitions. However, finding the result of a transition $S\tau$ requires scanning through the entire adjacency list of the vertex corresponding to state S.

10.2.4 Remark. It is also possible to consider a representation in which we do not explicitly store the graph structure, and instead keep only a list of the vertex labels described above as part of the vertex-labeled adjacency list representation. The graph structure implicit in this list of labels can then be recovered by finding pairs of labels that differ in a single bit. An approximate dictionary structure of Brodal and Gąsieniec (1996) allows this recovery of the graph structure to be performed efficiently: in time $O(|\mathcal{S}| \cdot |\mathcal{T}|)$ one can compute a data structure of size $O(|\mathcal{S}| \cdot |\mathcal{T}|)$ which allows the neighbors of each vertex to be listed in time $O(|\mathcal{T}|)$ per vertex, so the total time to convert this representation to a vertex-labeled adjacency list is $O(|\mathcal{S}| \cdot |\mathcal{T}|)$.

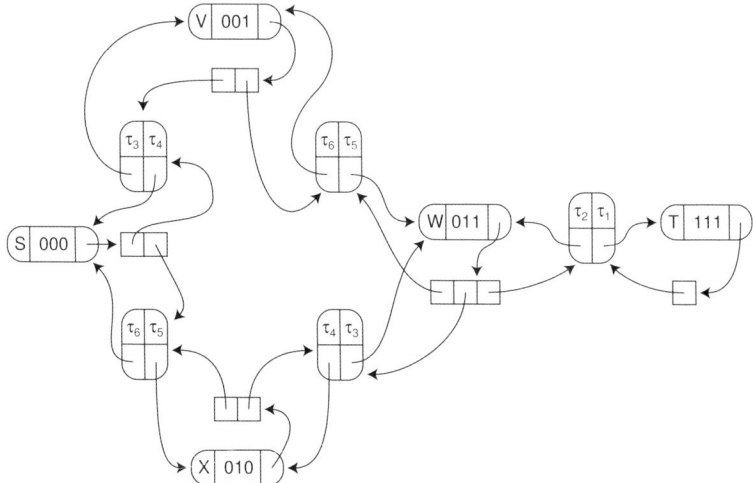

Figure 10.1. Labeled adjacency list for the medium of Figure 2.1.

10.2.5 Example. Figure 10.1 depicts an adjacency list representation for the medium depicted in Figure 2.1. This representation has one vertex object (drawn as a wide rounded rectangle) for each of the five states of the medium, one edge object (drawn as a tall narrow rounded rectangle) for each of the medium's five pairs of effective transitions and their reverses, and a list object (drawn as a sequence of small squares) for the list of edges incident to each vertex. Pointers from one object to another are drawn as arrows, and we have included labels both on the vertices (binary numbers) and the edges (two tokens associated with the two pointers to the endpoints of the edge).

10.2.6 Remark. If we desire a space-efficient data structure that allows us both to look up transition results in constant time and to find all effective transitions in time proportional to the number of such transitions, we may combine the hashed transition table and the edge-labeled adjacency list, for a total space bound of $O(|E|)$, and time $O(1)$ per transition or per effective transition listed.

In some cases it may also be of use to add to our representations an auxiliary table from which we can look up the reverse of each token. Such a table requires storage $O(|\mathcal{T}|)$, and may be constructed easily from the edge-labeled adjacency list, as it stores the tokens with their reverses at each edge.

We now discuss algorithms for converting among these input representations. A chart of these representations and their conversions is shown in Figure 10.2.

10.2.7 Theorem. *We may convert the transition table representation to the hashed transition table, and vice versa, in time* $O(|\mathcal{S}| \cdot |\mathcal{T}|)$.

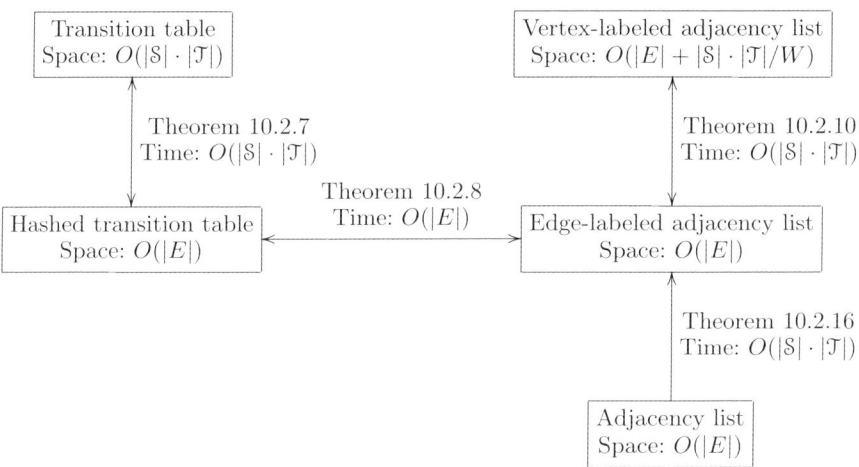

Figure 10.2. Media representations and conversions between them.

Proof. To convert a transition table into a hashed transition table, initialize H to be an empty hash table, loop through all state-token pairs S, t, for each pair use the transition table to look up the result of the transition St, compare this result to S, and if different add key S, t and value St to H. To convert a hashed transition table into a transition table, initialize matrix $M[S, t] = S$ for all state-token pairs S, t, then loop through all key-value pairs in the hash table. For each key S, t and value St, change $M[S, t]$ to St. □

10.2.8 Theorem. *We may convert the hashed transition table representation to the edge-labeled adjacency list representation, and vice versa, in expected time $O(|E|)$.*

Proof. To convert a hashed transition table into an edge-labeled adjacency list, create a vertex object for each state, and create an empty hash table of edge objects keyed by their endpoints. Loop through all key-value pairs in the hash table. For each key S, t and value St in the hash table, create (if it does not exist already) an edge object connecting S and St, and add this object to the adjacency lists of S and St and to the hash table of edge objects. Also, whether the edge object was created or already existed, store t as one of the two tokens associated with that edge. When the conversion is complete, the hash table of edge objects may be discarded.

To convert an edge-labeled adjacency list into a hashed transition table, simply create an empty hash table and add to it two entries for each edge of the graph. □

In order to efficiently convert the vertex-labeled adjacency list to the edge-labeled adjacency list, we need some bit-manipulation data structures.

10.2.9 Lemma. *Suppose we are given as input a collection of n B-bit binary words. Then in time $O(nB)$ we can construct a data structure that allows us to look up the position of the first bit at which any two words of our input differ, in time $O(1)$ per lookup.*

Proof. We build a *trie*. A trie is a binary tree that has a node for each prefix of each input word. The parent of a nonempty prefix is the prefix with one fewer bit, formed by removing the last bit from w, and this parent relation defines the tree structure of the trie. The root of the trie is the zero-length word, and its leaves are the n input words. A trie can be built in time $O(nB)$ by adding words to it one at a time. Each time we add a word we may step down a path from the root to the new leaf in time $O(B)$ per added word.

Then, the position of the first difference between two words w_1 and w_2 equals the length of the word forming the least common ancestor of w_1 and w_2 in this trie. We may process any tree, in time linear in the number of tree nodes, so that least common ancestor queries in that tree may be answered in constant time per query (see, e.g., Bender and Farach-Colton, 2000). Applying such a least common ancestor data structure to our trie gives the result. □

10.2.10 Theorem. *We may convert the vertex-labeled adjacency list representation to the edge-labeled adjacency list, and vice versa, in time $O(|\mathcal{S}| \cdot |\mathcal{T}|)$.*

Proof. In one direction, suppose we have a vertex-labeled adjacency list. We use the data structure of Lemma 10.2.9 to determine the position i at which the labels of the two endpoints of each edge differ, and we label that edge with the two tokens $t_{i,0}$ and $t_{i,1}$ defined in our description of the vertex-labeled adjacency list representation. It takes time $O(|\mathcal{S}| \cdot |\mathcal{T}|)$ to initialize the data structure of Lemma 10.2.9, after which we can label all edges in total time $O(|E|)$.

In the other direction, suppose we have an edge-labeled adjacency list, from which we have extracted the tokens to form a sequence of $|\mathcal{T}|/2$ token-reverse pairs. We initialize a label L to the all-zero bitvector, and perform a depth-first traversal of the graph starting from an arbitrarily chosen vertex v_0. Each time our traversal first reaches a vertex v, we assign it the label L, and each time our traversal crosses an edge labeled with the tokens from the ith token-reverse pair in the sequence, we change the ith bit of L from 0 to 1 or vice versa. In this way, each vertex v is assigned a label in which the nonzero bits correspond to the tokens on a concise message from v_0 to v. Clearly, such a labeling of the graph must be distance-preserving. □

10.2.11 Remark. Converting an edge-labeled adjacency list to an adjacency list is a trivial matter of omitting the labels, and may be performed in time $O(|E|)$. The sequence of results in the remainder of this section, in which we follow Eppstein (2007b), describes a slower conversion in the other direction. Recall that the Djoković-Winkler relation Θ defined by the equivalence (7.2) holds for two edges $e = \{x, y\}$ and $= \{u, v\}$ of G, thus $e\Theta f$, if and only if $\delta(x, u) + \delta(y, v) \neq \delta(x, v) + \delta(y, u)$.

10.2.12 Lemma. *Let pq be an edge in a bipartite graph G. Then $pq\Theta rs$ if and only if, for exactly one of r and s, there is a shortest path to p that passes through q.*

Proof. If neither r nor s has such a path, then $\delta(q, r) = \delta(p, r) + 1$ and $\delta(q, s) = \delta(p, s) + 1$, so $\delta(p, r) + \delta(q, s) = \delta(p, r) + 1 + \delta(p, s) = \delta(q, r) + \delta(p, s)$ by associativity of addition, and it is not the case that $pq\Theta rs$. Similarly, if both r and s have such paths, then $\delta(q, r) = \delta(p, r) - 1$ and $\delta(q, s) = d(p, s) - 1$, so $\delta(p, r) + \delta(q, s) = \delta(p, r) - 1 + \delta(p, s) = \delta(q, r) + \delta(p, s)$. Thus in neither of these cases can pq and rs be related. If, on the other hand, exactly one of r and s has such a path, we may assume (by swapping r and s if necessary) that it is r that has the path through q. Then $\delta(q, r) = \delta(p, r) - 1$ while $\delta(q, s) = \delta(p, s) + 1$, so $\delta(p, r) + \delta(q, s) = \delta(p, r) + \delta(p, s) + 1 \neq \delta(p, r) - 1 + \delta(p, s) = \delta(q, r) + \delta(p, s)$, so in this case $pq\Theta rs$. □

10.2.13 Remark. Thus, to find the equivalence class of edge pq, we may perform a breadth first search rooted at p, maintaining an extra bit of information for each vertex v traversed by the search: whether v has a shortest path to p that passes through q. This bit is set to false initially for all vertices except for q, for which it is true. Then, when the breadth first search traverses an edge from a vertex v to a vertex w, such that w has not yet been visited by the search (and is therefore farther from p than v), we set the bit for w to be the disjunction of its old value with the bit for v. As we will show, we can apply this technique to find several edge classes at once. Specifically, we will find the equivalence classes of each edge pq incident to a single vertex p, by performing a single breadth first search rooted at p. We observe that no two such edges pq and pr can be related to each other by Θ.

10.2.14 Remark. Our algorithm will need efficient data structures for storing and manipulating bit vectors, which we now describe. Our algorithms depend on a model of computation in which integers of at least $\log |V|$ bits may be stored in a single machine word, and in which addition, bitwise Boolean operations, comparisons, and table lookups can be performed on $\log |V|$-bit integers in constant time per operation. The constant-time assumption is standard in the analysis of algorithms, and any machine model that is capable of storing an address large enough to address the input to our problem has machine words with at least $\log |V|$ bits.

We store a bitvector with k bits in $\lceil 1 + k/\log |V| \rceil$ words, by packing $\log |V|$ bits per machine word. The disjunction operation on two such bitvectors, and the symmetric difference operation on two such bitvectors, may be performed by applying single-word disjunction or symmetric difference operations to each

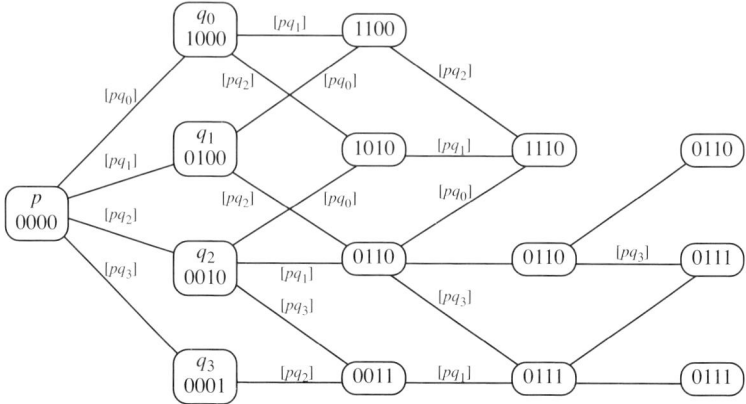

Figure 10.3. Data for the algorithm of Lemma 10.2.15. The left-to-right order of the vertices indicates their distance from the leftmost vertex p, and therefore their position in the breadth first traversal of the graph. Each vertex is labeled with a bitvector indicating which vertices q_i belong to shortest paths to p, and the edges of the graph are labeled $[pq_i]$ if they are part of the equivalence class of edge pq_i.

of the words in this representation. To test whether a bitvector is nonzero, we use a comparison operation to test whether each of its words is nonzero. To test whether a bitvector has exactly one nonzero bit, and if so find out which bit it is, we again use comparisons to test whether there is exactly one word in its representation that is nonzero, and then look up that word in a table that stores either the index of the nonzero bit (if there is only one) or a flag value denoting that there is more than one nonzero bit. Thus, all these operations may be performed in time $O(1 + k/\log|V|)$ per operation.

10.2.15 Lemma. *Let G be any partial cube graph with n vertices and m edges, and let p be a vertex in G that is incident to the largest number of edges among all vertices of G. Then there is an algorithm which finds the equivalence classes of Θ for each edge in G incident to p, in time $O(|V|)$ per equivalence class.*

Proof. Let k denote the number of incident edges of p, which must be at least $2m/n$. We denote the k neighbors of p in G by q_i, for an index i satisfying $0 \le i < k$. We create, for each vertex of G, a bitvector D_v with k bits; we denote the ith bit of this structure as $D_v[i]$. Bit $D_v[i]$ will eventually be 1 if v has a shortest path to p that passes through q_i; initially, we set all of these bits to 0, and then set we set $D_{q_i}[i] = 1$.

Next, we perform a breadth first traversal of G, starting at p. When this traversal visits a vertex v, it considers all edges vw such that w has not yet been visited; due to the breadth first ordering of the vertices, each such edge must be part of a shortest path from p to w. The algorithm then sets all bits

$D_w[i]$ to be the disjunction of their previous values with the corresponding bits $D_v[i]$.

Finally, once the breadth first search is complete and all data structures D_v have reached their final values, we examine each edge vw in the graph. If $D_v = D_w$, we ignore edge vw, as it will not be equivalent to any edge pq_i. Otherwise, we compute a bitvector B as the symmetric difference of D_v and D_w, and assign vw to the equivalence class of the edge pq_i for which $B[i]$ is nonzero. Figure 10.3 shows this assignment of edges to classes for a simple example graph.

All stages of the algorithm perform $O(|E|)$ steps, each one of which involves $O(1)$ of the bitvector operations described by Remark 10.2.14, so the total time is $O(|E|(1 + k/\log|V|)) = O(k|V|)$, as claimed. □

10.2.16 Theorem. *We may convert from the adjacency list representation into the edge-labeled adjacency list representation in time $O(|\mathcal{S}| \cdot |\mathcal{T}|)$.*

Proof. Given a graph G, we find the maximum degree vertex p in G and apply the algorithm of Lemma 10.2.15 to identify the equivalence classes of all adjacent edges, thus labeling those edges. Next, we find a graph G' representing the projection (as per Theorem 2.11.6) of the symmetric set of edge labels created by the application of Lemma 10.2.15. In graph-theoretic terms, this projection may be found in linear time by contracting all labeled edges. We apply the conversion algorithm recursively to label the edges of G', undo the edge contraction, and copy the labels of the contracted graph's edges to the corresponding edges of G.

The time for contracting and uncontracting is dominated by the time to apply Lemma 10.2.15, which is $O(|V|) = O(|\mathcal{S}|)$ per equivalence class found. Therefore, the total time is $O(|\mathcal{S}| \cdot |\mathcal{T}|)$. □

10.2.17 Remark. Unless stated otherwise, we will assume for the rest of our algorithms that the input medium is represented by an edge-labeled adjacency list, as this representation is space-efficient, efficiently convertable to the other representations, and (unlike the hashed transition table) deterministic.

10.3 Finding Concise Messages

We now follow Eppstein and Falmagne (2002) in describing an efficient algorithm for finding concise messages and distances between all pairs of states in a medium. This could be done in total time $O(|V| \cdot |E|)$ by performing a sequence of breadth first traversals in the graph representing the medium, starting a different traversal at each state, but the following algorithm is more efficient. Together with the $O(|\mathcal{S}| \cdot |\mathcal{T}|)$-time algorithm for labeling the edges of an unlabeled partial cube graph, described in Theorem 10.2.16, this fast distance computation algorithm forms a key step in the recent algorithm of Eppstein (2007b) for efficiently recognizing partial cube graphs.

10.3.1 Lemma. *For any medium, and any starting state S of the medium, there is a message $\boldsymbol{m}$ of at most $2|V| - 3$ tokens, such that applying $\boldsymbol{m}$ to S produces a sequence of states that includes every state of the medium. Message $\boldsymbol{m}$ may be constructed algorithmically in time $O(|E|)$ from a labeled adjacency list representation of the medium.*

Proof. We apply a standard depth-first search algorithm to the adjacency list representation. This algorithm is recursive, and begins by a recursive call having the initial state S as an argument; it maintains as it progresses a bit for each state of the medium, representing whether or not that state has already been included in the traversal. When called with a state Q as argument, the algorithm marks Q as having been included in the traversal; then, it loops in sequence through the states that can be reached from Q by a single transition. If this loop reaches a state R that has not yet been included in the traversal, the algorithm is called recursively with R as argument.

To turn this search procedure into a message, we build $\boldsymbol{m}$ by concatenating one token onto it at a time from an initial empty message. When the loop through the neighbors of Q finds an untraversed state R and calls the algorithm recursively with argument R, it also concatenates onto $\boldsymbol{m}$ a token τ such that $Q\tau = R$, and when it returns from the recursive call to R, it also concatenates onto $\boldsymbol{m}$ the reverse of τ. In this way, $S\boldsymbol{m}$ is always equal to the state currently being considered by the algorithm, and the sequence of states produced by applying $\boldsymbol{m}$ to S is always equal to the sequence of states that have been traversed so far by the algorithm.

At the end of this depth first search procedure, $\boldsymbol{m}$ is a stepwise effective but ineffective message for S. It has length $2|V| - 2$, since each state is the argument for one recursive call and since the recursive call for each state other than S is associated with the addition of two tokens onto $\boldsymbol{m}$. To produce the message of length $2|V| - 3$ described in the statement of the Lemma, we may remove the final token from $\boldsymbol{m}$. □

10.3.2 Remark. If S is the root of a star medium (cf. Definition 2.9.4), the bound of $2|V| - 3$ on the length of $\boldsymbol{m}$ cannot be improved; see Problem 10.9.

10.3.3 Theorem. *Given any medium, in time $O(|V|^2)$ we can build a table of dimensions $|V| \times |V|$ that stores, for each pair of states S and Q, two pieces of information: the distance d_{SQ} from S to Q, and a token τ_{SQ} such that τ_{SQ} is effective for S and τ_{SQ} belongs to a concise message transforming S to Q.*

Proof. We apply the conversions of Theorem 10.2.8 and Theorem 10.2.7 so that we have both the input edge-labeled adjacency list and a state-transition table allowing us to test in constant time whether a token is effective for a given state. We use Lemma 10.3.1 to find a sequence of all the states in the medium, connected by transitions of the medium. We then traverse the medium by stepping from state to state in this sequence, maintaining as we

10.3 Finding Concise Messages

do a data structure that allows us to compute the table entries d_{SQ} and τ_{SQ} for each traversed state Q.

The data structure that we maintain as we traverse the medium consists of the following components:

- A doubly-linked list L of pairs (τ, Λ_τ), where each τ is a token in the content of Q and Λ_τ is a pointer to a linked list described below.
- A pointer from each state $S \ne Q$ to the first pair (τ, Λ_τ) in L for which $S\tau$ is effective.
- Linked lists Λ_τ for each pair (τ, Λ_τ), listing the states pointing to that pair.

The pointers from each state S to the associated pair (τ, Λ_τ) provide half of the information we are trying to compute: τ is an effective token for S that is in the content of Q, and therefore can be stored as τ_{SQ}. We record this information for Q when the traversal first reaches Q. The other information, the numeric distances d_{SQ} to Q, can easily be computed in time $O(|V|)$ for each Q by traversing the tree formed by the effective transitions $S\tau_{SQ}$. Thus, it remains to show how to initialize and maintain the data structures described above as we perform our traversal.

To initialize the data structures, we determine the content of the initial state Q by performing a depth-first traversal of the graph and recording the tokens labeling the depth first search tree edges. We then create an empty list Λ_τ for each token τ in the content of the initial state Q. We make a list L of the pairs (τ, Λ_τ) in an arbitrary order. Then, for each state S, we find a pointer to the first (τ, Λ_τ) in this list such that τ is effective for S, by sequentially searching L, using our transition table to test whether each token in the sequence is effective for S in constant time per test.

It remains to describe how to update the data structure as we perform each step from a state Q to a new state $Q\tau = Q'$ of the depth first traversal. The token $\tilde\tau$ reversing this transition belongs to the content of Q, so prior to the transition it is listed as part of some pair $(\tilde\tau, \Lambda_{\tilde\tau})$ in L. We remove this pair from L, and append a new pair (τ, Λ_τ) to the end of L, where Λ_τ is a new empty list. We must then recompute the pointers from each state S that had previously been pointing to $(\tilde\tau, \Lambda_{\tilde\tau})$. We do so by sequentially searching list L, starting at the position of the deleted pair $(\tilde\tau, \Lambda_{\tilde\tau})$, for the first pair $(\tau', \Lambda_{\tau'})$ such that τ' is effective for S; once we find τ', we append S to $\Lambda_{\tau'}$. We set the pointer for Q to the pair (τ, Λ_τ) without searching, since τ will be the only token in $\hat{Q}'$ that is effective for Q.

We finish the proof by analyzing the time used by this algorithm. Finding the content of Q takes time $O(|E|)$, initializing L takes time $O(|V|)$, and initializing the pointers for each state takes time $O(|\mathcal{T}|)$ per state, or $O(|V|\cdot|\mathcal{T}|)$ overall. List L initially contains $|\mathcal{T}|/2$ pairs, and each traversal step appends a new pair to L. Therefore, the total number of pairs added to L over the course of the algorithm is at most $|\mathcal{T}|/2+2|V|-3 = O(|V|)$. The most expensive part of the algorithm is the sequential searching to find a new effective pair. For each state S, the sequence of sequential search steps never revisits a position in L, so the total number of steps of sequential searching over the course of the algorithm is at most $|V|(|\mathfrak{T}|/2+2|V|-3) = O(|V|^2)$. Computing numeric distances also takes a total of $O(|V|^2)$ time, and the other data structure update steps take only constant time per traversal step. Therefore, the total time for the algorithm is $O(|V|^2)$. □

From the table constructed above, one can construct a concise message m such that $Sm = Q$ for any two states S and Q, in time $O(|m|)$, by repeatedly using the table to find effective tokens in the message.

10.3.4 Example. We describe step-by-step the algorithm above as it finds concise messages in the medium of Figure 2.1. Recall that the adjacency list and transition table data structures used by the algorithm are depicted in Figure 10.1 and Table 10.1 respectively. In the depth first traversal used by the algorithm, we will test the edges out of each vertex in order by their position in the adjacency lists depicted in Figure 10.1.

To make visually apparent the distinction between lists that happen to contain a single item and the items within the lists, in this example, we represent a list of items using a syntax borrowed from the Python programming language, in which the list is enclosed in square brackets or parentheses and list items are separated by commas. Thus, [1] represents a list with a single item in it, which should be thought of as a different type of object than the number 1 that is the first and only item in the list [1]. Similarly, we use a Python-like syntax to represent the pointers that map states to items in L: each pointer is represented as a colon-separated pair $state : item$, and the set of these pairs is enclosed in curly brackets. To avoid confusion over whether it is appropriate to use mathematical equality to describe a relationship that is contingent on the step at which it is tested, we also represent the value of the program's variables by following the variable name by a colon and a value, rather than using the equal sign that would be more typical in computer science writing.

1. We begin the traversal at the state S. The tokens in the content of S are τ_2, τ_4, and τ_6. We order these arbitrarily (in our example, we will use the numeric order of the subscripts) and initialize list L to

$$L : [(\tau_2, \varLambda_2), (\tau_4, \varLambda_4), (\tau_6, \varLambda_6)].$$

The initial value of the set of pointers from states to L assigns each state (other than our initial state S) to the first token in L that is effective for that state and belongs to a concise message that transforms that state into S:

$$\{T : (\tau_2, \varLambda_2), V : (\tau_4, \varLambda_4), W : (\tau_4, \varLambda_4), X : (\tau_6, \varLambda_6)\}.$$

The edges corresponding to these effective transitions form a spanning tree of G, from which we can calculate the distances from each state to V:

V and X are at distance one, W is at distance two, and T is at distance three. Finally, each Λ_i stores a sequence of the states pointing to it:

$$\Lambda_2 : [T]$$
$$\Lambda_4 : [V, W]$$
$$\lambda_6 : [X].$$

2. The depth first traversal of G follows the first edge out of S, leading to the transition $S\tau_3 = V$. We remove τ_4, which is the reverse of τ_3, from L, and append τ_3:

$$L : [(\tau_2, \Lambda_2), (\tau_6, \Lambda_6), (\tau_3, \Lambda_3)].$$

For each state in Λ_4 other than the new position of the traversal (that is, only state W) we search sequentially through the positions of L after the deleted pair for the next effective token. The first position tested is the one for τ_6, which is effective. So we update the pointers from states to L, including as well a new pointer from S and deleting the one from V:

$$\{S : (\tau_3, \Lambda_3), T : (\tau_2, \Lambda_2), W : (\tau_6, \Lambda_6), X : (\tau_6, \Lambda_6)\}.$$

These pointers give effective tokens for each state on a concise message to V. After updating the lists Λ_i to reflect these changes, they stand at

$$\Lambda_2 : [T]$$
$$\Lambda_3 : [S]$$
$$\Lambda_6 : [X, W].$$

3. The depth first traversal attempts to use the first edge out of V, corresponding to the transition $V\tau_4 = S$, but this leads to a state that has already been visited, so the transition is ignored and our data structures remain unchanged.

4. The second edge out of V leads to the transition $V\tau_5 = W$. Updating the data structures for this transition gives

$$L : [(\tau_2, \Lambda_2), (\tau_3, \Lambda_3), (\tau_5, \Lambda_5)].$$

The pointers from states to L are

$$\{S : (\tau_3, \Lambda_3), T : (\tau_2, \Lambda_2), V : (\tau_5, \Lambda_5), X : (\tau_3, \Lambda_3)\}$$

(giving effective tokens on concise messages from each state to W), and the lists Λ_i are

$$\Lambda_2 : [T]$$
$$\Lambda_3 : [S, X]$$
$$\Lambda_5 : [V].$$

5. The first edge out of W leads back to the already-visited state V. The second edge corresponds to the transition $W\tau_4 = X$. Updating our data structures for this transition leads to a situation with

$$L : [(\tau_2, \Lambda_2), (\tau_5, \Lambda_5), (\tau_4, \Lambda_4)],$$

pointers from states

$$\{S : (\tau_5, \Lambda_5), T : (\tau_2, \Lambda_2), V : (\tau_5, \Lambda_5), W : (\tau_4, \Lambda_4)\}$$

(giving effective tokens on concise messages from each state to X), and the lists are

$$\Lambda_2 : [T]$$
$$\Lambda_4 : [W]$$
$$\Lambda_5 : [T, V].$$

6. Both edges out of X lead to already-visited states, so the depth first traversal returns to W to try the remaining edge out of it, via the transition $X\tau_3 = W$. Updating our data structures for this transition leads to a situation with

$$L : [(\tau_2, \Lambda_2), (\tau_5, \Lambda_5), (\tau_3, \Lambda_3)].$$

The pointers from states are

$$\{S : (\tau_5, \Lambda_5), T : (\tau_2, \Lambda_2), V : (\tau_5, \Lambda_5), X : (\tau_3, \Lambda_3)\},$$

and the lists are

$$\Lambda_2 : [T]$$
$$\Lambda_3 : [X]$$
$$\Lambda_5 : [S, V].$$

Note that our data structures, while still valid, are in a different configuration than they were the first time our traversal reached W: the tokens in L appear in a different order, and this leads to other changes in the rest of our data structures.

7. The final edge out of W corresponds to the effective transition $W\tau_1 = T$. We update our data structures to

$$L : [(\tau_5, \Lambda_5), (\tau_3, \Lambda_3), (\tau_1, \Lambda_1)],$$

with pointers from states

$$\{S : (\tau_5, \Lambda_5), V : (\tau_5, \Lambda_5), W : (\tau_1, \Lambda_1), X : (\tau_3, \Lambda_3)\}$$

(giving effective tokens on concise messages from each state to T), and lists

$$\Lambda_1 : [W]$$
$$\Lambda_3 : [X]$$
$$\Lambda_5 : [S, V].$$

8. The remaining steps backtrack through the depth first traversal without reaching any additional unexplored states, and we omit them from our example (cf. Problem 10.11).

10.4 Recognizing Media and Partial Cubes

As an application of our algorithm for finding concise messages efficiently, we use it as part of an algorithm for verifying that an input in one of the formats described earlier is a correct description of a medium.

10.4.1 Theorem. *If we are given as input an object purported to be a medium, in the transition table, hashed transition table, edge-labeled adjacency list, or vertex-labeled adjacency list format, we can test whether the input is in fact a medium in time $O(|\mathcal{S}|^2)$.*

Proof. Our algorithm works most directly for the vertex-labeled adjacency list format. If the input is an edge-labeled adjacency list, we use Theorem 10.2.10 to construct from it a vertex-labeled adjacency list. This algorithm uses only the assumption that the edge labels consistently group the tokens into token-reverse pairs; if not, we terminate the algorithm and report that the input is invalid. Once we have constructed a vertex-labeled adjacency list, we use Lemma 10.2.9 to verify that, for each effective transition St of the edge-labeled adjacency list, the first differing bit in the labels of S and of St is the one corresponding to token t. If this is so, and the vertex-labeled adjacency list validly describes a medium, then the edge-labeled adjacency validly describes the same medium. If the input is a transition table (hashed or not), we use Theorem 10.2.8 to convert it to an edge-labeled adjacency list and then proceed as above. If this conversion process attempts to label the same orientation of the same edge with multiple tokens, or leaves some orientation of an edge unlabeled, we terminate the algorithm and report that the input is invalid; otherwise, the reverse conversion will result in a transition table identical to our input, so if the resulting adjacency list validly describes a medium, then the transition table validly describes the same medium.

With our input in vertex-labeled format, we apply Lemma 10.2.9 twice, once on the labels of the vertices and once on the reverses of those labels (that is, the labeled graph formed by replacing the token labeling each edge by its reverse). With these two data structures, we may verify that, for each edge of the adjacency list structure, the two endpoints of that edge differ in exactly one bit of their labels. From this we may conclude (using the triangle inequality for Hamming distance) that the graph distances between any two vertices are at least equal to the Hamming distances between their labels.

Finally, we apply Theorem 10.3.3 to find, for each vertex v of the graph representing the purported medium, a tree of paths from each other vertex to v, such that if the graph is indeed a medium the paths will be shortest paths to v. If the algorithm of this theorem ever fails to find an effective transition from some vertex w towards the current node v of its depth first traversal, and hence fails to find the tree of paths to v, we terminate the algorithm and report that the input is invalid. For each v, once the tree of paths is found, we test for each edge of the tree that the vertex label on the endpoint nearer to v is closer to the label of v than is the vertex label on the endpoint farther than v. If this is the case, then along any path from another vertex to v in this tree, each label bit can change at most once, so the path length equals the Hamming distance between labels.

If our input passes all of these tests, it describes a labeled graph in which graph distances are always greater than or equal to the Hamming distance of the labels, and in which any two vertices can be connected by a path with path length equal to the Hamming distance of their labels. Such a graph must be a partial cube, and the labeling defines a valid medium structure on the graph.

The most time consuming steps of the algorithm reside in the application of Theorem 10.3.3 and the testing of each edge of each shortest path tree; these steps take total time $O(|S|^2)$. □

10.4.2 Remark. Eppstein (2007b) describes a simplified procedure for testing whether an edge-labeled graph describes a medium, that is also based on applying Theorem 10.3.3. However, the simplification is only guaranteed to work for labels produced by a modified version of the algorithm of Theorem 10.2.16. By combining Theorem 10.2.16 and Theorem 10.4.1, we can determine whether an unlabeled input graph is a partial cube in time $O(|V|^2)$, as was shown by Eppstein (2007b).

10.5 Recognizing Closed Media

In this section, we consider the problems of determining whether an oriented medium is closed, and of finding a closed orientation for an unoriented medium. A naive algorithm for testing whether an orientation is closed would test each triple S, τ, and τ', where S is any state and τ and τ' are any two different positive tokens, to determine whether $S \neq S\tau$ and $S \neq S\tau'$ but $S\tau = S\tau\tau'$. If any triple is found for which all three of these inequalities and equalities are true, then the medium is not closed. However, there are $O(|S| \cdot |\mathcal{T}|^2)$ triples to test, leading to a high running time. Our closedness testing algorithm reduces this time bound by using the adjacency list structure of our representation and the following lemma to reduce the number of triples to be tested.

10.5.1 Definition. We use the notation $S^{\mathcal{E}^+}$ to refer to the set of positive tokens of an oriented medium that are effective for S.

10.5.2 Lemma. *In any closed oriented medium* $(S, \mathcal{T})$, *for any state* S, $|S^{\mathcal{E}^+}| \leq \log_2 |S|$.

Proof. For each $P \subset S^{\mathcal{E}^+}$, let $\boldsymbol{m}_P$ be a message formed by concatenating the tokens in P. Then $\boldsymbol{m}_P$ is stepwise effective for S (by the assumption that the medium is closed), so the positive content of S differs from that of $S\boldsymbol{m}_P$ by exactly the tokens in P, from which it follows that two subsets $P \neq P'$ of $S^{\mathcal{E}^+}$ have $S\boldsymbol{m}_P \neq S\boldsymbol{m}_{P'}$. As each subset of $S^{\mathcal{E}^+}$ gives rise to a different state in this way, there can be at most $|S|$ possible subsets of $S^{\mathcal{E}^+}$, from which the bound follows. □

10.5.3 Remark. It is natural to define a closedness testing algorithm based on this property: first check whether the input medium satisfies Lemma 10.5.2 and, if so, test only those triples S, τ, τ' where τ and τ' belong to $S^{\mathcal{E}^+}$. There are two drawbacks to this approach: first, for media not satisfying the lemma, we can be sure that they are not closed but we do not find an explicit triple for which closure is violated. And second, to perform the tests efficiently we apparently require a hashed transition table representation of the medium, and the hash table operations of that representation force us to use expected-case analysis instead of giving a deterministic worst-case time bound. We remediate both of these issues with a more sophisticated technique, described below, that is based on a second property about the number of effective positive tokens in closed media.

10.5.4 Lemma. *In any closed oriented medium, let S be any state and let τ be a positive token. Then* $|S\tau^{\mathcal{E}^+}| \geq |S^{\mathcal{E}^+}| - 1$.

Proof. By the definition of a closed medium, every positive effective token for S other than τ itself must also be effective for $S\tau$ (whether or not $S\tau \neq S$). □

10.5.5 Remark. The bound of this lemma need not be tight; see Problem 10.12. For the case when τ is a negative token, see Problem 10.13.

10.5.6 Theorem. *Given an edge-labeled adjacency list representation of a medium, and an orientation on that medium, we can determine whether the medium is closed, and if not find a triple S, τ, τ' such that τ and τ' are effective on S but τ' is not effective on $S\tau$, in time $O(|E| \cdot \log |V|)$.*

Proof. We can compute $|S\tau^{\mathcal{E}^+}|$ for each state S in total time $O(m)$. If some state S and positive token τ are such that $p(S\tau) < p(S) - 1$, then the medium is not closed. In this case, we can compare the lists of effective tokens for S and St in time $O(|\mathcal{T}|)$ and find a token τ' that is effective for S but ineffective for $S\tau$; S, τ, and τ' form the desired triple.

Next, if $|S\tau^{\mathcal{E}^+}| \geq |S^{\mathcal{E}^+}| - 1$ is true for all S and positive τ, but some state has more than $\log_2 |\mathcal{S}|$ positive effective tokens, we can follow a sequence of positive effective transitions from that state until we find a state S with exactly $1 + \lfloor \log_2 |\mathcal{S}| \rfloor$ positive effective tokens. We will apply the remainder of our algorithm to the submedium formed by the states reachable from S via the positive tokens that are effective on S. This submedium still violates Lemma 10.5.2, so searching within this submedium will not prevent us from finding the triple we desire. This submedium can be constructed algorithmically in linear time by applying a depth first search from S in the adjacency list representation, following only transitions for token pairs the positive token of which is effective on S, and copying all searched vertices and edges to an edge-labeled adjacency list representation of the submedium. By restricting our attention to this submedium, we can assume without loss of generality that the number of positive effective tokens at each node is at most $\log_2 |\mathcal{S}| + 1$.

Finally, we make a sorted list of the $O(\log |\mathcal{S}|)$ positive tokens effective at each state of the (restricted) medium, in total time $O(|E| \log \log |V|)$; the additional doubly-logarithmic factor in this time bound comes from applying a sorting routine to a collection of lists of logarithmic size, the total length of which is $O(|E|)$. For each state S, and each positive token τ effective for S, we merge the two sorted lists for S and $S\tau$ to determine whether the list for S contains a token τ' that is missing from the list for $S\tau$. If so, we report S, τ, and τ' as the triple that witnesses the fact that the medium is not closed.

Finally, if our medium passes all the tests above, we report that it is closed. The time for this algorithm is dominated by the times to merge sorted lists; there are $O(|E|)$ such merges performed (one per positive effective transition) and each takes time $O(\log |V|)$, so the total time is $O(|E| \log |V|)$. $\square$

Next, we consider the problem of finding closed orientations for unoriented media.

10.5.7 Remark. Not every medium has a closed orientation; for instance, there is no closed orientation of the medium having six states in a cycle.

10.5.8 Lemma. *In any medium the number of triples S, τ, τ' where $\tau \neq \tau'$ and where both τ and τ' are effective for S is at most $\min(\frac{1}{2}|E| \cdot |\mathcal{T}|, |V|^2)$.*

Proof. Exactly $2|E|$ pairs S, τ have τ active for S. Each such pair can form a triple with at most $|\mathcal{T}|/2$ tokens τ' (because only one of each token-reverse pair can be effective for S), and each triple comes from two pairs (one for τ and one for τ'), so the total number of triples is at most $\frac{1}{2}|E| \cdot |\mathcal{T}|$.

Each triple S, τ, τ' can be associated with the pair of states $S\tau$, $S\tau'$. There are fewer than $\frac{1}{2}|V|^2$ pairs of states, and each possible pair of states can only be associated with two triples: the other triple that can be associated with the pair $S\tau$, $S\tau'$ is $S\tau\tau'$, $\tilde{\tau}$, $\tilde{\tau}'$. Therefore, the number of triples is at most twice the number of pairs of states, $|V|^2$.

Since we have separately proved that the number of triples is at most $\frac{1}{2}|E| \cdot |\mathcal{T}|$ and that it is at most $|V|^2$, it must be at most the minimum of these two quantities. □

10.5.9 Remark. The example of the hypercube shows that the bound of Lemma 10.5.8 cannot be simplified to $O(|S| \cdot |\mathcal{T}|)$, while the example of the star medium shows that it cannot be simplified to $O(|E| \log |V|)$.

10.5.10 Theorem. *If we are given an unoriented medium, we can find a closed orientation, if one exists, or determine that no such orientation exists, in time $O(\min(|E| \cdot |\mathcal{T}|, |V|^2))$.*

Proof. We translate the problem to an instance of 2-SAT (see, e.g., Aspvall et al., 1979); that is, a problem in which we are given a boolean formula in the form of a conjunction, each clause of which is a disjunction of two variables or negated variables, and must determine whether all variables can be assigned truth values in such a way as to make the overall formula true.

We do so by creating a Boolean variable for each token in the medium, where the truth of variable t in a truth assignment for our instance will correspond to the positivity of token τ in an orientation of the medium. By adding clauses $t \vee \tilde{t}$ and $\neg t \vee \neg \tilde{t}$, we guarantee that exactly one of the two variables t and $\tilde{t}$ corresponding to tokens τ and $\tilde{\tau}$ is true, so that the truth assignments of the instance correspond exactly to orientations of the medium. We then add further clauses $\neg t \vee \neg t'$ for each triple S, τ, τ' such that τ and τ' are effective for S but $\tau\tau'$ is not stepwise effective. In an orientation corresponding to a truth assignment satisfying these clauses, τ and τ' cannot both be positive, so triple S, τ, τ' will not violate the definition of a closed orientation. Thus, our 2-SAT instance will have a satisfying truth assignment if and only if the medium has a closed orientation, and the closed orientation, if it exists, can be determined easily from the truth assignment.

We can construct the 2-SAT instance by using a transition table to test each triple S, τ, and τ', where τ and τ' belong to the adjacency list of tokens effective for S. By Lemma 10.5.8 the time for this construction is $O(\min(|E| \cdot |\mathcal{T}|, |V|^2))$. As it has $|\mathcal{T}|$ variables, the 2-SAT instance has size $O(|\mathcal{T}|^2)$, and can be solved in time proportional to its size as shown by Aspvall et al. (1979). □

10.5.11 Remark. The solvable 2-SAT instances can be characterized in terms of the nonexistence of a cyclic chain of implications that includes both a variable and its negation; it would be possible to translate this back into a characterization of the media having closed orientations, but the resulting characterization is unsatisfactory. It would be of interest to find a more natural characterization of the media having closed orientations.

10.5.12 Example. We apply the algorithm of Theorem 10.5.10 to the medium depicted in Figure 2.1. The resulting 2-SAT instance consists of the formula

$$(\tau_1 \vee \tau_2) \wedge (\neg\tau_1 \vee \neg\tau_2) \wedge$$
$$(\tau_3 \vee \tau_4) \wedge (\neg\tau_3 \vee \neg\tau_4) \wedge$$
$$(\tau_5 \vee \tau_6) \wedge (\neg\tau_5 \vee \neg\tau_6) \wedge$$
$$(\neg\tau_1 \vee \neg\tau_4) \wedge (\neg\tau_1 \vee \neg\tau_6),$$

the first six clauses of which enforce that the orientation of each token's reverse is the reverse of the token's orientation. The last two clauses prevent the pairs τ_1, τ_4 and τ_1, τ_6 from being positive in the orientation, as if they were the triples W, τ_1, τ_4 or W, τ_1, τ_6 would violate the definition of a closed medium.

This 2-SAT instance can be satisfied, for instance, by a truth assignment in which τ_1, τ_4, and τ_6 are false and τ_2, τ_3, and τ_5 are true. This truth assignment corresponds to the orientation $\{\mathcal{T}^+, \mathcal{T}^-\}$ with $\mathcal{T}^+ = \{\tau_2, \tau_3, \tau_5\}$ and $\mathcal{T}^- = \{\tau_1, \tau_4, \tau_6\}$). The oriented medium formed by this orientation is closed.

10.5.13 Remark. An alternative method for finding closed orientations is based on the observation (Theorem 4.1.11) that any closed orientation of a finite medium has a unique apex S, a state having no positive effective transitions, and that an orientation with apex S must be the of the form $\{\widehat{S}, \mathcal{T} \setminus \widehat{S}\}$. We can use Theorem 10.5.6 to test each orientation of this form and determine whether one is closed. However this does not seem to lead to a method more efficient than the one described above.

10.6 Black Box Media

A medium may have exponentially many states, relative to its number of tokens, making storage of all states and transitions infeasible. However, for many media, a single state can easily be stored, and state transitions can be constructed algorithmically. The relative ease of storing a single state and performing transitions versus finding the whole medium can be seen, for instance, in the medium of linear extensions of a partial order described in Problem 9.3: one can represent a state as a permutation of elements of a partial order, and apply a token to a state by attempting the swap of two elements and by testing whether the result is still a linear extension of the partial order, without computing or storing the set of all linear extensions. Similarly, for the medium of acyclic orientations described in Example 9.1.16, one can represent a state as an orientation for each edge of the given graph, and apply a token to a state by attempting the reversal of an edge and by testing whether the result is still an acyclic orientation, without computing or storing the set of all acyclic orientations. More generally, we have discussed in Definition 4.5.1 the concept of a *base*, which may be used to summarize succinctly any closed medium; one may represent a state in a medium summarized in this way as a union of base sets, and one may apply a token by adding or removing an element from the set and testing whether the result is again a union of base sets.

Thus, we would like to design algorithms for media that do not depend on storing an explicit list of states and state transitions. However, we would like to do this in as general way as possible, not depending on details of the state representation. This motivates an oracle-based model of media, similar to the black box groups from computational group theory (Babai and Szemerédi, 1984). We are interested here in the limits that a severely restricted input model such as we give below may place on our ability to compute with media. Developing algorithms for such a model allows them to be applied very generally, in cases where it is inconvenient to construct ahead of time an explicit list of the states and transitions of the medium.

10.6.1 Definition. We define a *black box medium* to be a representation of a medium consisting of the following parts:

- A list of the tokens of the medium.
- A procedure `transitionFunction` that takes as input the representation of a state S and a token t, and produces as output the representation of the state St.
- The representation of a single state S_0.

We require each state to have a unique representation, so that states can be compared for equality. However the details of the state representation and transition function of a black box medium are unavailable to an algorithm that processes a medium of this form.

10.6.2 Example. Media formed by a hyperplane arrangement (cf. Chapter 9) can be described as black box media as follows. We represent a state (that is, a chamber of the arrangement) as the set of halfspaces containing it, with one halfspace bounded by each hyperplane of the arrangement. We represent a token as an individual halfspace. The transition function takes as input a state S and a halfspace h, and returns the state formed by replacing the complement of h by h in the set of halfspaces representing S, if such a state exists, or it returns S itself if the modified state would not be part of the arrangement. To implement the transition function, we need to determine whether a set of halfspaces has a nonempty intersection; this can be done in polynomial time by linear programming. Therefore, we can represent this medium as a black box medium without explicitly constructing all chambers of the arrangement.

This linear programming test can also be viewed as an example of another implicit model of a medium, an *independence oracle* (Karp et al., 1985) in which we represent a well-graded set family (in our example, a family of halfspaces) by using a bitmap to represent a set, and perform a transition by changing a single bit of the bitmap and calling another function (in our example, the linear programming feasibility test) to determine whether the modified set belongs to the family. However, the independence oracle model provides greater computational power than the black box model; for instance, given two bitmaps representing well-graded sets S and Q, one can quickly find an effective transition from S that is in the content of Q, by testing bits in the symmetric difference of the two bitmaps until finding one that leads to another set within the family. In contrast, for the black box model, finding a token that gives an effective transition from a state S and is in the content of another state Q appears to require a much lengthier exploration process.

We can bound the amount of memory required to store a single state by a parameter s. We denote by T the amount of time required per call to `transitionFunction`. In this context, we consider "polynomial time" to mean time that is polynomial in $|\mathcal{T}|$ and T, and "polynomial space" to mean a space bound that is polynomial in $|\mathcal{T}|$ and s.

One of the most basic computational problems for a black box medium is to find all of its possible states. As with any deterministic finite automaton, one could list all states by performing a depth first traversal of the state transition graph. However, performing a depth-first search efficiently requires keeping track of the set of already-visited states, so that the search can recognize when it has reached a state that it has already explored, and backtrack rather than repeatedly exploring the same states. If the medium is large enough to make an adjacency list representation infeasibly large, then it would also likely be infeasible to store this set of already-visited nodes. As we now show, it is possible instead to traverse all states of a black box medium, using an amount of storage limited to only a constant number of states.

The main idea behind our traversal algorithm is to use a modified version of the *reverse search* procedure of Avis and Fukuda (1996). The reverse search procedure allows one to list all the states of a general state transition system, as long as one can define *canonical paths*: sequences of transitions from each state to an initially given state S_0 such that the union of the transitions on all the paths forms a spanning tree of the state space, the *canonical path tree*. If one can quickly determine whether a transition belongs to the canonical path tree, then one can list the states by traversing this tree, ignoring the transitions that do not form edges in the tree. In our case, canonical paths can be constructed from the orientation $(\widehat{S_0}, \mathcal{T} \setminus \widehat{S_0})$. If we have an ordered list of the positive tokens of this orientation, then a step in a canonical path from state S can be found by taking the first token in the list that is effective on S. The difficulty, however, is that we are not given the orientation, and cannot construct it without searching the medium, the task we are trying to solve. Fortunately, we can interleave the construction of the orientation and the enumeration of states, in a way that allows the reverse search algorithm to proceed, taking polynomial time per state and using polynomial space.

10.6.3 Theorem. *If we are given a black box medium, we can list all states of the medium in time $O(|S| \cdot |\mathcal{T}|^2 T)$ and space $O(s + |\mathcal{T}|)$.*

Proof. We perform the reverse search procedure described above, building a list of positive tokens in the orientation defined by S_0 and simultaneously searching the tree of canonical paths; the canonical path from a state to S_0 is found by applying the first effective token stored in the list being built. The data stored by the algorithm consists of the current state S in the search, the list of positive tokens discovered so far, and a pointer to the first token τ in this list for which we have not yet searched transition $S\tau$. At any point in the algorithm, the tokens on the canonical path from the current state S to S_0 will have already been included in our list of positive tokens. Initially, the list of positive tokens is empty, $S = S_0$, and τ is the first token in the list.

At each step of the algorithm, we test transition $S\tau$ as follows: if τ is already listed in the list of positive tokens, it cannot be the reverse of a step in a canonical path, and we set τ to the next token in the list of all tokens. Similarly, if $S\tau$ is ineffective we set τ to the next token. If τ is effective, then we search the list of tokens for the reverse token $\tilde{\tau}$ satisfying $S\tau\tilde{\tau} = S$. The path formed by composing the transition from $S\tau$ back to S with the canonical path from S to S_0 must correspond to a concise message (otherwise it would include τ and τ would have been already included in the list of positive tokens) so $\tilde{\tau}$ must be positive; we include it in the list of positive tokens. We then check whether $\tilde{\tau}$ is the first listed positive token that is effective for $S\tau$. If it is the first effective positive token, then the transition from S to $S\tau$ is the reverse of a canonical step, so we set S to $S\tau$, reset τ to the token at the beginning of the token list, output state $S\tau$, and continue searching at the new state. If not, we set τ to the next token in the list and continue searching from t.

Whenever the token τ advances past the end of the list of tokens in our search, this indicates that we have exhausted all transitions from the state S. So, we return to the parent of S in the canonical path tree, by searching the list of positive tokens for the first one that is effective on S. We must then also reset the token τ to the appropriate position in the list of tokens. We do this by searching sequentially through the list of tokens for the one that caused the parent of S to transition to S. When we advance past the end of the list of tokens for state S_0, the search is complete.

Each step maintains the invariants discussed above. By induction on the lengths of canonical paths, each state is output exactly once, when it is found by the reverse of a canonical step from its parent in the canonical path tree. The space bound is easy to compute from the set of data stored by the algorithm. For each effective transition τ from each state S, we perform $O(|\mathcal{T}|)$ calls to the transition function: one scan of the token list to find the reverse $\tilde{\tau}$ and determine whether $S\tau$ is canonical, a second scan of the token list to find the effective transitions from $S\tau$ (if $S\tau$ was canonical), and a third scan of the token list to find the correct position of τ after returning from $S\tau$ to S. Therefore, the total number of calls to the transition function is $O(|\mathfrak{T}|^2)$ per state, which dominates the total running time of the algorithm. □

Table 10.3. Python implementation of reverse search for black box media states.

```
def mediumStates(s0, tokens, transitionFunction):
    """
    List all states of a medium.
    The input is a starting state s0, a list of tokens, and the transition
    function (state,token)->state of a medium. The output is the sequence of
    states of the medium. We do not assume that the tokens are organized into
    reverse pairs. The memory usage consists of a constant number of states,
    together with additional data linear in the number of tokens. The algorithm
    is a stackless version of the reverse search technique of Avis and Fukuda.
    """
    tokens = list(tokens)  # make sure we can reuse the list of tokens
    positiveTokens = []  # ordered list of tokens on concise messages to s0

    def step(state, default=None):
        """
        Find the first token on a concise message from state to s0,
        and return the next state on that path. If no token is found, return
        default after searching for a token that takes the given state to it.
        """
        for t in positiveTokens:
            x = transitionFunction(state, t)
            if x != state:
                return x
        positiveTokens.extend([t for t in tokens
                               if transitionFunction(state, t) == default])
        return default

    state = s0
    yield state
    tokenSequence = iter(tokens)
    while True:
        try:
            x = transitionFunction(state, tokenSequence.next())
            if x != state and x != s0 and step(x,state) == state:
                state = x
                yield state
                tokenSequence = iter(tokens)

        except StopIteration:
            # We reach here after exhausting all tokens for the current state.
            # Backtrack to the parent and reset the token sequence pointer.
            if state == s0:
                return
            parent = step(state)
            tokenSequence = iter(tokens)
            for token in tokenSequence:
                if transitionFunction(parent, token) == state:
                    break
            state = parent
```

Table 10.3 displays an implementation of our reverse search procedure in the Python programming language. The `yield` keyword triggers Python's *simple generator protocol*, which creates an iterator object suitable for use in `for`-loops and similar contexts and returns it from each call to `mediumStates`.

Problems

10.1 Let $G = (V, E)$ be a partial cube graph. Prove that $|E| = \frac{1}{2}|V|\log_2 |V|$ if and only if G is a hypercube.

10.2 For each of the following pairs of functions, determine which of the two would be preferable as the running time bound of an algorithm.

  (i) $|\mathcal{S}| \log |\mathcal{S}|$ versus $|\mathcal{S}| \cdot |\mathcal{T}|$.
  (ii) $|\mathcal{S}| \cdot |\mathcal{T}|$ versus $|\mathcal{S}|^2$.
  (iii) $|V|^2 \log |V|$ versus $|V| \cdot |E|$.

10.3 Draw the transition table for the medium shown in Figure 6.2(A).

10.4 List the key-value pairs for the hashed transition table representation of the medium shown in Figure 6.2(A).

10.5 Draw the labeled adjacency list representation for the medium shown in Figure 6.2(A).

10.6 Draw the transition table for the medium shown in Figure 6.2(B).

10.7 List the key-value pairs for the hashed transition table representation of the medium shown in Figure 6.2(B).

10.8 Draw the labeled adjacency list representation for the medium shown in Figure 6.2(B).

10.9 Let S be the root of a star medium $(\mathcal{S}, \mathcal{T})$, and let $\boldsymbol{m}$ be a message such that applying $\boldsymbol{m}$ to S produces a sequence of states that includes all states of $\mathcal{S}$. Prove that the number of tokens in $\boldsymbol{m}$ is at least $2|\mathcal{S}| - 3$.

10.10 Let S be a state in a medium $\mathcal{M} = (\mathcal{S}, \mathcal{T})$, such that either $\mathcal{M}$ is not a star or S is not its root. Prove that there exists a message $\boldsymbol{m}$ of fewer than $2|\mathcal{S}| - 3$ tokens, such that applying $\boldsymbol{m}$ to S produces a sequence of states that includes all states of $\mathcal{S}$.

10.11 Complete Example 10.3.4 by describing the configuration of the algorithm's data structures after each of the remaining traversal steps until the depth first traversal returns to the original starting vertex.

10.12 Give an example of a closed oriented medium, a state S in that medium, and a positive token τ, such that τ is effective for S and $|S\tau^{\varepsilon^+}| > |S^{\varepsilon^+}| - 1$.

10.13 Let S be a state of a closed oriented medium and let τ be a negative token. Prove that $|S\tau^{\varepsilon^+}| \leq |S^{\varepsilon^+}| + 1$.

11
Visualization of Media

In this chapter, following Eppstein (2005a), we describe methods for the planar layout of finite media and partial cube graphs in a way that makes the medium structure apparent to the human viewer. Our focus is on graph drawing techniques that may be efficiently implemented in a computer algorithm. The algorithms we consider are based on two principles: embedding the state transition graph in a low-dimensional integer lattice (as described in Chapter 8) and projecting the lattice onto the plane, or drawing the medium as a *zonotopal tiling*. After this material, we describe some more specialized algorithms for drawing learning spaces, as in Eppstein (2006).

The definition of a graph drawing, below, is essentially the same as that in Di Battista et al. (1999).

11.0.1 Definition. Let f be a continuous one-to-one function from the closed interval $[0,1]$ to the plane $\mathbb{R}^2$. Then the image under f of the open interval $(0,1)$ is a *simple curve*, with *endpoints* $f(0)$ and $f(1)$. A *drawing* of a graph G is an assignment of a point for each vertex of G, and a simple curve for each edge of G, such that an edge (u,v) is assigned to a curve that has the points assigned to u and v as its endpoints. As a shorthand, we refer to a drawing of the graph of a medium $\mathcal{M}$ as a drawing of $\mathcal{M}$.

A *straight-line drawing* is a drawing in which each curve is an open line segment. A *planar drawing* is a drawing in which the points assigned to the vertices and the curves assigned to the edges are all disjoint from each other. A classical result in this area is Fáry's theorem (e.g., see section 4.10 of Di Battista et al., 1999) which states that any finite graph that has a planar drawing has a planar straight-line drawing.

If U is the union of the points and curves in a planar drawing, then a *face* of the drawing is a connected component of the complement of U. We will need to distinguish between *bounded faces* of drawings (that is, faces contained within a bounded region of the plane) and *unbounded faces*. Any planar drawing of a finite graph will have exactly one unbounded face, but drawings of infinite graphs may have multiple unbounded faces, or no unbounded faces. An *exterior edge* or *exterior vertex* of a drawing is an edge or vertex that lies on the boundary of an unbounded face, and an *interior edge* or *interior vertex* is an edge or vertex that is not exterior.

11.1 Lattice Dimension

As we describe in Chapter 8, the minimum dimension of an integer lattice into which a given medium may be isometrically embedded has a simple characterization involving graph matching, from which it follows that the dimension (and a minimum-dimension embedding) may be constructed efficiently. If the lattice dimension of a medium is low, we may be able to use the embedding directly as part of an effective drawing algorithm. For instance, if a medium $\mathcal{M}$ can be embedded isometrically onto the planar integer lattice $\mathbb{Z}^2$, then we can use the lattice positions as vertex coordinates of a drawing in which each edge is a vertical or horizontal unit segment. As we describe below, it is sometimes also possible to find planar drawings from three-dimensional lattice embeddings.

11.1.1 Definition. The *isometric projection* of a three dimensional lattice is formed by mapping each point (x, y, z) onto the point

$$\left(\frac{2x - y - z}{3}, \frac{2y - x - z}{3}, \frac{2z - x - z}{3} \right)$$

on the plane $x + y + z = 0$; that is, it is the vector projection perpendicular to the vector $(1, 1, 1)$. The name, "isometric projection," indicates that the unit vectors of the lattice are mapped to vectors that have the same lengths as each other; it has little to do with isometry or projection in the media-theoretic senses.

11.1.2 Remark. If $\mathcal{M}$ can be embedded isometrically onto the cubic lattice $\mathbb{Z}^3$, in such a way that the isometric projection of $\mathbb{Z}^3$ maps the states of $\mathcal{M}$ to distinct planar points, then the straight-line drawing with these point placements is planar and has unit length edges meeting each other at 60° and 120° angles. For instance, Figure 9.4 can be viewed in this way as a drawing formed by an isometric projection of a three-dimensional lattice embedding.

We now briefly describe our algorithm for finding low-dimensional lattice embeddings more generally.

11.1.3 Theorem. *Suppose that we are given a finite medium $\mathcal{M} = (\mathcal{S}, \mathcal{T})$, and a hypercube isometry $f : G \mapsto \{0, 1\}^\tau$, where G is the graph of $\mathcal{M}$. Then we can compute in time $O(|\mathcal{S}| \cdot |\mathcal{T}|^2)$ the lattice dimension d of G, and in the same time construct a lattice isometry $g : G \mapsto \mathbb{Z}^d$.*

Proof. We construct the semicube graph $\mathrm{Sc}(G)$ as defined in Chapter 8 directly from the definition: there is one semicube per token in $\mathcal{T}$, and we can test whether any two semicubes should be connected by an edge in the semicube graph in time $O(|\mathcal{S}|)$, so the total time for this step is $O(|\mathcal{S}| \cdot |\mathcal{T}|^2)$. We then use a maximum matching algorithm to find a matching with the largest possible number of edges in $\mathrm{Sc}(G)$; as Micali and Vazirani (1980) show, this can be done in time $O(|\mathcal{T}|^{2.5})$. We can then transform the matching in $\mathrm{Sc}(G)$ to a collection of d paths in time $O(|\mathcal{S}| \cdot |\mathcal{T}|)$ by adding to the matching an edge from each semicube to its complement. The dth coordinate of a vertex in the lattice embedding equals the number of semicubes that contain the vertex in even positions along the dth path. The total time is dominated by the $O(|\mathcal{S}| \cdot |\mathcal{T}|^2)$ bound for finding $\mathrm{Sc}(G)$. □

It is essential for the efficiency of this technique that the embedding of the medium into $\mathbb{Z}^d$ be isometric. Even for a tree (a very special case of a partial cube) it is NP-complete to find an embedding into $\mathbb{Z}^2$ with unit length edges that maps distinct vertices to distinct lattice points but is not otherwise required to be distance-preserving (Bhatt and Cosmodakis, 1987).

We can use this embedding algorithm as part of a graph drawing system, by embedding our input medium in the lattice of the lowest possible dimension and then projecting that lattice onto the plane. For two-dimensional lattices, no projection is needed, and we have already discussed in Remark 11.1.2 the projection of certain three-dimensional integer lattices onto two-dimensional planar drawings. We discuss more general techniques for lattice projection in the next section.

11.2 Drawing High-Dimensional Lattice Graphs

Two-dimensional lattice embeddings of media, and some three-dimensional embeddings, lead to planar graph drawings with all edges short and all angles bounded away from zero. However, we are interested in drawing media that may not have such well-behaved lattice embeddings. We describe here a method for transforming any lattice embedding of any medium into a drawing with the following properties:

[V1] All vertices are assigned distinct integer coordinates in $\mathbb{Z}^2$.

[V2] All edges are drawn as straight line segments.

[V3] No edge passes closer than unit distance to a vertex that is not one of its endpoints.

[V4] The line segments representing two edges of the drawing are translates of each other if and only if the two edges are parallel in the lattice embedding.

232   11 Visualization of Media

[V5] The medium corresponding to a Cartesian product of intervals of $\mathbb{N}$, $[a_0, b_0] \times [a_1, b_1] \times \cdots [a_{d-1}, b_{d-1}]$ is drawn in area $O(n^2)$, where n is the number of its states.

Because of Property [V4], the lattice embedding and hence the medium structure of the state transition graph can be read from the drawing. To draw a lattice-embedded medium, we map $\mathbb{Z}^d$ to $\mathbb{Z}^2$ linearly, by choosing two particular vectors X and $Y \in \mathbb{Z}^d$, and using these vectors to map any point $p \in \mathbb{Z}^d$ to the point $(X \cdot p, Y \cdot p) \in \mathbb{Z}^2$ (here "·" denotes the inner product of vectors). We now describe how to choose these vectors X and Y in order to achieve the desired properties of our drawing. If $L \subseteq \mathbb{Z}^d$ is the set of vertex placements in the lattice embedding of our input medium, define a *slice* $L_{i,j} = \{p \in L \,|\, p_i = j\}$ to be the subset of vertices having ith coordinate equal to j. We choose the coordinates X_i sequentially, from smaller i to larger, so that all slices $L_{i,j}$ are separated from each other in the range of x-coordinates they are placed in. Specifically, set $X_0 = 0$. Then, for $i > 0$, define

$$X_i = \max_j \left(\min_{p \in L_{i,j}} \sum_{k=0}^{i-1} X_k p_k - \max_{q \in L_{i,j-1}} \sum_{k=0}^{i-1} X_k q_k \right),$$

where the outer maximization is over all j such that $L_{i,j}$ and $L_{i,j-1}$ are both nonempty. We define Y similarly, but we choose its coordinates in the opposite order, from larger i to smaller: $Y_{d-1} = 0$, and

$$Y_i = \max_j \left(\min_{p \in L_{i,j}} \sum_{k=i+1}^{d-1} X_k p_k - \max_{q \in L_{i,j-1}} \sum_{k=i+1}^{d-1} X_k q_k \right).$$

11.2.1 Theorem. *The projection method described above satisfies properties [V1]–[V5]. The method's running time is $O(|\mathcal{S}| \cdot |\mathcal{T}|^2)$.*

Proof. Property [V2] and Property [V4] follow immediately from the fact that our drawing is formed by projecting $\mathbb{Z}^d$ linearly onto $\mathbb{Z}^2$, and from the fact that the formulas used to calculate X and Y assign different values to different coordinates of these vectors.

All vertices are assigned distinct coordinates (Property [V1]): for, if vertices p and q differ in the ith coordinates of their lattice embeddings, they belong to different slices $L_{i,j}$ and $L_{i,j'}$ and are assigned X coordinates that differ by at least X_i (unless $i = X_i = 0$ in which case their Y coordinates differ by at least Y_i).

The separation between vertices and edges (Property [V3]) is almost equally easy to verify: consider the case of three vertices p, q, and r, with an edge $\{p,q\}$ to be separated from r. Since p and q are connected by an edge, their lattice embeddings must differ in only a single coordinate i. If r differs from p and q only in the same coordinate, it is separated from edge $\{p,q\}$ by a multiple of (X_i, Y_i). Otherwise, there is some coordinate $i' \neq i$

in which r differs from both p and q. If $i' > i$, the construction ensures that the slice $L_{i',j}$ containing pq is well separated in the x-coordinate from the slice $L_{i',j'}$ containing r, and if $i' < i$ these slices are well separated in the y coordinate.

Finally, we consider Property [V5]. For Cartesian products of intervals, in the formula for X_i, the value for the subexpression $\min_{p \in L_{i,j}} \sum_{k=0}^{i-1} X_k p_k$ is the same for all j considered in the outer maximization, and the value for the subexpression $\max_{q \in L_{i,j-1}} \sum_{k=0}^{i-1} X_k q_k$ is also the same for all j considered in the outer maximization, because the slices are all just translates of each other. Therefore, there is no gap in x-coordinates between vertex placements of each successive slice of the medium. Since our drawings of these media have vertices occupying contiguous integer x coordinates and (by a symmetric argument) y coordinates, the total area is at most n^2.

The time for implementing this method is dominated by that for finding a minimum-dimension lattice embedding of the input graph, which is bounded as stated in Theorem 11.1.3. □

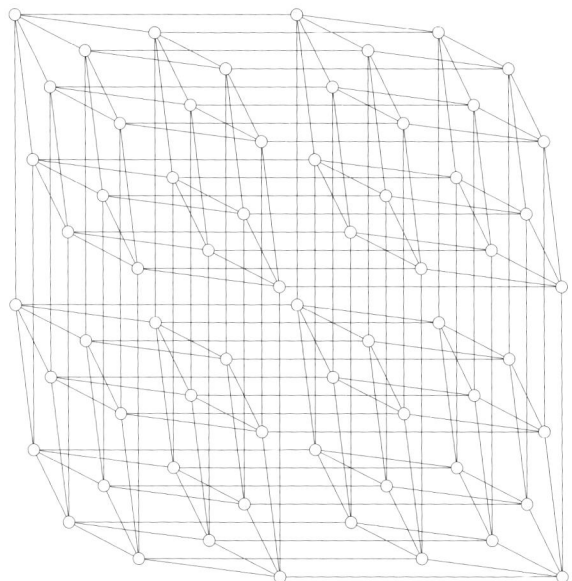

Figure 11.1. Lattice drawing of six-dimensional hypercube.

11.2.2 Example. When applied to a hypercube, the coordinates X_i become powers of two, and this vertex placement algorithm produces a uniform placement of vertices (Figure 11.1) closely related to the Hammersley point set commonly used in numerical computation and computer graphics in view of its low discrepancy properties (Wong et al., 1997).

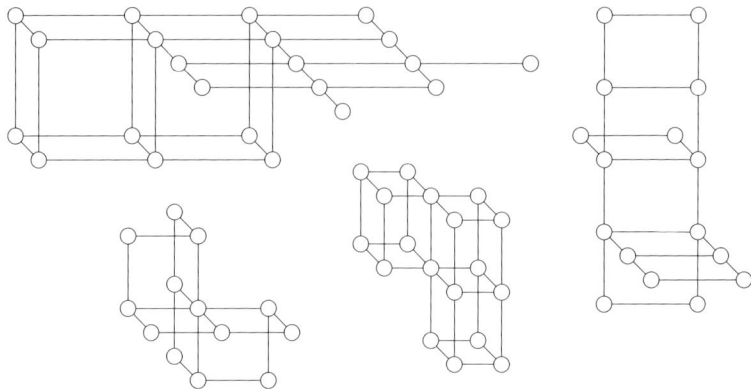

Figure 11.2. Lattice drawings of four irregular media with three-dimensional lattice embeddings (from Eppstein, 2005b). The bottom left drawing is of a medium isomorphic to the weak ordering medium previously shown in a more symmetrical form in Figure 9.9.

11.2.3 Example. Figure 11.2 provides several additional examples of media drawn by this lattice projection method, including another view of the medium of weak orders from Example 9.4.2.

11.3 Region Graphs of Line Arrangements

We now turn our attention from drawing methods that apply to all media, and instead consider methods that work for specific types of media. In this section, we describe a form of planar graph duality that can be applied to find straight-line planar drawings for the region graphs of line arrangements in $\mathbb{R}^2$.

11.3.1 Definition. Suppose D is a planar drawing of a graph, in which each edge separates two distinct faces, and R is the union of the closures of the bounded faces of D. In this case we say that D is a *tiling* of R, and that the closure of each face of D is a *tile* of the tiling. A *mosaic* is a tiling of $\mathbb{R}^2$ in which all faces are regular polygons.

A polygon P is *centrally symmetric* if it is a translate of its pointwise reflection through the origin. For example a regular n-gon is centrally symmetric if and only if n is even. A *zonotopal tiling* is a planar straight-line graph drawing in which each bounded face is a centrally symmetric polygon.

For additional tiling terminology, results, and further references, see, for instance, Grünbaum and Shephard (1987) and Senechal (1995). For more on finite zonotopal tilings see e.g. Felsner and Weil (2001).

11.3 Region Graphs of Line Arrangements

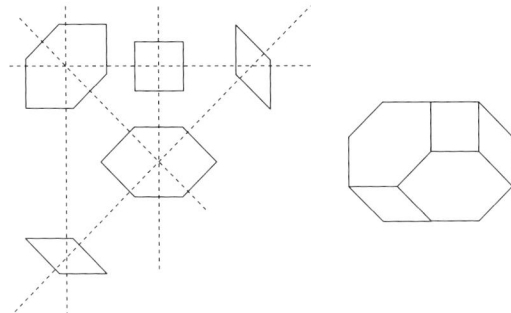

Figure 11.3. A line arrangement (left), with a symmetric convex polygon corresponding to each of its vertices; joining these polygons along their edges forms a zonotopal tiling (right).

We now describe a construction for zonotopal tilings as drawings of the region graphs of certain arrangements of lines. Subsequently, in Section 11.4, we will use a similar construction to precisely characterize the graphs that can be drawn as zonotopal tilings, as region graphs of more general kinds of arrangements. In Section 11.5, we will use zonotopal tilings constructed in this way as part of algorithms for visualizing a class of media more general than those formed from line arrangements, and in Section 11.6 we will examine a special subclass of these graphs that correspond to learning spaces.

Let $\mathcal{A}$ be a locally finite line arrangement. Form a set of centrally symmetric convex polygons, one per vertex of the arrangement, with unit length sides perpendicular to the lines meeting at that vertex. Then, form a cell complex by gluing together two such polygons along an edge whenever they correspond to adjacent vertices along the same line of the arrangement. It can be seen that this cell complex has cells which meet at interior vertices (corresponding to bounded cells of the arrangement) at angles totaling 2π, while each boundary vertex of the complex (corresponding to an unbounded cell of the arrangement) has angles totaling at most π. Therefore, it forms a drawing of the region graph of $\mathcal{A}$ that covers a convex subset of the plane by tiles, each of which is a centrally symmetric and strictly convex polygon.

This construction, which is shown in Figure 11.3, can also be performed in the following equivalent way. Choose a vertex of the region graph of the arrangement to be placed at the origin of the plane. Then, for each other vertex v, find a straight path in the graph from the origin to v, and set the coordinates of v to be the sum of unit vectors perpendicular to the arrangement lines corresponding to the edges in the straight path. The axioms defining a medium can be used to show that these coordinates are independent of the choice of straight path to v, and that the choice of base point affects the overall vertex placement only by a translation.

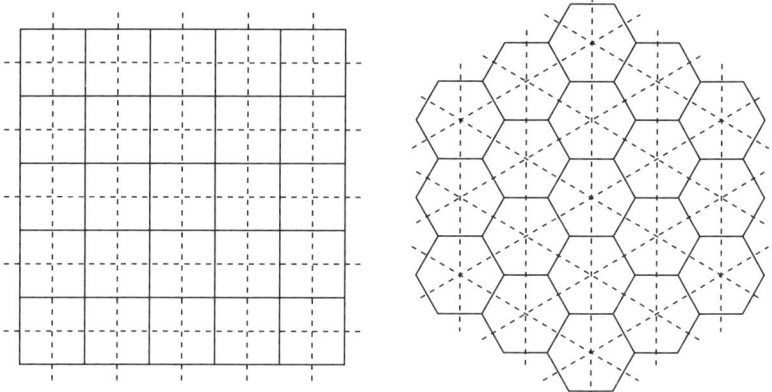

Figure 11.4. A 2-grid (left, dashed), and a 3-grid (right, dashed), superimposed on their dual zonotopal tilings (solid).

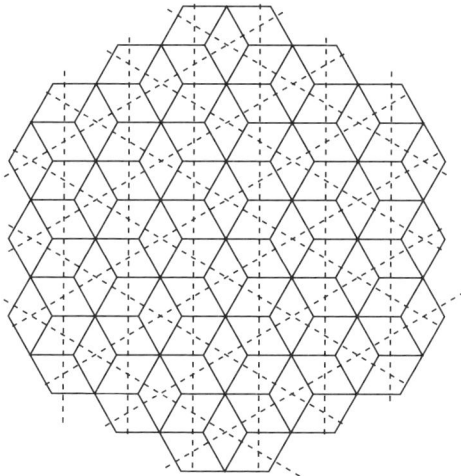

Figure 11.5. Another 3-grid and its dual zonotopal tiling.

11.3.2 Example. A line arrangement consisting of k infinite families of parallel lines, spaced at unit distance within each family, is called a k *multigrid*. The zonotopal tiling formed by drawing the region graph of a 2-multigrid is just the usual periodic tiling of the plane by edge-to-edge rhombi or squares (Figure 11.4, left). The tiling of the plane by edge-to-edge equilateral triangles can be viewed as a 3-multigrid, the drawing of the region graph of which is a zonotopal tiling of the plane by regular hexagons (Figure 11.4, right). By Theorem 9.1.9, this hexagonal lattice is a drawing of an infinite partial cube. In fact, this lattice is isometrically embeddable into the graph of the lattice $\mathbb{Z}^3$ (Deza and Shtogrin, 2002).

11.3 Region Graphs of Line Arrangements   237

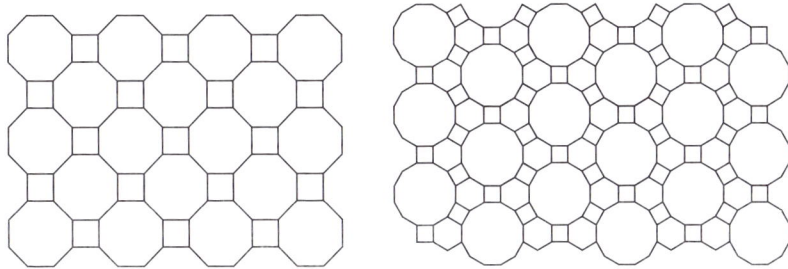

Figure 11.6. The mosaic of octagons and squares (left), and the mosaic of dodecagons, hexagons, and squares (right).

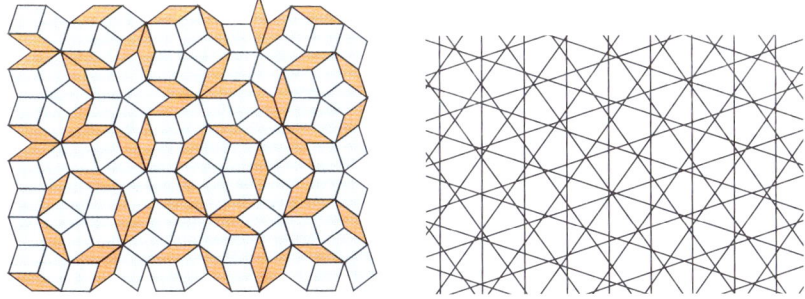

Figure 11.7. A portion of a Penrose rhombic tiling (left), and a pentagrid (right).

11.3.3 Example. Another 3-grid is shown in Figure 11.5. The line arrangement itself defines a mosaic of hexagons and triangles, but the drawing of its region graph shown in this figure has faces that are rhombs. It is almost obvious from the drawing that the region graph is isometrically embeddable into $\mathbb{Z}^3$. Note that graphs shown in Figures 1.5 and 9.3 are finite isometric subgraphs of the graph in Figure 11.5.

11.3.4 Example. Two more mosaics, of octagons and squares and of dodecagons, hexagons, and squares, are shown in Figures 11.6. These are again drawings of partial cubes, as they are drawings of region graphs of line arrangements with regularly spaced lines having four and six different slopes respectively. However, these arrangements are not multigrids, because lines of different slopes are spaced at different distances.

11.3.5 Example. A more sophisticated example of a tiling that is the drawing of an infinite partial cube is given by a Penrose rhombic tiling shown in Figure 11.7(left). A construction suggested by de Bruijn (1981) demonstrates that these kind of tilings can be formed from the region graphs of a particular class of line arrangements known as *pentagrids*: 5-grids such as the one in Figure 11.7(right), having equal angles between each pair of parallel line

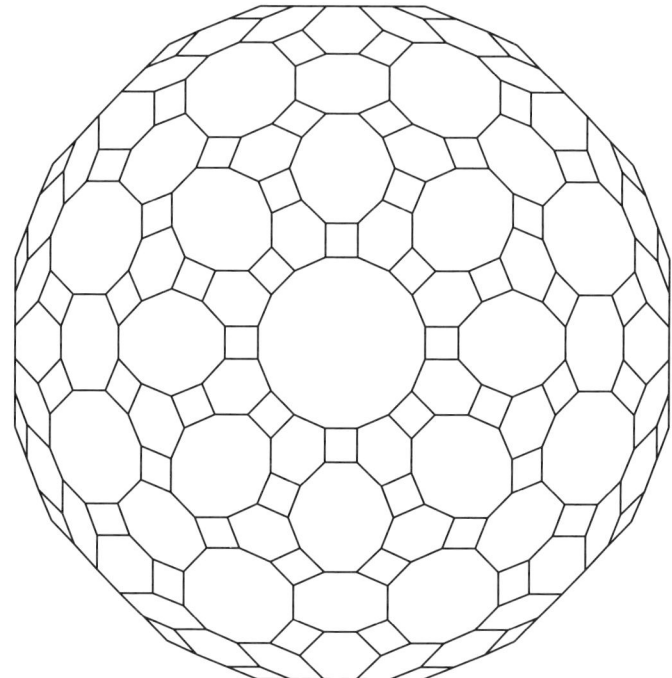

Figure 11.8. A zonotopal tiling that is not the region graph of a line arrangement.

families, and satisfying certain additional constraints on the placement of the grids relative to each other. It follows implicitly from de Bruijn (1981) that these rhombic tilings are isometrically embeddable in $\mathbb{Z}^5$ (see also Deza and Shtogrin, 2002).

11.4 Pseudoline Arrangements

As described in the previous section, every line arrangement leads to a zonotopal tiling of a convex polygon. The reverse is not true; some zonotopal tilings, even of convex polygons, are not drawings of region graphs of line arrangements (e.g., Figure 11.8). In general, such tilings arise from *nonstretchable pseudoline arrangements*; see, e.g., Grünbaum (1972). However, Eppstein (2005a) (who we follow here) characterized finite zonotopal tilings in terms of a generalization of line arrangements, known as weak pseudoline arrangements. The significance of these results for media is that it will lead to techniques for visualizing media by drawing them in the plane as zonotopal tilings: the characterization we give in this section for these tilings is the basis for the visualization algorithms described later in this chapter.

11.4 Pseudoline Arrangements

11.4.1 Definition. Following Shor (1991), we define a *pseudoline* to be the image of a line under a homeomorphism of the plane (see Remark 11.4.2). That is, if f is a continuous one-to-one mapping of the plane to itself, and ℓ is a line in the plane, then $f(\ell)$ is a pseudoline. It follows from this definition that a pseudoline is a non-self-crossing curve in the plane that extends to infinity in both directions and partitions the plane into two parts. A *weak pseudoline arrangement* is a collection of finitely many pseudolines, such that any pair of pseudolines has at most one point of intersection, and such that if any two pseudolines intersect then they cross properly at their intersection point. Weak arrangements were studied by de Fraysseix and Ossona de Mendez (2004); they generalize both pseudoline arrangements (in which each pair of pseudolines must cross; see, e.g., Grünbaum, 1972) and hyperbolic line arrangements, and are a special case of *extendible pseudosegment arrangements* (Chan, 2003).

A *region* of a weak pseudoline arrangement $\mathcal{A}$ is a connected component of $\mathbb{R}^2 \setminus \mathcal{A}$. The *region graph* of a weak pseudoline arrangement $\mathcal{A}$ is defined analogously to that of a line arrangement, as having a vertex for each region and an edge between the vertices corresponding to any two regions that are separated by a single pseudoline.

11.4.2 Remark. There is considerable disagreement in the mathematical literature over the proper definition of a pseudoline. Two common definitions are, first, that a pseudoline is an uncontractible simple closed curve in the projective plane (e.g., Grünbaum, 1972), or, second, that a pseudoline is the *monotone curve* formed as the graph of a continuous function $y = f(x)$ in the (x, y) plane (e.g., Edelsbrunner, 1987). Berman (2007) suggests as an alternative definition that a pseudoline is the result of replacing a bounded segment of a line by a simple curve that does not cross the rest of the line. None of these definitions is suitable for defining the weak pseudoline arrangements considered here, as they do not allow regions of the plane to be bounded by more than two nonintersecting pseudolines. Instead, de Fraysseix and Ossona de Mendez (2004) define a pseudoline as a subset of the plane homeomorphic to the real line, but their definition is overbroad as it allows curves that do not separate the plane into two components.

11.4.3 Lemma. *Every zonotopal tiling with finitely many tiles is a drawing of the region graph of a weak pseudoline arrangement.*

Proof. Draw a collection of line segments connecting the opposite pairs of edge midpoints in each bounded face of the drawing, as shown in Figure 11.9. These segments meet up at their endpoints to form piecewise-linear curves in the plane; in the unbounded face, extend these curves to infinity without adding any additional crossings.

Each of these curves passes through a family of parallel edges of the zonotopal tiling. Any crossing between these curves occurs within a tile of the

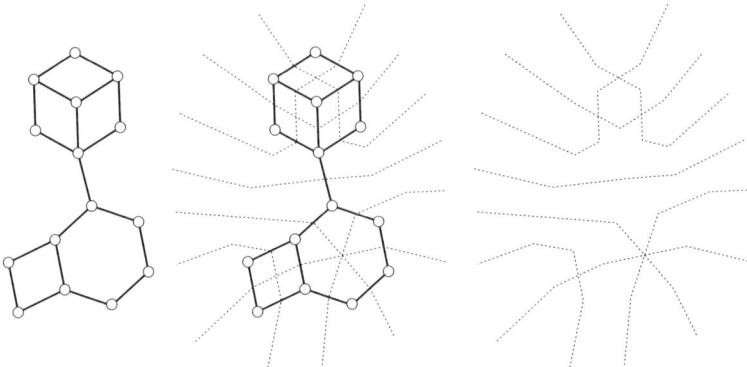

Figure 11.9. Converting a zonotopal tiling to a weak pseudoline arrangement (from Eppstein (2005a)).

tiling, between two line segments that pass through pairs of tiling edges with different slopes; in particular, this implies that no curve can cross itself, so each curve is a pseudoline. However, we must still show that any two pseudolines cross at most once, so that they form a weak pseudoline arrangement.

We may rotate the plane if necessary so no tiling edge is vertical. Each pseudoline then partitions the plane into a left side (the side containing the leftmost vertices of the tiles it passes through) and a right side (the side containing the rightmost vertices of the tiles it passes through). The boundary of each tile can be partitioned at the leftmost and rightmost vertices into a top boundary and bottom boundary. We view each pseudoline as being oriented from the point it enters the tile on the bottom boundary towards the point it exits the tile on the top boundary; this orientation is consistent with our definition of the left and right sides of the pseudoline.

With this choice, suppose that a curve c_1, passing through a family of tiling edges with slope s_1, crosses curve c_2, passing through a family of tiling edges with slope s_2. Necessarily, $s_1 \neq s_2$; assume $s_1 > s_2$. The points of c_1 that occur earlier than the crossing point in the orientation of c_1 belong to the right side of c_2, and the points of c_2 that occur later than the crossing point belong to the left side of c_2. Similarly, the points of c_2 earlier than the crossing point belong to the left side of c_1, and the points of c_2 later than the crossing point belong to the right side of c_1. There is no way for these curves to switch sides in this way other than at a crossing, so c_1 and c_2 can cross only once. This argument applies to any pair of pseudolines in the arrangement, so each pair crosses at most once and the result is a weak pseudoline arrangement. Our tiling is the region graph of this arrangement. □

To prove a converse result, we need some technical material on cyclic orders.

11.4 Pseudoline Arrangements

11.4.4 Definition. A *cyclic order* on a set S is an assignment of a value ± 1 to every ordered triple of distinct elements from S, such that:

1. The value assigned to (a, b, c) equals that assigned to (b, c, a) and (c, a, b), and is the negation of the value assigned to (a, c, b), (b, a, c), and (c, b, a).
2. For any four distinct elements a, b, c, and d, exactly one of (a, d, b), (b, d, c), and (c, d, a) is assigned the same value as (a, b, c).

If (a, b, c) is assigned $+1$, we say that this triple is in *clockwise order*, while if it is assigned -1, we say that it is *counterclockwise*.

11.4.5 Example. Let S be any subset of points on a unit circle in the Cartesian plane, and assign (a, b, c) the value $+1$ if the arc of the circle extending clockwise from a to c contains b. Then this is assignment is a cyclic order on S.

For finite sets, cyclic orders may equivalently be defined in terms of directed graphs:

11.4.6 Lemma. *Let S be a finite set with a cyclic order, with $|S| \geq 3$, and form a graph G having the elements of S as vertices, with a directed edge from a to b if there is no c in the order for which (a, c, b) is assigned $+1$. Then G is a directed cycle, and any triple (a, b, c) is assigned $+1$ in the cyclic order if and only if there is a simple directed path from a to c in G that passes through b.*

Proof. If $|S| = 3$, with (a, b, c) assigned $+1$, the axioms defining a cyclic order show that the graph contains edges ab, bc, and ca, and has the property described. Otherwise, let x be any member of S; by induction, there exists a graph G' describing the cyclic order on the smaller set $S' = S \setminus \{x\}$. It follows from the axioms defining a cyclic order that there is a unique edge yz of G' for which (y, x, z) is assigned $+1$; replacing edge yz by a path of two edges yx and xz forms the required graph G for S. □

11.4.7 Lemma. *Let S be any countable set with a cyclic order. Then there exists a function f that maps elements of S to distinct points on the unit circle, such that the values assigned to (a, b, c) and $(f(a), f(b), f(c))$ are equal for every a, b, and c.*

Proof. As S is countable, we can assume that its elements form a sequence x_i, $i = 0, 1, 2, \ldots$. We will fix the values of $f(x_i)$ in the ordering given by this sequence. Thus, suppose that we have already fixed all $f(x_j)$ for $j < i$, and now wish to choose a value for $f(x_i)$. The points $f(x_j)$, $j < i$, form a polygon, the edges of which form the graph G for these points described by Lemma 11.4.6. As in the proof of that lemma, there is a unique edge $f(a)f(b)$ of this polygon such that (a, x_i, b) is assigned $+1$ in the cyclic order. We set $f(x_i)$ to equal the midpoint of the circular arc from $f(a)$ to $f(b)$; from

Lemma 11.4.6, this can be seen to preserve the cyclic ordering of all triples involving points (x_i, x_j, x_k) with $j, k < i$. □

11.4.8 Lemma. *Let $\mathcal{A}$ be a weak pseudoline arrangement, and partition each pseudoline of $\mathcal{A}$ into two rays. Then there exists a cyclic ordering on the rays, such that two pseudolines ℓ and ℓ' cross if and only if the rays r_1, r_2, r_1', and r_2' into which they are partitioned occur in the cyclic order (r_1, r_1', r_2, r_2') or (r_1, r_2', r_2, r_1').*

Proof. To determine the cyclic ordering of a triple of rays (a, b, c), we consider the subarrangement $\mathcal{A}'$ formed by the (at most three) pseudolines containing these rays. $\mathcal{A}'$ forms a drawing with at most six rays; each ray of a, b, and c coincides except for a finite portion of its length with one of the rays of the drawing. So we may determine the cyclic ordering of (a, b, c) from the order in which the rays of the drawing extend to infinity. □

11.4.9 Lemma. *The region graph of any weak pseudoline arrangement can be drawn as a zonotopal tiling.*

Proof. We mimic the construction of a zonotopal tiling from a line arrangement, by choosing arbitrarily a vertex r_0 of the region graph of the arrangement to place at the origin, forming a unit vector v_ℓ for each line ℓ of the arrangement, and placing each vertex v of the region graph at the point formed by summing the vectors corresponding to edges of the shortest path from v to v_0 in the region graph. We then draw a line segment between each pair of vertex placements corresponding to adjacent regions in the region graph.

To finish the description of the construction, we need to describe how to choose the vector v_ℓ. By Lemma 11.4.8, the infinite ends of the pseudolines extend to infinity in a consistent circular ordering, so by Lemma 11.4.7 we may assign to each pseudoline ℓ two points $p_{\ell,0}$ and $p_{\ell,1}$ on the unit circle, consistent with that ordering, such that no two pseudolines are assigned the same point, and such that two pseudolines ℓ and ℓ' cross if and only if their assigned points appear in the cyclic order $p_{\ell,0}, p_{\ell',0}, p_{\ell,1}, p_{\ell',1}$ or its reverse. We number these points in such a way that, if ℓ is oriented from the end corresponding to $p_{\ell,0}$ to the end corresponding to $p_{\ell,1}$, the region corresponding to v_0 is on the right side of this oriented curve. We then let v_ℓ be a unit vector formed by scaling the vector $p_{\ell,1} - p_{\ell,0}$.

As we now show, this placement of vertices corresponds to a planar straight-line drawing of the region graph in which each face has as its boundary a non-self-intersecting polygon and in which the bounded faces are centrally symmetric polygons; thus, it is a zonotopal tiling. First, we describe the bounded faces. If c is a point where two or more pseudolines of the arrangement meet, the drawing constructed as above contains a polygon having vertices corresponding to the regions meeting at c. The exterior angle at each successive vertex of this polygon equals the angle between successive lines

$p_{\ell,0} p_{\ell,1}$, from which it follows that this polygon is convex and centrally symmetric; it will form a face in our drawing. A similar argument shows that the angles of adjacent pairs of edges around each vertex in our drawing of the region graph add to 2π. Thus, our drawing is locally planar. However, we still must verify that the unbounded face of the drawing is drawn in a non-self-crossing way, as that does not follow from the above assertions.

The unbounded face of the drawing corresponds to the cyclic sequence of unbounded regions in the arrangement, which are separated by unbounded arcs of pseudolines. This sequence of unbounded regions corresponds to a polygonal chain of vertices in our drawing; we must verify that this polygonal chain f does not cross itself. To verify this, we prove a stronger property: that, for each point q on an edge e of f, there is a ray r_q with q as its starting point that extends to infinity without crossing any other edge or vertex of the drawing. To find r_q, suppose that e belongs to an edge of the region graph corresponding to a pair of regions separated by an unbounded arc of pseudoline ℓ, corresponding to the point $p_{\ell,0}$. We let r be a ray starting at $p_{\ell,0}$, tangent to the unit circle in a clockwise direction, and form r_q by translating r so that it starts at q instead of at $p_{\ell,0}$. Then no other edge e' of f, corresponding to a pair of regions separated by pseudoline ℓ', can block r_q, because to do so it would have to have a slope inconsistent with our choice of $p_{\ell',0}$ and $p_{\ell'_1}$: the most extreme slopes possible for e' are slightly to the clockwise of r_q, for an edge e' counterclockwise of e around f for which $p_{\ell',0}$ and $p_{\ell'_1}$ are both clockwise of and near $p_{\ell,0}$, or slightly to the counterclockwise of r_q, for an edge e' clockwise of e around f for which $p_{\ell',0}$ and $p_{\ell'_1}$ are both counterclockwise of and near $p_{\ell,0}$.

As our placement of the vertices of the region graph can be represented as a collection of faces, each faces non-self-crossing and glued together in a non-self-crossing way, with each bounded face being a centrally symmetric polygon, the result is a zonotopal tiling. □

11.4.10 Theorem. *A partial cube G can be drawn as a zonotopal tiling if and only if it the region graph of a weak pseudoline arrangement.*

Proof. The implication in one direction is Lemma 11.4.3, and the other direction is Lemma 11.4.9. □

11.4.11 Theorem. *Let G be a graph drawn as a zonotopal tiling. Then G is a partial cube.*

Proof. By Theorem 11.4.10, we may instead assume that G is the region graph of a weak pseudoline arrangement. We may arbitrarily choose a positive side for each pseudoline, and label the regions of the arrangement by the set of positive sides of pseudolines containing them. We must show that this labeling is distance-preserving.

In one direction, each edge in G separates two regions whose labels differ by a single element. By repeated application of the triangle inequality for the symmetric difference distance, any path in G must be at least as large as the distance between the regions corresponding to its endpoints.

In the other direction, suppose we are given any two regions c_1 and c_2 in a weak pseudoline arrangement $\mathcal{A}$ bounded by a simply connected subset of the plane. We must show that there exists a path in G connecting them with length at most equal to the distance between the regions. We prove this by induction on the distance; as a base case, if $c_1 = c_2$ there clearly exists a zero-length path in G having the corresponding vertices as endpoints.

If $\mathcal{A}$ contains a pseudoline that does not separate c_1 from c_2, we consider the subarrangement formed by removing from $\mathcal{A}$ the region of the plane opposite c_1 and c_2 from that pseudoline, and removing the pseudoline itself. The resulting subset of the plane is again simply connected by the Jordan Curve Theorem, and (as each pseudoline crosses the removed one at most once) we again have a weak pseudoline arrangement in this simply connected subset. The region graph of this arrangement is a subgraph of G, and the labels of c_1 and c_2 in this smaller arrangement are at the same distance as they were in G, as the only removed pseudolines are those that do not separate c_1 from c_2. By a sequence of such removals we reach a situation where c_1 and c_2 belong to a weak arrangement $\mathcal{A}'$ of pseudolines in which every pseudoline separates c_1 from c_2, and in which the distance between the labels of c_1 and c_2 is equal to their distance in the original arrangement. Let c_3 be any region neighboring c_1 in $\mathcal{A}'$. Then c_3 has a label that is closer to c_2's by one unit, compared to c_1's label, so we may find the desired path by concatenating a path from c_3 to c_2 (which must exist by the induction hypothesis) to an edge from c_1 to c_3. □

11.4.12 Example. Some finite subgraphs of the hexagonal lattice play important roles in chemistry (see, for instance Gutman and Cyvin, 1989). Let C be a simple cycle of the hexagonal lattice. A *benzenoid graph* B_C is formed by the vertices and edges of this lattice lying on and in the interior of C. Figure 11.10 illustrates this definition (interior edges of the graph B_C are shown by dashed lines).

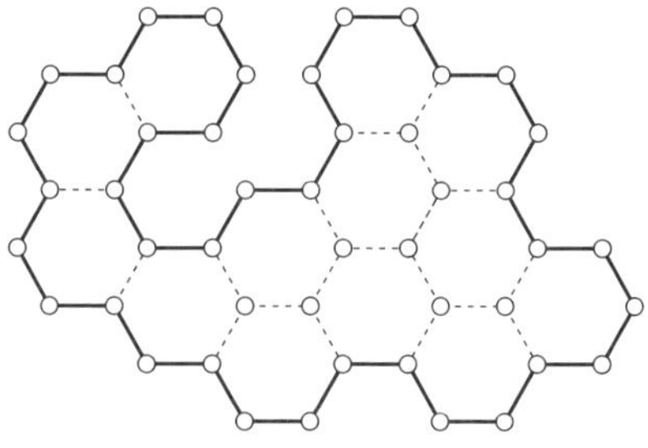

Figure 11.10. A benzenoid graph B_C.

It is clear from this example that a benzenoid graph is not, generally speaking, an isometric subgraph of the hexagonal lattice. Nevertheless, a benzenoid graph is a partial cube itself. This was proven using a geometric argument by Imrich and Klavžar (2000), but it also follows immediately from Theorem 11.4.11 since every benzenoid graph is a zonotopal tiling.

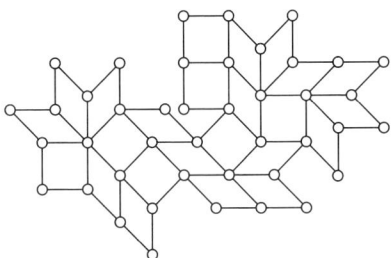

Figure 11.11. A squaregraph.

11.4.13 Definition. A *squaregraph* (Chepoi et al., 2002) is a graph drawn in the plane with quadrilateral faces in such a way that each interior vertex of the drawing has four or more incident edges. Squaregraphs are a special case of the median graphs discussed in Chapter 4 (Bandelt and Chepoi, 2005). An example of a squaregraph is shown in Figure 11.11.

11.4.14 Lemma. *Every squaregraph is the region graph of a weak pseudoline arrangement.*

Proof. As in the proof of Lemma 11.4.3, we form a system of curves by connecting opposite edge midpoints within each quadrilateral of the squaregraph (keeping the connecting curve interior to the quadrilateral even in cases where the quadrilateral is drawn as concave) and by extending each boundary edge midpoint to infinity without further crossings. We must show that this curve system is a weak pseudoline arrangement.

Suppose, for a contradiction, that this curve system is not a weak pseudoline arrangement. Then we have one of three types of feature that prevent it from being a weak pseudoline arrangement: some curve is closed, some curve crosses itself, or some two curves cross twice. In all three cases, the quadrilaterals that the curve or curves forming the feature pass through surround a bounded region of the plane, and contribute three edges to each vertex on the boundary of the region, with the exception of at most two vertices which have four edges from these quadrilaterals.

The set of quadrilaterals inside this region must themselves form a squaregraph. But it can be shown by Euler's formula (cf. Problem 11.11) that any squaregraph has at least four boundary vertices with degree two. When adjoined to the quadrilaterals surrounding the region in question, these vertices would form interior vertices with degree three in the overall graph, contradicting the assumption that it is a squaregraph. This contradiction shows that the curve system we formed must be a weak pseudoline arrangement. □

11.4.15 Corollary. *Every squaregraph can be drawn as a zonotopal tiling.*

11.4.16 Corollary. *Every squaregraph is a partial cube.*

11.4.17 Example. A polyomino (Golomb, 1994) is a simply connected union of squares in the plane, meeting edge to edge; Figure 11.12 shows the distinct polyominos formed from up to five squares. The vertices and edges of most of these examples form isometric subsets of $\mathbb{Z}^2$, but this is not true e.g. for the U-shaped pentomino in the bottom left of the figure. Nevertheless, every pentomino forms a squaregraph and is therefore a partial cube.

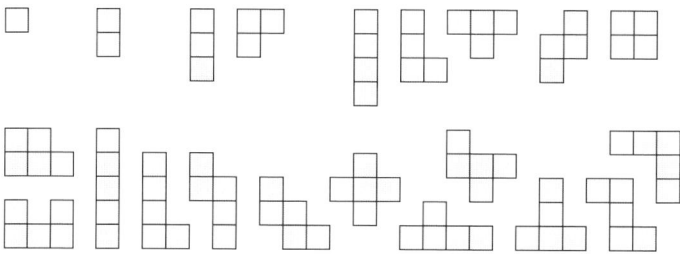

Figure 11.12. The polyominos of up to five squares.

11.5 Finding Zonotopal Tilings

In the previous section, we characterized the media that can be drawn as zonotopal tilings: they are the media formed from weak pseudoline arrangements. We now describe algorithms for efficiently constructing drawings of this type.

Note that not every medium, and not even every medium with a planar state transition graph, can be drawn as a zonotopal tiling; see for instance Figure 11.13 for media that have planar mediatic graphs but that cannot be drawn as a zonotopal tiling.

We now describe how to efficiently construct a zonotopal tiling corresponding to a given medium, if one exists.

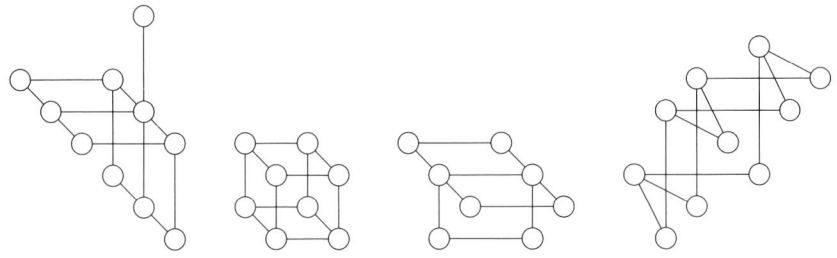

Figure 11.13. Media with planar mediatic graphs that cannot be drawn as a zonotopal tiling.

11.5.1 Definition. A graph is *k-vertex-connected* if there is no set of $k-1$ vertices the removal of which disconnects the remaining graph. In particular a 2-vertex-connected graph is called *biconnected* and a 3-vertex-connected graph is called *triconnected*.

11.5.2 Definition. An *SPQR tree* (Di Battista and Tamassia, 1989; Mutzel, 2003) is a tree structure that represents the triconnected components of a graph, and is a standard tool in graph drawing and planar embedding algorithms. Each node v in the SPQR tree of G has associated with it a multigraph G_v consisting of some subset of vertices of G, edges of G, and *virtual edges* representing contracted parts of the remaining graph that can be separated from the edges of G_v by a *split pair* of vertices (the endpoints of the virtual edge). Each non-virtual edge of G occurs exactly once as an edge in some G_v. If two SPQR tree nodes are connected by an edge in the tree, each has a virtual edge connecting two vertices shared by both nodes.

The SPQR tree for a graph contains four types of nodes:

- An S node v is associated with a graph G_v that is a simple cycle (or *polygon*) of virtual edges.
- A P node v is associated with a graph G_v that consists of two vertices connected by three or more parallel virtual edges.
- A Q node v is associated with a graph G_v formed by two vertices connected by a single non-virtual edge.
- An R node v is associated with a graph G_v that is triconnected and in which all edges are virtual.

An SPQR tree representing a given graph may be formed recursively by the following procedure:

- If the graph contains a non-virtual edge, replace it by a virtual edge, form the SPQR tree of the replaced graph, and form a Q node for the replaced edge, connecting it to a node in the SPQR tree that contains the corresponding virtual edge.

- If the graph consists of a cycle of virtual edges, form an S node for it, and if the graph consists of two vertices connected by three or more parallel edges, form a P node for it.
- If the graph is not of either of the two types described above, but contains a split pair, separate the graph into two smaller graphs (both containing copies of a virtual edge connecting the two vertices of the split pair), form SPQR trees for these two graphs, and connect the two trees by an edge connecting nodes containing the two virtial edges corresponding to the split pair.
- If none of the previous cases applies, the remaining graph must be triconnected. Form an R node for it.

Mutzel (2003) describes a more efficient linear-time procedure for constructing SPQR trees.

11.5.3 Definition. If D is a planar graph drawing, we define the *type* of the embedding to consist of the clockwise ordering of the edges around each face of the embedding. We say that two embedding types are *equivalent* either when they are consist of the same set of cyclic orderings, or when each cyclic ordering in one embedding type is the reverse of the corresponding cyclic ordering in the other embedding type.

The interior and exterior edges of an embedding type may be determined from the embedding type: an interior edge will be listed as part of the cyclic orderings of two faces of the embedding, and those orderings will pass in opposite directions through the edge, whereas the two cyclic orderings of faces that contain an exterior edge will both traverse the edge in the same direction. For the same reason we may determine the bounded and unbounded faces of an embedding directly from its embedding type.

Observe that the region graph of a weak pseudoline arrangement can be given an embedding type by listing the edges in clockwise order around each region of the arrangement, and that this type is equivalent to the embedding type of the drawing we have constructed of the region graph of the arrangement as a zonotopal tiling. As a shorthand, we call this embedding type the embedding type of the arrangement.

We call a graph *ambiguously embeddable* if it has multiple inequivalent embedding types that are the embedding types of weak pseudoline arrangements.

11.5.4 Lemma. *Let $G = (V, E)$ be a biconnected planar partial cube. Then G cannot be ambiguously embeddable. If G is the region graph of an arrangement, its embedding type can be found in time $O(|V|)$.*

Proof. We form the SPQR tree of G, and root it arbitrarily; let (s_v, t_v) denote the split pair connecting a non-root node v to its parent, and let H_v denote the graph (with one virtual edge) represented by the SPQR subtree rooted at v. We work bottom up in the rooted tree, showing by induction on tree size that the following properties hold for each node of the tree:

1. Each graph H_v has at most one equivalence class of embedding types that can include the embedding type of a weak pseudoline arrangement.
2. If v is a non-root node, and H_v has the embedding type of a weak pseudoline arrangement, then edge $\{s_v, t_v\}$ is an exterior edge of that embedding type.
3. If v is a non-root node, and H_v has the embedding type of a weak pseudoline arrangement, form the path p_v by removing virtual edge $s_v t_v$ from the unbounded face of H_v. Then p_v must be part of the unbounded face of any embedding type of G that is an embedding type of a weak pseudoline arrangement.

SPQR trees are divided into four different cases (represented by the initials S, P, Q, and R) and our proof follows the same case analysis, below, in each case showing that the properties at each node follow from the same properties at the descendant nodes.

Trivial case: If G_v consists of a single graph edge and a single virtual edge (a Q-node), then clearly there can only be one planar embedding (up to reflection) of G_v.

Parallel case: If G_v consists of three or more edges connecting (s_v, t_v) (a P-node). In this case, G can only have the embedding type of a weak pseudoline arrangement if G_v has three edges, one of which corresponds to an edge of G. For, if all three edges of G_v are virtual edges, each of which corresponds to a nontrivial subgraph of G, the two faces of any drawing of G that separate the middle subgraph from the other two subgraphs could not both be drawn strictly convexly.

Since our initial graph G is not a multigraph, it must have exactly one edge connecting s_v to t_v, and the other two edges of G_v must be virtual. Further, the argument above implies that the edge connecting s_v to t_v must be in the interior of any drawing of G.

If v is the root of the SPQR tree, it has two children u and w. In this case, the embedding of $H_v = G$ must be formed by placing H_u and H_w on opposite sides of the edge $\{s_v, t_v\}$, with the paths p_u and p_v facing outwards. If these conditions are satisfied, we have found as desired a unique embedding for G. If v is not the root, it has one child u, as the other virtual edge connects v to its parent; H_v differs from H_u by the addition of a single non-virtual edge $\{s_v, t_v\}$. As before, the non-virtual edge must be sandwiched between the two other parts of G, so the only possible embedding of H_v is to place the non-virtual edge $\{s_v, t_v\}$ parallel to the virtual edge of H_u connecting the same two vertices, on the internal side of this virtual edge.

Series case: If G_v is a polygon (an S-node) then the embedding of H_v is formed by orienting the graph H_u for each child node u so that p_u is placed on the outside of the polygon. If v is the root of the SPQR tree, this completes the proof that the embedding of G is unique. Otherwise, $\{s_v, t_v\}$ must be on the unbounded face of H_v (since it is an edge of the polygon. Path p_v must lie along the unbounded face of any embedding of G, because (if any child of v is nontrivial) it contains vertices already required to lie along the unbounded face from lower levels of the SPQR tree. If all children of v are trivial, then H_v is just the same polygon as G_v, and separates two faces in any planar embedding of G; in this case p_v must lie along the unbounded face because it is not possible for two strictly convex internal faces to share a path of three or more vertices.

Rigid case: In the final case, G_v is a triconnected graph, which must be planar (else G has no planar drawing). Such graphs have a unique planar embedding up to the choice of unbounded face. By the same reasoning as in the parallel case, each virtual edge must lie on the unbounded face, or else it would be sandwiched between two internal faces leading to a nonconvexity in the drawing. We divide into subcases according to the number of virtual edges.

- If there are no virtual edges, then G is itself 3-connected. If G is to have the embedding type of a pseudoline arrangement with L lines, then the unbounded face of G must have $2L$ edges. No other face of G could have so many edges, because G has at least four faces and any internal face with k edges would correspond to crossings between $(k/2)(k/2-1)/2$ pairs of pseudolines, leaving no crossings for the other faces. So in this case the unbounded face can be uniquely identified as the face with the largest number of edges. (In fact we can prove that no 3-connected graph can be drawn as a zonotopal tiling, but the proof is more complex, and we reuse this subcase's reasoning in the next subcase.)
- If there is a single virtual edge, it must be on the unbounded face, so this narrows down the choice of the unbounded face to two possibilities, the two faces of G_v containing the virtual edge. By the same reasoning as for the subcase with no virtual edges, these two faces must have differing numbers of edges and the unbounded face must be the one with the larger number of edges. If v is not the root, it has no children and $H_v = G_v$; otherwise, the embedding of H_v is formed from that of G_v by orienting the child of v with p_v along the unbounded face of G_v.
- If there are two or more virtual edges, there can only be one face in G_v containing these edges, which must be the unbounded face of G_v. The embedding of H_v is fixed by placing the graph H_u for each child u of v so that the unbounded face of H_u (minus the virtual edge connecting it to G_v) lies along the unbounded face of G_v.

□

11.5.5 Theorem. *Given a partial cube graph $G = (V, E)$, we can determine whether G is the region graph of a weak pseudoline arrangement, and if so construct a drawing of G as a zonotopal tiling, in time $O(|V|)$.*

Proof. If G is biconnected, we choose a planar embedding of G by the method of Lemma 11.5.4. Otherwise, each articulation point of G must be on the unbounded face of any embedding. We find biconnected components of G, embed each component by Lemma 11.5.4, and verify that these embeddings place the articulation points on the unbounded faces of each component. We then connect the embeddings together into a single embedding having as its unbounded face the edges that are exterior in each biconnected component; the choice of this embedding may not be unique but does not affect the correctness of our algorithm.

Certain graphs, such as odd cycles, may be embedded by the method of Lemma 11.5.4 even though they do not have the embedding type of a weak pseudoline arrangement. Thus, once we have an embedding of G, we must still verify that it has the embedding type of a weak pseudoline arrangement, and construct a face-symmetric planar drawing. We first make sure all faces of G are even, and construct an arrangement of curves $\mathcal{A}$ dual to G. We verify that $\mathcal{A}$ has no closed curves, a necessary condition for it to be a weak pseudoline arrangement. We defer the verification that each curve in $\mathcal{A}$ crosses each other such curve at most once until later in the algorithm.

We then produce vertex placements for a drawing of the region graph of $\mathcal{A}$ as described in the proof of Lemma 11.4.9. We test for each edge of the resulting drawing that the endpoints of that edge are placed at unit distance apart with the expected slope, and we test that each internal face of G is drawn as a correctly oriented strictly convex and centrally symmetric polygon. If so, our algorithm returns the drawing it has constructed; otherwise, it returns a failure result indicating that no drawing as a zonotopal tiling exists.

If $\mathcal{A}$ were not a weak pseudoline arrangement, either due to a curve self-crossing or to two curves crossing each other with the wrong orientation, the face of G dual to that crossing point would be drawn as a nonconvex polygon or as an incorrectly oriented convex polygon. However, our algorithm tests for both of these possibilities. Thus if our input passes all these tests we have determined that it is the region graph of a weak pseudoline arrangement and found a drawing as a zonotopal tiling. □

We note that the seemingly very closely related algorithmic problem of determining whether a planar graph is the region graph of a two-dimensional line arrangement is NP-hard (Shor, 1991) and therefore unlikely to have an efficient solution.

11.5.6 Example. Examples of drawings produced by the face-symmetric planar drawing algorithm of Theorem 11.5.5 are shown in Figure 11.14.

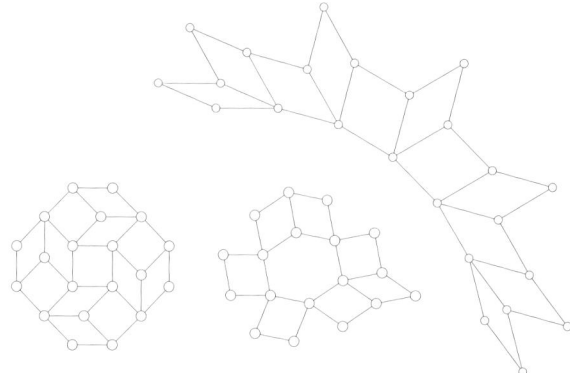

Figure 11.14. Face-symmetric planar drawings of three irregular media.

11.6 Learning Spaces

Recall (Section 1.6 and Theorem 4.2.1) that a *learning space* is a medium formed by a well-graded set family that is closed under pairwise unions and includes the empty set. As established by Theorem 4.2.2, any learning space can be represented by a closed, rooted medium. We describe algorithms for drawing such media. For some omitted proofs and details, see Eppstein (2006).

11.6.1 Definition. Learning spaces are equipped with a natural orientation coming from the set membership relation. With this orientation, they can be viewed as directed acyclic graphs having a unique source (vertex without incoming edges) at the empty set, and a unique sink (vertex without outgoing edges) at the set corresponding to the domain of the learning space. In the graph theory literature, such graphs are known as *st-oriented* or as having a *bipolar orientation* (Ebert, 1983; Even and Tarjan, 1976).

It is natural, then, to consider learning spaces for which the associated bipolar orientation is compatible with a planar drawing of the graph, in that the source and sink can both be placed on the unbounded face of a planar drawing: bipolar oriented graphs with such an embedding are known as *st-planar*, so the learning spaces with this property can be called *st-planar learning spaces*.

As we show, *st*-planar learning spaces can be characterized by drawings of a very specific type: Every *st*-planar learning space has a drawing in which all bounded faces are convex quadrilaterals with the bottom side horizontal and the left side vertical. We call such a drawing an *upright-quad drawing*, and we describe linear time algorithms for finding an upright-quad drawing of any *st*-planar learning space. Conversely, every upright-quad drawing comes from an *st*-planar learning space in this way.

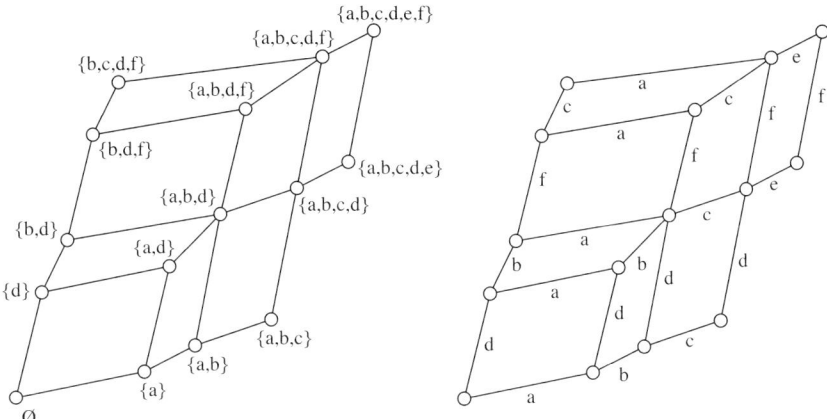

Figure 11.15. An *st*-planar learning space, from Eppstein (2006).

11.6.2 Example. An example of an *st*-planar learning space is shown in Figure 11.15; in the left view, the vertices of a drawing of the graph are labeled by the corresponding sets in a well-graded ∪-closed set family, while on the right view, each edge is labeled by the name of a token in the corresponding medium. Positive tokens are drawn as edges oriented from lower left to upper right.

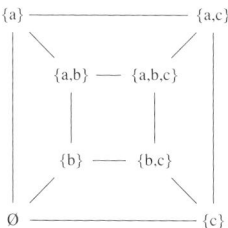

Figure 11.16. A learning space that is planar but not *st*-planar, from Eppstein (2006). There is no drawing of this learning space in which the source and sink are both exterior vertices; in the drawing shown, the sink is interior.

11.6.3 Example. Figure 11.16 shows an example of a learning space that is planar but not *st*-planar: the power set on three elements, forming a graph with the structure of a cube. In any planar embedding of this graph, the vertex representing the empty set and the vertex representing the whole domain are on different faces, so they cannot both be on the unbounded face of any drawing.

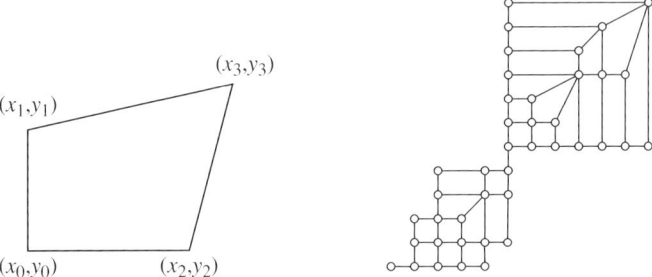

Figure 11.17. Left: An upright quadrilateral. Right: An upright-quad drawing. From Eppstein (2006).

11.6.4 Definition. For any subset S of points in the plane, we say that (x, y) is *minimal* if no other point (x', y') in the set has $x' \le x$ and $y' \le y$, and *maximal* if no other point (x', y') in the set has $x' \ge x$ and $y' \ge y$. Any finite set S has at least one minimal point and at least one maximal points, but these points may not be unique.

Let D be a drawing of a graph G, v be a vertex in G, and S_v be the set of points assigned by D to v and to the neighbors of v in G. We say that v is a *minimal vertex* of the drawing if the point assigned to v is minimal in S_v, and we say that v is a *maximal vertex* of the drawing if the point assigned to v is maximal in S_v. Note that any drawing has at least one minimal vertex and at least one maximal vertex: if S is the set of all points assigned by D to vertices in G, then the minimal and maximal points in S correspond to minimal and maximal vertices in G (respectively). However, a drawing may have additional minimal or maximal vertices that do not correspond to minimal or maximal points of the larger set S (see Problem 11.13).

Let Q be a convex quadrilateral having the four points in a set S as its corners. We say that Q is an *upright quadrilateral* if S has a unique minimal point and a unique maximal point, such that the edges of Q incident to the minimal point are horizontal and vertical. That is, it is the convex hull of four points $\{(x_i, y_i) \mid 0 \le i < 4\}$ where $x_0 = x_1 < x_2 \le x_3$ and $y_0 = y_2 < y_1 \le y_3$ (Figure 11.17, left). We define the *bottom edge* of an upright quadrilateral to be the horizontal edge incident to the minimal point, the *left edge* to be the vertical edge incident to the minimal point, and the *top edge* and *right edge* to be the edges opposite the bottom and left edges respectively.

We define an *upright-quad drawing* of a graph G to be planar straight-line drawing D of G with the following properties:

[U1] There is exactly one minimal vertex of D and exactly one maximal vertex of D.
[U2] Every bounded face of the drawing is an upright quadrilateral, the sides of which are edges of the drawing.

In an upright-quad drawing, all edges connect a pair of points (x, y) and (x', y') with $x' \le x$ and $y' \le y$; if we orient each such edge from (x', y') to (x, y) then the resulting graph is directed acyclic with a unique source and sink.

As we now show, upright-quad drawings may be produced from a certain special type of weak pseudoline arrangement.

11.6.5 Definition. A *right angle* is a geometric figure in $\mathbb{R}^2$ formed by two perpendicular rays that share a common endpoint; this endpoint is usually called the *vertex* of the angle, but to avoid confusion with graph-theoretic terminology we instead call it the *corner* of the right angle. If A is a right angle, $\mathbb{R}^2 \setminus A$ has two components, the *interior* of the right angle (the convex set between the two rays) and the *exterior* of the right angle (the complementary concave set).

Two right angles are *parallel* if the intersection of their interiors is also the interior of a right angle. Note that parallelism is an equivalence relation. Two parallel right angles are *coincident* if one of the rays forming one of the angles is a subset of one of the rays forming the other angle. A finite collection of co-planar parallel right angles, no two of which are coincident is called an *arrangement of parallel right angles*.

By convention, we consider for the remainder of this section arrangements of parallel right angles in which each right angle is parallel to the negative quadrant of the Cartesian coordinate system. That is, the two rays bounding each angle run parallel to the x- and y-axes of the coordinate system, and the corner of each angle is its unique maximal point.

We may view any right angle as a (non-smooth) curve in the plane. It is straightforward to observe that an arrangement of parallel right angles is thus a special type of arrangement of pseudolines. As with any pseudoline arrangement, we may define a *region graph* that has one vertex per region of the arrangement, with two vertices adjacent whenever the corresponding regions are separated by a single curve of the arrangement. For our arrangements of parallel right angles, it is convenient to draw the region graph by assigning each region's vertex to the unique maximal point of its region, except for the upper right region which has no maximal point. We draw the vertex for the upper region at any point with x and y coordinates strictly larger than those of any curve in our arrangement.

11.6.6 Theorem. *The assignment of vertices to points in the plane described above produces an upright-quad drawing for the region graph of any arrangement of parallel right angles.*

Proof. The drawing's edges consist of finite segments of the rays forming the angles, together with diagonal segments within each region that connect corners of angles to the region's maximal point; none of these edges cross, so

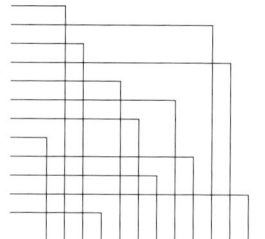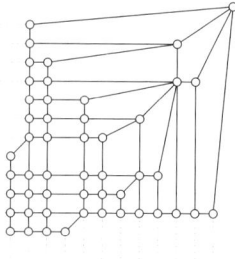

Figure 11.18. Left: an arrangement of parallel right angles. Right: the region graph of the arrangement, drawn with each vertex (except the top right one) at the maximal point of its region. From Eppstein (2006).

we have a planar straight-line drawing. Each finite region of the arrangement is bounded above and to the right by the rays of a right angle, either one of the angles of the given arrangement or the angle that forms the boundary of the intersection of two of the interiors of given angles. Each finite region is also bounded below and to the left by a staircase formed by a union of interiors of angles of the arrangement; the drawing's edges subdivide this region into upright quadrilaterals by diagonals connecting the concave corners of the region to its maximal point. A similar sequence of upright quadrilaterals connects the staircase formed by the union of interiors of all given angles to the point representing the upper right region, which is the unique maximal vertex of the drawing. The unique minimal vertex of the drawing represents the region formed by the intersection of the interiors of all given angles. Thus, all requirements of an upright-quad drawing are met. □

11.6.7 Example. Figure 11.18 shows an arrangement of parallel right angles (left), and the region graph of the arrangement drawn as described above (right). As can be seen from the figure, the drawing of the region graph is an upright-quad drawing.

11.6.8 Theorem. *The region graph of any arrangement $\mathcal{A}$ of parallel right angles is the mediatic graph of an st-planar learning space.*

Proof. As $\mathcal{A}$ is a weak pseudoline arrangement, its region graph is the mediatic graph of a medium by Theorem 11.4.11. When the tokens represented by each angle of $\mathcal{A}$ are oriented from the lower left of the arrangement to the upper right, the resulting medium is rooted at the most extreme region to the lower left of the drawing. Thus, by Theorem 4.2.2, if we show that this orientation of the medium is also closed, it will follow that it is a learning space.

If a region of $\mathcal{A}$, associated with set S, has a single angle c of $\mathcal{A}$ as its upper right boundary, there can be no two distinct states $S\tau$ and $S\rho$ for positive tokens τ and ρ in the associated medium, and closure is satisfied

11.6 Learning Spaces   257

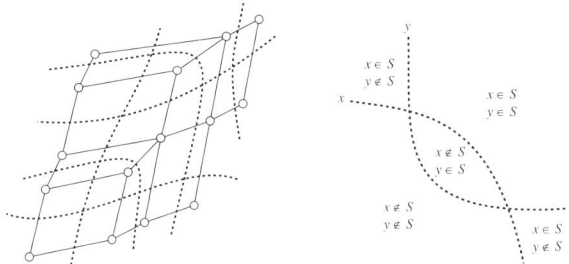

Figure 11.19. Left: A curve arrangement $\mathcal{A}(G)$ corresponding to an *st*-planar learning space. Right: Two crossings between the same two curves lead to a contradiction, so $\mathcal{A}$ must be a weak pseudoline arrangement (Lemma 11.6.10). From Eppstein (2006).

vacuously. On the other hand, if a region r of $\mathcal{A}$, associated with set S, has a ray bounding the angle associated with token τ as its upper boundary and a ray bounding the angle associated with token ρ as its right boundary, then the only successors of S in the associated medium can be $S\tau$ and $S\rho$. In this case, the region diagonally opposite r across the vertex where x and y meet is associated with the state $S\tau\rho = S\rho\tau$, so again the definition of closure is met. Thus, the region graph represents a learning space, and by Theorem 11.6.6 it is *st*-planar. □

11.6.9 Definition. Define a *zone* of a token τ in an *st*-planar learning space $\mathcal{M}$ to be the set of bounded faces of the *st*-planar drawing of $\mathcal{M}$ that contain edges labeled by τ. It can be shown that a zone consists of a sequence of faces, in which consecutive faces in the sequence each share an edge labeled by x. Further, the two edges labeled by τ in any face of the zone partition the boundary of the face into two paths with equal numbers of edges.

We may form a curve arrangement $\mathcal{A}(\mathcal{M})$ from an *st*-planar drawing of $\mathcal{M}$ as follows. For each token τ, we draw a curve C_τ that crosses each edge labeled by τ and passes only through the bounded faces of the zone of τ. Each bounded face of the drawing is crossed by two such curves, which we draw in such a way that they cross exactly once within the face. We extend these curves to infinity past the exterior edges of the drawing without any crossings in the unbounded face (Figure 11.19, left).

11.6.10 Lemma. *If $\mathcal{M}$ is an st-planar learning space, then $\mathcal{A}(\mathcal{M})$ is a weak pseudoline arrangement.*

Proof. The curves in $\mathcal{A}(\mathcal{M})$ are topologically equivalent to lines and meet only at crossings. Suppose for a contradiction that two curves labeled τ and ρ in $\mathcal{A}(\mathcal{M})$ cross more than once. Then (Figure 11.19, right) there would exist two or more disjoint regions between the curves, and there could be no concise path between states in the upper left region to states in the lower right region, contradicting the defining axioms of a medium. □

11.6.11 Definition. We are now ready to define the vertex coordinates for our upright-quad drawing algorithm. Let D be a given st-planar drawing of a learning space. Consider the sequence of labels $x_0, x_1, \ldots x_{\ell-1}$ occurring on the right path from the bottom to the top vertex of the external face of the drawing. For any vertex v of our given st-planar learning space, let $X(v) = \min\{i \mid x_i \notin v\}$. If v is the topmost vertex of the drawing, define instead $X(v) = \ell$. Similarly, consider the sequence of labels $y_0, y_1, \ldots y_{\ell-1}$ occurring on the left path from the bottom to the top vertex of the external face of the drawing. For any vertex v of our given st-planar learning space, let $Y(v) = \min\{i \mid y_i \notin v\}$. If v is the topmost vertex of the drawing, define instead $Y(v) = \ell$. We call this system of coordinates the *upright-quad coordinates* for D.

11.6.12 Lemma. *Let D be an st-planar drawing of a learning space. Then the upright-quad coordinates for D form a drawing of the region graph of $\mathcal{A}(\mathcal{M})$ in which the lower left boundary of each zone follows a right angle, parallel to the negative quadrant of the Cartesian coordinate system, and no two of these parallel right angles share a boundary line.*

Specifically, if token τ appears in position x in the right path of the drawing, and position y in the left path, then the lower left boundary of its zone lies on the right angle with apex (x, y).

11.6.13 Corollary. *The coordinates described above form a drawing of a region graph of translated parallel right angles, and therefore form an upright-quad drawing of $\mathcal{M}$.*

11.6.14 Example. A drawing produced by the technique described above, for the st-planar learning space from Example 11.6.2 is shown on the left of Figure 11.20. As in standard st-planar dominance drawing algorithms (Di Battista et al., 1999) we may compact the drawing by merging coordinate values $X(v) = i$ and $X(v) = i+1$ whenever the merge would preserve the dominance ordering of the vertices; a compacted version of the same drawing is shown on the right of Figure 11.20.

We summarize the results of this section:

11.6.15 Theorem. *Every st-planar learning space $\mathcal{M} = (\mathcal{S}, \mathcal{T})$ has an upright-quad drawing in an integer grid of area $O(|\mathcal{T}|^2)$ that may be found in time $O(|\mathcal{S}|)$.*

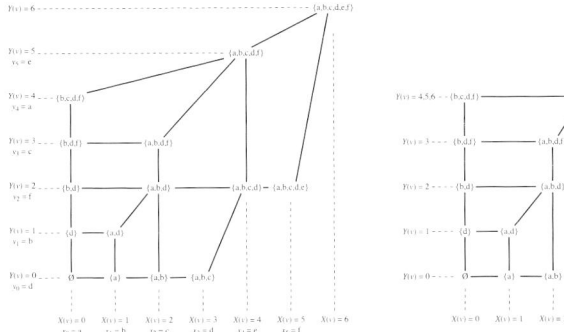

Figure 11.20. Left: coordinates for conversion of *st*-planar learning space to upright-quad drawing. Right: the same drawing with compacted coordinates. From Eppstein (2006).

Proof. We construct an *st*-planar embedding for $\mathcal{M}$, form from it the parallel right angle arrangement $\mathcal{A}(\mathcal{M})$, and use the indices of the curves to assign coordinates to vertices as above. The coordinates of the vertices in a face f may be assigned by referring only to the labels of edges in f, in time $O(|f|)$; therefore, all coordinates of G may be assigned in linear time. The area bound follows easily. □

11.6.16 Corollary. *Any st-planar learning space* $\mathcal{M} = (\mathcal{S}, \mathcal{T})$ *has* $|\mathcal{S}| \leq 1 + (|\mathcal{T}|/2 + 1)|\mathcal{T}|/4$.

Proof. Our drawing technique assigns each vertex (other than the topmost one) a pair of coordinates associated with a pair of elements $\{x_i, y_j\} \subset \mathcal{T}^+$ (possibly with $x_i = y_j$), and each pair of elements can supply the coordinates for only one vertex. Thus, there can only be one more vertex than subsets of one or two members of $\mathcal{T}^+$. □

11.6.17 Example. Figure 11.21 shows an *st*-planar learning space formed from the well-graded family of sets that are the unions of a prefix and a suffix of a totally ordered set. Learning spaces formed in this way have a number of states exactly matching the bound of Corollary 11.6.16, showing that this bound is tight.

We have shown that every *st*-planar learning space can be represented as the region graph of an arrangement of parallel right angles, every arrangement of parallel right angles represents an *st*-planar learning space in this way, and every region graph of an arrangement of parallel right angles or *st*-planar learning space has an upright-quad drawing. Eppstein (2006) additionally shows that every upright-quad drawing represents an *st*-planar learning space, although not all such drawings have vertices placed at the maximal points of regions in an arrangement of parallel right angles; we omit the details.

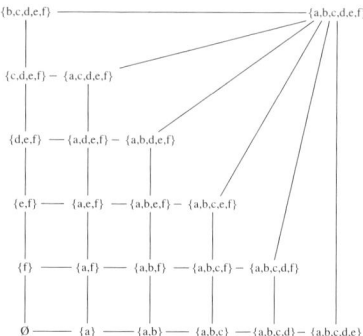

Figure 11.21. The family of sets formed by the union of a prefix and a suffix of some ordered universe forms an st-planar learning space with the maximum possible number of states. From Eppstein (2006).

Problems

11.1 Let M be embedded isometrically onto the cubic lattice $\mathbb{Z}^3$, in such a way that the isometric projection of that lattice (see Definition 11.1.1) projects different vertices to distinct positions in the plane. Show that all faces of the resulting drawing must be regular hexagons or 60°-120° rhombi.

11.2 Find a planar drawing of a medium, using only regular hexagons and 60°-120° rhombi as faces, that is not the isometric projection of any three-dimensional lattice embedding of the medium.

11.3 Find a drawing of a biconnected graph G as a zonotopal tiling, such that the edges of G have only three distinct slopes, that cannot be drawn using only regular hexagons or 60°-120° rhombi.

11.4 Find a medium such that the projection method of Section 11.2 produces a drawing that has no edge crossings but does not form a zonotopal tiling.

11.5 Describe the coordinates of the drawing produced by the projection method of Section 11.2 on a graph in the form of a cycle of $2n$ vertices. What is the area of the minimal axis-aligned rectangle containing the drawing, as a function of n?

11.6

(i) Show that a cycle of $2n$ vertices can be drawn as a symmetric convex polygon, with integer vertex coordinates, within a rectangle of area $O(n^3)$.

(ii) Show that a cycle of $2n$ vertices can be drawn as a symmetric convex polygon, with integer vertex coordinates, within a square of area $O(n^3)$.

(iii) Show that, if a cycle of $2n$ vertices is drawn as a symmetric convex polygon, with integer vertex coordinates, then the minimal axis-aligned rectangle containing the cycle has area $\Omega(n^3)$.

11.7 Figure 11.22 below shows cycles of 4, 6, 8, and 10 vertices, drawn as non-convex polygons with integer coordinates using only unit-length edges. The drawings are symmetric under $180°$ rotation, and therefore can be formed as the projection of a lattice embedding of the cycles.

(i) Prove that, for any n, a $2n$-vertex cycle can be embedded in this way into a square of area $O(n)$.

(ii) In a drawing of a $2n$-vertex cycle be drawn as a rotationally-symmetric non-self-crossing polygon with integer vertex coordinates, and unit-length edges, let there be v_1 vertices with convex angles, v_2 vertices with concave angles, and v_3 vertices with $180°$ angles. Prove that $v_1 = v_2 + 4$ and that $v_3 \le v + 1 + v + 2$.

(iii) The drawings for $2n = 4$ and $2n = 10$ have no $180°$ angles. More generally, for which n can a $2n$-vertex cycle be drawn as a rotationally-symmetric non-self-crossing polygon with integer vertex coordinates, unit-length edges, and no $180°$ angles?

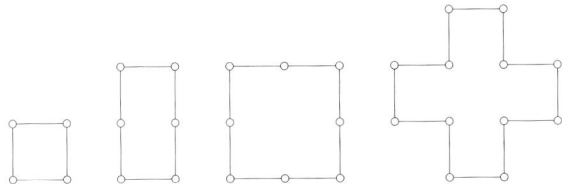

Figure 11.22. Figure for Problem 11.7

11.8 Find a zonotopal tiling that is a drawing of the region graph of the line arrangement in Figure 1.3.

11.9 Find an example of a tiling of a non-simply-connected region of the plane, by strictly convex centrally symmetric tiles, that is not the drawing of the mediatic graph of a partial cube.

11.10 Find an example of a tiling of a convex region of the plane, by convex but not necessarily strictly convex centrally symmetric tiles (that is, some internal angles can equal pi), that is not the drawing of the mediatic graph of a partial cube.

11.11 Find a planar straight line drawing of a finite graph G, such that all bounded faces of the drawing are convex quadrilaterals and G is not a partial cube.

11.12 Use Euler's formula $E - V + F = 2$ for planar graphs to prove that every squaregraph has at least four degree-two exterior vertices.

11.13 Find a graph G, a drawing D of G, and a vertex v, such that v is a minimal vertex of the drawing, but such that the point assigned to v is not a minimal point among the set of all points assigned to vertices by D.

11.14 Find a drawing of the st-planar learning space shown in Figure 11.6.2 in which all bounded faces are drawn as parallelograms.

11.15 Prove that any st-planar learning space has a drawing in which all bounded faces are drawn as parallelograms.

11.16 Find a drawing of a planar graph in which all bounded faces are drawn as parallelograms, but for which the graph does not represent an st-planar learning space.

11.17 Find a planar straight line drawing of a graph that satisfes property [U2] of an upright-quad drawing, but that does not satisfy [U1], and that is not the drawing of an st-planar learning space.

12
Random Walks on Media

In some finite or countable situations, a medium $(\mathcal{S}, \mathcal{T})$ can serve as the algebraic component of a random walk model whose states coincide with those of the medium[1]. The basic idea is that there exists a probability distribution

$$\vartheta : \mathcal{T} \to [0,1] \;:\; \tau \mapsto \vartheta_\tau$$

on the set of all tokens, such that, if a token-event occurs, then the probability of a transition from some state S to some state $T \neq S$ is equal to ϑ_τ if $S\tau = T$ and vanishes otherwise. We also suppose that there is an 'initial' distribution

$$\eta : \mathcal{S} \to [0,1] \;:\; S \mapsto \eta_S$$

on the set of all states. This distribution governs the choice of first state. The pair (η, ϑ) defines a stochastic process which is properly called a 'random walk' because the transitions can occur only between adjacent states. A realization of the process is a sequence of states, which is induced by first sampling the distribution η on $\mathcal{S}$, and then sequentially sampling the distribution ϑ on $\mathcal{T}$.

12.0.1 Example. An illustration is given in Figure 12.1 which displays the beginning of a sequence of states in the permutohedron medium Π_2 of Example 3.5.9. This type of 'cyclical random walk' was already considered by Feller (1968). However, we use it here as an exemplary case of a general type of random walk on the states of a finite medium. We shall refer to this example on several occasions in our presentation of the theory.

We recall that the states are the six strict linear orders on the set $\{1,2,3\}$. We use on the graph the abbreviation ijk to mean the strict linear order $\{ij, ik, jk\}$. Such abbreviations will be used in the sequel whenever convenient. We denote by $\mathcal{L}_3$ the set of those six strict linear orders, and by $\mathcal{T}_3$ the

---

[1] Some of the material in this chapter follows Chapter 22 in Falmagne (2003) and Falmagne et al. (2007). The introduction to Markov chains is included for completeness. These concepts and results can be found in any standard text such as Chung (1967), Barucha-Reid (1974), or Parzen (1994).

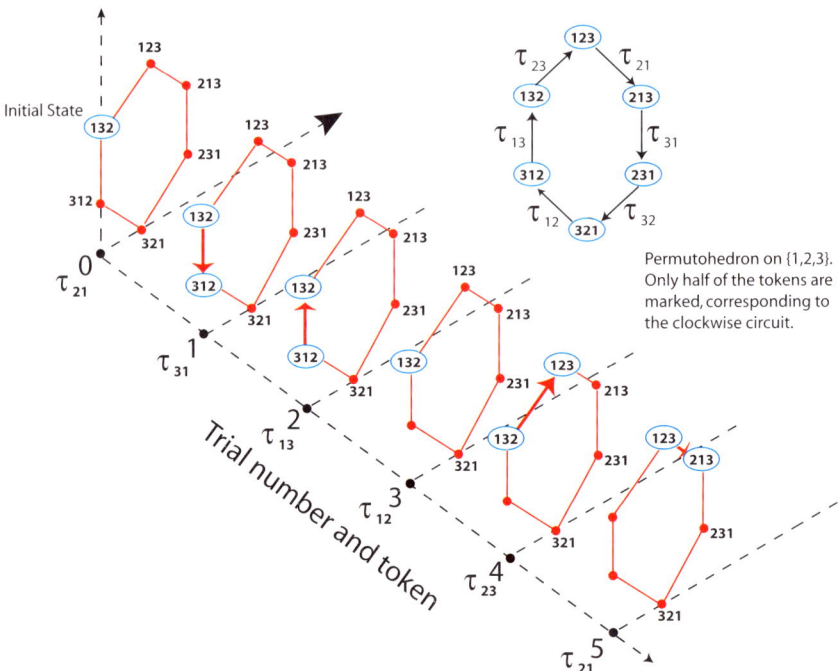

Figure 12.1. A realization of a random walk on the permutohedron medium Π_2 of Example 3.5.9. The initial state on trial 0 is 132, presumably selected from sampling the initial probability distribution η on the set of states. The sequence of ensuing states, namely 321, 132, 321, 312, 123, is obtained by sampling the probability distribution ϑ on the set of tokens. These states are produced by the sequence of tokens $\tau_{21}, \tau_{31}, \tau_{13}, \tau_{12}, \tau_{23}, \tau_{21}$. Note that the tokens τ_{21} and τ_{12} are ineffective for the state 132 on two occasions: no transition takes place on trials 0 and 3.

corresponding set of tokens τ_{ij}, $i, j \in \{1, 2, 3\}$, $i \neq j$, which are defined by the equation

$$\tau_{ij} : \mathcal{L}_3 \to \mathcal{L}_3 \ : \ L \mapsto L\tau_{ij} = \begin{cases} (L \setminus \{ji\}) \cup \{ij\} & \text{if } j \text{ covers } i \\ L & \text{otherwise.} \end{cases}$$

In this example the states observed on the successive trials are:

trials	0	1	2	3	4	5	...
states	132	312	132	132	123	213	...

The symmetry inherent in the axioms defining a medium, together with some hypotheses on the probability distribution ϑ, ensures that the asymptotic distribution on the set of states exists. It is in fact easy to describe: the asymptotic probability of some state $S \in \mathcal{S}$ is proportional to the product of the probabilities ϑ_τ of all the tokens τ occuring in any of the concise messages producing S, that is, all the tokens belonging to $\widehat{S}$ (cf. Definition 2.4.1; see Theorem 12.4.1).

Such a mechanism may provide a suitable model for the evolution of a physical, biological, or social system subjected to a constant probabilistic barrage of 'token-events', each of which may induce a 'mutation' of its state taking the form of an addition or a removal of some constituent feature. We develop the theory for two variants of this idea in this chapter. In one of them, we assume that the token-events occur on discrete times numbered $0, 1, \ldots, n, \ldots$, giving rise to what is called a 'discrete parameter' random walk (cf. Definition 12.2.1). This case is illustrated in Figure 12.1. The other variant deals with a phenomenon in which a token-event can occur at any instant of a potentially infinite interval of time. This leads to describe a 'continuous parameter' random walk. We only consider random walks on the states of finite media. Random walks for infinite media—for example, random walks on a tessellation of the plane (cf. Coxeter, 1973, Section IV)—can be defined by standard methods. We do not cover such material here because, as far as we know, no relevant application exists. We begin by briefly reviewing some standard facts on regular Markov chains[2].

12.1 On Regular Markov Chains

Loosely speaking a Markov chain is a precise description, in the language of probability theory, of a system with a very limited memory of its past. Consider a system (a random walk on a medium, say) whose states evolve over time. Suppose that the states have been observed on j discrete moments or 'trials' numbered $n_1 < n_2 < \ldots < n_j$. (These trials are not necessarily consecutive and we know nothing about the states of the system on any other trials.) Suppose also that the state of the system on any later trial n_{j+k} only depends upon that last observed state on trial n_j. A system satisfying such a property for any choice of trials numbers $n_1, n_2, \ldots, n_j, n_{j+k}$ is said to be 'Markovian[3].' In words: a system is Markovian if the prediction of its future state only depends upon the last recorded state. Let us state this idea formally.

12.1.1 Definition. Let $\mathbf{X}_0, \mathbf{X}_1, \ldots, \mathbf{X}_n, \ldots$ be a sequence of random variables taking their values on the same finite set $\mathcal{S}$. We write $\mathbb{P}$ for the probability measure of the common sample space of these random variables. Suppose that, for any positive integer n and any event $\mathcal{E}$ depending only on the values of the random variables $\mathbf{X}_0, \mathbf{X}_1, \ldots \mathbf{X}_{n-1}$, we have, for any $S, V \in \mathcal{S}$,

$$\mathbb{P}(\mathbf{X}_{n+1} = V \mid \mathbf{X}_n = S, \mathcal{E}) = \mathbb{P}(\mathbf{X}_{n+1} = V \mid \mathbf{X}_n = S). \tag{12.1}$$

---

[2] For Markov chain terminology in this chapter, we follow mostly Kemeny and Snell (1960).
[3] From the Russian mathematician Andrei Andreïevich Markov (1856-1922).

Then, the sequence $(\mathbf{X}_n)$ is a *(finite) Markov chain.* Note that in this case we also have

$$\mathbb{P}(\mathbf{X}_{n+1} = V \mid \mathbf{X}_n = S, \ldots, \mathbf{X}_1, \mathbf{X}_0) = \mathbb{P}(\mathbf{X}_{n+1} = V \mid \mathbf{X}_n = S). \tag{12.2}$$

The formulations of the Markovian property in terms of (12.1) and (12.2) are actually equivalent (see Problem 12.1.) The set $\mathcal{S}$ is the *state space* of the chain, and the elements of $\mathcal{S}$ are called the *states*. Note in passing the fortunate coincidence that the term 'state' is used consistently for a medium and the corresponding Markov chain on that medium. We shall use the abbreviations

$$p_n(S) = \mathbb{P}(\mathbf{X}_n = S) \tag{12.3}$$

$$t_{n,m}(S, V) = \mathbb{P}(\mathbf{X}_m = V \mid \mathbf{X}_n = S). \tag{12.4}$$

Thus, Eq. (12.1) can be rewritten as

$$\mathbb{P}(\mathbf{X}_{n+1} = V \mid \mathbf{X}_n = S, \mathcal{E}) = t_{n,n+1}(S, V).$$

The quantities $t_{n,m}(S, V)$ are referred to as the *transition probabilities* of the chain $(\mathbf{X}_n)$. Defining the chain requires thus specifying the *initial probabilities* $p_0(S)$ of the states and the *one-step transition probabilities* $t_{n,n+1}(S, V)$. When these transition probabilities do not depend upon the trial number, that is, when

$$t_{n,n+1}(S, V) = t_{m,m+1}(S, V) \tag{12.5}$$

for all $S, V \in \mathcal{S}$ and $n, m \in \mathbb{N}_0$, then the Markov chain is called *homogeneous* or, equivalently, is said to have *stationary transition probabilities*. In this case, there are numbers $\mathfrak{t}(S, V)$ defined for all states S and V in $\mathcal{S}$ such that, for all trial numbers $n \in \mathbb{N}_0$,

$$\mathfrak{t}(S, V) = \mathbb{P}(\mathbf{X}_{n+1} = V \mid \mathbf{X}_n = S). \tag{12.6}$$

We have thus

$$\mathfrak{t}(S, V) \geq 0, \quad \text{for all } S, T \in \mathcal{S}; \tag{12.7}$$

$$\sum_{V \in \mathcal{S}} \mathfrak{t}(S, V) = 1, \quad \text{for all } S \in \mathcal{S}. \tag{12.8}$$

This case is of special interest here because we suppose that the transitions between states result from the tokens-events, which occur with constant probabilities.

The following result is fundamental, and holds whether or not the chain is homogeneous.

12.1.2 Theorem. *For any trial numbers $0 \leq n < m < r$ and any states S and V in $\mathcal{S}$,*

$$t_{n,r}(S, V) = \sum_{W \in \mathcal{S}} t_{n,m}(S, W) t_{m,r}(W, V). \tag{12.9}$$

Equation (12.9) is known as the *Chapman-Kolmogorov Equation.*

Proof. We have by definition

$$
\begin{aligned}
t_{n,r}(S, V) &= \mathbb{P}(\mathbf{X}_r = V \mid \mathbf{X}_n = S) \\
&= \sum_{W \in \mathcal{S}} \mathbb{P}(\mathbf{X}_r = V \mid \mathbf{X}_m = W, \mathbf{X}_n = S) \mathbb{P}(\mathbf{X}_m = W \mid \mathbf{X}_n = S) \\
&= \sum_{W \in \mathcal{S}} \mathbb{P}(\mathbf{X}_r = V \mid \mathbf{X}_m = W) \mathbb{P}(\mathbf{X}_m = W \mid \mathbf{X}_n = S) \\
&= \sum_{W \in \mathcal{S}} t_{n,m}(S, W) t_{m,r}(W, V).
\end{aligned}
$$

□

12.1.3 Corollary. *For any trial numbers $0 \leq n < m < r$, and any states i, j, and k in $\mathcal{S}$,*

$$t_{n,r}(i, j) \geq t_{n,m}(i, k) t_{m,r}(k, j). \tag{12.10}$$

12.1.4 Matrix Notation. Results concerning finite Markov chains are conveniently written in the notation of vectors and matrices. Let the elements of a state space $\mathcal{S}$ be numbered $1, 2, \ldots, q$. Thus, the variables i, k, and j in Eqs. (12.10) run in the set $\{1, 2, \ldots, q\}$. Let

$$T_{n,m} = \begin{pmatrix} t_{n,m}(1,1) & t_{n,m}(1,2) & \ldots & t_{n,m}(1,q) \\ t_{n,m}(2,1) & t_{n,m}(2,2) & \ldots & t_{n,m}(2,q) \\ \ldots & \ldots & \ldots & \ldots \\ \ldots & \ldots & \ldots & \ldots \\ t_{n,m}(q,1) & t_{n,m}(q,2) & \ldots & t_{n,m}(q,q) \end{pmatrix} \tag{12.11}$$

denote a square matrix, the cell (i, j) of which contains the probability of a transition from state i on trial n to state j on trial m. Note that the probabilities of each row of the matrix $T_{n,m}$ sum to one. Such a matrix is often called a *transition matrix*, or a *stochastic matrix*. Extending the notation introduced in (12.3), we also write

$$\mathbf{p}_n = \bigl(p_n(1), \ldots, p_n(t)\bigr) \tag{12.12}$$

for the vector of the state probabilities on trial n. In the notation of (12.11) and (12.12), the Chapman-Kolmogorov Equation (cf. Theorem 12.1.2) has the compact expression

$$T_{n,r} = T_{n,m} T_{m,r}, \tag{12.13}$$

for any positive integers $n < m < r$. A simple expression is available for the vector of the state probabilities on trial $n + 1$ as a function of the state probabilities on trial n and of the transition probabilities. We clearly have, for any trial number n and any state i in $\mathcal{S}$,

$$\mathbf{p}_{n+1}(i) = \sum_{k \in \mathcal{S}} p_n(k) t_{n,n+1}(k, i).$$

Using the standard notation for the product of a vector by a matrix, this becomes
$$\mathbf{p}_{n+1} = \mathbf{p}_n T_{n,n+1}.$$
By induction, it follows that
$$\mathbf{p}_{n+m} = \mathbf{p}_m T_{m,m+1} T_{m+1,m+2} \cdots T_{n-1,n}.$$
Thus, in particular,
$$\mathbf{p}_n = \mathbf{p}_0 T_{0,1} T_{1,2} \cdots T_{n-1,n}. \tag{12.14}$$

Of particular interest in the sequel is the case in which the Markov chain is homogeneous. The transition probabilities are then constant (see Eq. (12.6)) and all the one-step transition matrices $T_{n,n+1}$ are identical to some basic transition matrix
$$\mathfrak{T} = (\mathfrak{t}(i,j)), \qquad (i,j \in \{1,\ldots,q\}).$$
In the theorem below, $\mathfrak{T}^k$ stands for the kth power of the matrix $\mathfrak{T}$.

12.1.5 Theorem. *In an homogeneous Markov chain $(\mathbf{X}_n)$ with a transition matrix $\mathfrak{T}$, the transition probabilities $t_{n,m}(i,j)$ only depend upon the difference $m - n$. Specifically, we have*
$$T_{n,m} = \mathfrak{T}^{m-n}. \tag{12.15}$$
Moreover, for any trial number $n \geq 1$,
$$\mathbf{p}_n = \mathbf{p}_0 \mathfrak{T}^n \tag{12.16}$$
where $\mathbf{p}_0$ is the initial vector of probabilities of the states, and $\mathbf{p}_n$ is the vector of probabilities of those states on trial n.

Proof. Equation (12.15) follows from Equation (12.13), which yields
$$T_{n,m} = T_{n,n+1} T_{n+1,n+2} \cdots T_{m-1,m} = \mathfrak{T}^{m-n},$$
since each of the factors in the product is equal to $\mathfrak{T}$.

Equation (12.16) is then obtained from the fact that, by definition of $T_{0,n}$ and by Eq. (12.15), we get $\mathbf{p}_n = \mathbf{p}_0 T_{0,n} = \mathbf{p}_0 \mathfrak{T}^n$. □

12.1.6 Definition. In view of this result, it makes sense in the homogeneous case to write $\mathfrak{t}_k(i,j)$ for the conditional probability of a transition from state i on trial n to state j on trial $n + k$, for any nonnegative integers n; that is,
$$\mathfrak{t}_k(i,j) = \mathbb{P}(\mathbf{X}_{n+k} = j \,|\, \mathbf{X}_n = i). \tag{12.17}$$

Note that $\mathfrak{t}_1(i,j) = \mathfrak{t}(i,j)$, and that $\mathfrak{t}_k(i,j)$ is the entry in the cell (i,j) of the matrix $\mathfrak{T}^k$ and satisfies $\mathfrak{t}_k(i,j) = t_{n,n+k}(i,j)$ for every integer $n \geq 0$ (cf. Eq. (12.4)).

12.1 On Regular Markov Chains

We now focus on the special kind of homogeneous Markov chains that provided the title of this section.

12.1.7 Definition. A finite, homogeneous Markov chain $(\mathbf{X}_n)$ is *regular* if there is a positive integer N such that whenever $n \geq N$, then $\mathfrak{t}_n(i,j) > 0$ for all states i, j in $\mathcal{S}$. Thus, all entries of the matrix $\mathfrak{T}^n$ are positive for any $n \geq N$. By convention, we say that a Markov chain is regular to mean that it is also finite and homogeneous.

12.1.8 Remark. Actually, the definition of a regular Markov chain given here is often given in the form of as a theorem stating the equivalence of regularity and a collection of other properties, homogeneity being one of them. Our discussion is only intended to establish economically some key results regarding random walks on finite media. A different way of obtaining regularity is sketched in Problem 12.16.

The next theorem concerns the convergence of the transition probabilities for large n and is the fundamental result for regular Markov chains.

12.1.9 Theorem. *For any state j in a regular Markov chain, there is a number $\alpha_j > 0$ such that $\sum_{j \in \mathcal{S}} \alpha_j = 1$, and for any pair (i,j) of states, we have*

$$\lim_{n \to \infty} \mathfrak{t}_n(i,j) = \alpha_j.$$

In other terms, the powers $\mathfrak{T}^n$ of the transition matrix $\mathfrak{T}$ are converging to a stochastic matrix $\mathfrak{A}$, each of the $q = |\mathcal{S}|$ rows of which is the same vector $\boldsymbol{\alpha} = (\alpha_1, \ldots, \alpha_q)$.

We postpone the proof of this result to establish a preliminary lemma, in which we show that for $1 \leq j \leq q$, the jth column vector of the matrix $\mathfrak{T}^n$ tends to a vector containing all identical terms α_j; that is, as $n \to \infty$, we have

$$\begin{pmatrix} \mathfrak{t}_n(1,j) \\ \mathfrak{t}_n(2,j) \\ \ldots \\ \ldots \\ \mathfrak{t}_n(q,j) \end{pmatrix} \to \begin{pmatrix} \alpha_j \\ \alpha_j \\ \ldots \\ \ldots \\ \alpha_j \end{pmatrix}.$$

12.1.10 Lemma. *Let $\mathfrak{T} = (\mathfrak{t}(i,j))$, $i,j \in \{1,\ldots q\}$, be a stochastic matrix with positive entries. For any index j, let $M_n(j)$ be the maximum value in the jth column vector of $\mathfrak{T}^n$, and let $m_n(j)$ be the minimum value in that column vector. Then, the sequence $m_n(j)$ is nondecreasing and the sequence $M_n(j)$ is nonincreasing. Accordingly, the sequence*

$$r_n(j) = M_n(j) - m_n(j), \qquad (n \in \mathbb{N})$$

specifying the range of values in that column vector is nonincreasing and in fact tends to zero.

Proof. Let (i,j) be any pair of states, and let δ be the smallest entry in the matrix $\mathfrak{T}$. Witout loss of generality, suppose that $m_n(j) = \mathfrak{t}_n(1,j)$. Using the Chapman-Kolmogorov Equation (Theorem 12.1.2), we obtain successively

$$\begin{aligned}\mathfrak{t}_{n+1}(i,j) &= \sum_{k\in\mathcal{S}} \mathfrak{t}_1(i,k)\mathfrak{t}_n(k,j) \\ &\leq \mathfrak{t}_1(i,1)m_n(j) + (1-\mathfrak{t}_1(i,1))M_n(j) \\ &\leq \delta m_n(j) + (1-\delta)M_n(j) \\ &= M_n(j) - \delta(M_n(j) - m_n(j)) \ .\end{aligned}$$

We have in particular

$$M_{n+1}(j) \leq M_n(j) - \delta(M_n(j) - m_n(j)) \tag{12.18}$$

which shows that the sequence $M_n(j)$ is nonincreasing. A similar argument applied to the sequence $m_n(j)$ yields

$$m_{n+1} \geq m_n(j) + \delta(M_n(j) - m_n(j)) \ ,$$

or, equivalently,

$$-m_{n+1} \leq -m_n(j) - \delta(M_n(j) - m_n(j)) \ . \tag{12.19}$$

Adding (12.18) and (12.19) and grouping terms, we obtain

$$\begin{aligned}r_{n+1}(j) &= M_{n+1}(j) - m_{n+1}(j) \\ &\leq M_n(j) - m_n(j) - 2\delta(M_n(j) - m_n(j)) \\ &= (M_n(j) - m_n(j))(1 - 2\delta) \\ &= r_n(j)(1 - 2\delta) \ .\end{aligned}$$

Hence for $n > 1$,

$$r_n(j) \leq r_1(j)(1-2\delta)^{n-1} \ . \tag{12.20}$$

Since $0 < 2\delta < 1$, we have $r_n(j) \to 0$, as asserted. $\square$

Proof of Theorem 12.1.9. If the entries of the transition matrix of the Markov chain are all positive, the result follows from Lemma 12.1.10. Indeed, the two sequences $m_n(j)$ and $M_n(j)$ converge, since they are monotonic and bounded. Because the range $r_n(j)$ tends to zero, these two sequences must converge to the same limit, which we denote by α_j. By definition of $m_n(j)$ and $M_n(j)$, we have $0 < m_n(j) \leq \mathfrak{t}_n(i,j) \leq M_n(j)$ for any pair of states (i,j), yielding $\lim_{n\to\infty} \mathfrak{t}_n(i,j) = \alpha_j$, with necesssarily $\sum_{j=1}^q \alpha_j = 1$.

In general, the matrix $\mathfrak{T}$ may have some zero entries. However, by Theorem 12.1.7, there is a positive integer N such that all the entries of $\mathfrak{T}^N$ are positive. Applying Lemma 12.1.10 to the matrix $\mathfrak{T}^N$, it follows that, for any state j, the sequence $r_{Nn}(j)$ tends to zero as $n \to \infty$. Because $r_{Nn}(j)$ is nonincreasing, we must conclude that it tends to zero. The rest of the argument is as in the first paragraph of this proof. $\square$

The next result completes the picture.

12.1.11 Theorem. Let $\mathbf{p}_0 = (p_0(1), \ldots, p_0(q))$ be the initial vector of state probabilities, and suppose that all the conditions of Theorem 12.1.9 are satisfied, with in particular $\mathfrak{T}^n \to \mathfrak{A}$. Then:

(i) $\mathbf{p}_0 \mathfrak{T}^n$ tends to the vector $\boldsymbol{\alpha}$, for any arbitrarily chosen vector $\mathbf{p}_0$;
(ii) $\boldsymbol{\alpha}$ is the unique vector satisfying the equation $\boldsymbol{\alpha}\mathfrak{T} = \boldsymbol{\alpha}$;
(iii) $\mathfrak{A}\mathfrak{T} = \mathfrak{T}\mathfrak{A} = \mathfrak{A}$.

Proof. We have $\mathbf{p}_0 \mathfrak{A} = \boldsymbol{\alpha}$ because all the rows of the matrix $\mathfrak{A}$ are identical to $\boldsymbol{\alpha}$ and $\sum_{j \in S} p_0(j) = 1$. This leads to

$$\lim_{n \to \infty} \mathbf{p}_0 \mathfrak{T}^n = \mathbf{p}_0 \mathfrak{A} = \boldsymbol{\alpha},$$

which proves (i). Let $\mathbf{q}$ be any vector satisfying $\mathbf{q}\mathfrak{T} = \mathbf{q}$. By induction, we get $\mathbf{q}\mathfrak{T}^n = \mathbf{q}$. Condition (ii) follows from Condition (i) since

$$\lim_{n \to \infty} \mathbf{q}\mathfrak{T}^n = \boldsymbol{\alpha} = \mathbf{q}.$$

Finally, (iii) results from the string of equalities
$\mathfrak{A} = \lim_{n \to \infty} \mathfrak{T}^n = \lim_{n \to \infty} \mathfrak{T}^n \mathfrak{T} = \mathfrak{A}\mathfrak{T} = \lim_{n \to \infty} \mathfrak{T}\mathfrak{T}^n = \mathfrak{T}\mathfrak{A}$.
□

Thus, for a regular Markov chain $(\mathbf{X}_n)$, regardless of the initial vector $\mathbf{p}_0$, the long range—or *asymptotic*—probabilities of the states will be those specified by the vector $\boldsymbol{\alpha} = (\alpha_1, \ldots, \alpha_q)$. The vector $\boldsymbol{\alpha}$ is often referred to as the *stationary distribution* of the Markov chain $(\mathbf{X}_n)$. The following sufficient condition for a vector $\boldsymbol{\alpha} = (\alpha_1, \ldots, \alpha_q)$ to be a stationary distribution will be instrumental.

12.1.12 Lemma. Let $\mathfrak{T} = (\mathfrak{t}(i,j))$, $i, j \in \{1, \ldots, q\}$ be the transition matrix of a regular Markov chain, and let $\boldsymbol{\alpha} = (\alpha_1, \ldots, \alpha_q)$ be a vector defining a probability distribution. Suppose that

$$\alpha_i \mathfrak{t}(i,j) = \alpha_j \mathfrak{t}(j,i) \qquad (i, j \in \{1, \ldots, q\}).$$

Then, $\boldsymbol{\alpha}$ is the unique stationary distribution of the Markov chain.

Proof. It suffices to verify that $\sum_{i}^{q} \alpha_{i=1} \mathfrak{t}(i,j) = \alpha_j$ and use Theorem 12.1.11(ii).
□

12.2 Discrete and Continuous Stochastic Processes

There are important situations in which a Markov chain is only one of the components of a more complex stochastic processes. This potentially applies to the random walks on media. The two key components are the state of the random walk on trial n (or at time t—the two cases will be considered) and the last token-event having occurred before or on trial n (or time t). Other components may also be introduced which depend on those two. We introduce here the relevant terminology.

12.2.1 Definition. Let $(\mathbf{X}_{1,n},\ldots,\mathbf{X}_{r,n})_{n\in\mathbb{N}_0}$ be a sequence of r-tuples of jointly distributed random variables. Such a sequence of r-tuples is called a *discrete (parameter) process* because the index set is the countable set $\mathbb{N}_0$ of non–negative integers.

A partial example of a realization of such a process has been given in Figure 12.1, which displays graphically a sequence of pairs (state, token) in the course of first 6 trials. In some real life cases, it makes sense to suppose that the events occur in the real time interval $\mathbb{R}_+ = [0, \infty[$.

12.2.2 Definition. Let $(\mathbf{X}_{1,t},\ldots,\mathbf{X}_{r,t})_{t\in\mathbb{R}_+}$ be a sequence of r-tuples of jointly distributed random variables. Such a sequence is called a *continuous parameter stochastic process* in view of its index set $\mathbb{R}_+$.

Homogeneous Markov chains have been defined in 12.1.1 by the property

$$t_{n,n+1}(R,S) = t_{m,m+1}(R,S), \qquad (n,m \in \mathbb{N}_0)$$

which states that the one-step transition probabilities from state R to state S do not depend upon the trial number. Homogeneous Markov processes are conceptually similar, in the sense that the transition probabilities from some state at time t to some other state at time $t + \delta$ only depend upon the time difference δ.

12.2.3 Definition. We say that a continuous stochastic process $(\mathbf{X}_t)$ is a Markov process if

$$\mathbb{P}(\mathbf{X}_{t_{n+1}} = S_{n+1} \,|\, \mathbf{X}_{t_n} = S_n, \ldots, \mathbf{X}_{t_1} = S_1) = \mathbb{P}(\mathbf{X}_{t_{n+1}} = S_{n+1} \,|\, \mathbf{X}_{t_n} = S_n)$$
$$(t_{n+1} > t_n > \ldots > t_1).$$

In such a case, the stochastic process $(\mathbf{X}_t)$ is defined by the pairwise conditional probabilities

$$\mathbb{P}(\mathbf{X}_{t+\delta}) = S \,|\, \mathbf{X}_t = R). \qquad (12.21)$$

that the process is in state S at time $t + \delta$ given that it is in state R at time t. A Markov process $(\mathbf{X}_t)$ is *homogeneous* if the conditional probabilities in Eq. (12.21) do not depend upon t. In this case, we can define the transition function[4]

$$\mathfrak{t}_\delta(R,S) = \mathbb{P}(\mathbf{X}_{t+\delta}) = S \,|\, \mathbf{X}_t = R). \qquad (12.22)$$

There is no need for us to recall facts from the theory of continuous Markov processes because we shall derive all our results from a distinguished Markov chain embedded in[5] the continuous Markov process under consideration.

---

[4] Compare with (12.5) and (12.6).
[5] This can be read as "defined from."

12.3 Continuous Random Walks on a Medium

The results described here are based on Falmagne (1997). We consider some organism capable of evolving from one discrete state to some other one under the influence of a barrage of punctual token-events. Both the set $\mathcal{S}$ of feasible states of the organism and the set $\mathcal{T}$ of possible token-events are finite, and the pair $(\mathcal{S}, \mathcal{T})$ satisfies Axioms [Ma] and [Mb] of a medium. The state of the organism is observed continuously from time zero on and continuously after that, and so are the arrival times and the identities of the tokens. We suppose that the arrival times of the tokens are governed by a homogeneous Poisson process[6]. Whenever a Poisson event occurs, a token-event is sampled from a probability distribution ϑ on $\mathcal{T}$, which is invariant over time. The resulting continuous parameter stochastic process has thus three components:

(1) the state of the organism at any time $t \geq 0$;
(2) the number of tokens having occurred from time 0 up to and including time t;
(3) the last token to occur before or at time t.

We introduce these three concepts formally.

12.3.1 Stochastic Concepts. We assume that there exists some probability distribution

$$\eta : \mathcal{S} \to [0,1] : S \mapsto \eta_S$$

on the set of states, governing the choice of the initial state of the process, and a positive probability distribution

$$\vartheta : \mathcal{T} \to]0,1] : \tau \mapsto \vartheta_\tau$$

on the set of tokens. The evolution is described in terms of three collections of random variables corresponding to the three components of the stochastic process outlined above. For $t \in]0, \infty[$,

$\mathbf{S}_t$ denotes the state of the random walk at time t,

$\mathbf{N}_t$ specifies the number of tokens that have occurred in the half-open interval of time $]0, t]$, and

$\mathbf{T}_t$ is the last token to occur before or at time t. We set $\mathbf{T}_t = 0$ if no token has occurred, that is, if $\mathbf{N}_t = 0$.

Thus, $\mathbf{S}_t$ takes its values in the set $\mathcal{S}$ of states, $\mathbf{N}_t$ is a nonnegative integer, and $\mathbf{T}_t \in (\mathcal{T} \cup \{0\})$. The random variables $\mathbf{N}_t$ will turn out to be the counting random variables of a Poisson process governing the delivery of the token-events (see Axiom [T]).

---

[6] Some of our results would still hold under the weaker assumption of renewal process.

The following random variables will also be instrumental:

$$\mathbf{N}_{t,t+\delta} = \mathbf{N}_{t+\delta} - \mathbf{N}_t$$

denoting the number of tokens in the half open interval $]t, t+\delta]$.

Axioms

The three axioms below specify the stochastic process $(\mathbf{N}_t, \mathbf{T}_t, \mathbf{S}_t)_{t>0}$ up to the parameters $(\eta_S)_{S \in \mathcal{S}}$ and $(\vartheta_\tau)_{\tau \in \mathcal{T}}$ for the two probability distributions, and a parameter $\lambda > 0$ specifying the intensity of the Poisson process. A discrete random walk $(\mathbf{M}_n)_{n \in \mathbb{N}_0}$ describing the succession of states of $\mathcal{S}$ is then defined from $(\mathbf{N}_t, \mathbf{T}_t, \mathbf{S}_t)_{t>0}$ as an embedded process (cf. e.g. Parzen (1994)). The notation $\mathcal{E}_t$ stands for any arbitrarily chosen history of the process before time $t > 0$; $\mathcal{E}_0$ denotes the empty history.

[T] (OCCURRENCE OF THE TOKENS.) The occurrence times of the token-events are governed by a homogeneous Poisson process of intensity λ. When a Poisson event is realized, the token τ is selected with probability $\vartheta_\tau > 0$, regardless of past events. Thus, for any nonnegative integer k, any real numbers $t > 0$ and $\delta > 0$, and any history $\mathcal{E}_t$,

$$\mathbb{P}(\mathbf{N}_{t,t+\delta} = k \,|\, \mathcal{E}_t) = \frac{(\lambda \delta)^k e^{-\lambda \delta}}{k!} \quad (12.23)$$

$$\mathbb{P}(\mathbf{T}_{t+\delta} = \tau \,|\, \mathbf{N}_{t,t+\delta} = 1, \mathcal{E}_t)$$
$$= \mathbb{P}(\mathbf{T}_{t+\delta} = \tau \,|\, \mathbf{N}_{t,t+\delta} = 1) = \vartheta_\tau > 0. \quad (12.24)$$

[I] (INITIAL STATE.) Initially, the system is in state S with probability η_S. The system remains in that state at least until the occurrence of the first token-event. That is, for any $S \in \mathcal{S}$ and $t > 0$

$$\mathbb{P}(\mathbf{S}_t = S \,|\, \mathbf{N}_t = 0) = \eta_S.$$

[L] (LEARNING.) If R is the state of the system at time t, and a single token τ occurs between times t and $t + \delta$, then the state at time $t + \delta$ will be $R\tau$ regardless of past events before time t. Formally, we have

$$\mathbb{P}(\mathbf{S}_{t+\delta} = S \,|\, \mathbf{T}_{t+\delta} = \tau, \mathbf{N}_{t,t+\delta} = 1, \mathbf{S}_t = R, \mathcal{E}_t)$$
$$= \mathbb{P}(\mathbf{S}_{t+\delta} = S \,|\, \mathbf{T}_{t+\delta} = \tau, \mathbf{N}_{t,t+\delta} = 1, \mathbf{S}_t = R) = \eth(S, R\tau),$$

where $\eth$ is the Kronecker $\eth$ function on $\mathcal{S} \times \mathcal{S}$, that is

$$\eth(S, T) = \begin{cases} 1 & \text{if } S = T, \\ 0 & \text{otherwise}. \end{cases}$$

12.3.2 Definition. A 5-tuple $(\mathcal{S}, \mathcal{T}, \mathbf{N}_t, \mathbf{T}_t, \mathbf{S}_t)_{t>0}$, where $(\mathcal{S}, \mathcal{T})$ is a medium and $(\mathbf{N}_t, \mathbf{T}_t, \mathbf{S}_t)_{t>0}$ satisfies Axioms [I], [T] and [L] for some appropriate λ, η and ϑ, is called a *Poisson driven random walk* on the medium $(\mathcal{S}, \mathcal{T})$, or more briefly, a *Poisson walk* on $(\mathcal{S}, \mathcal{T})$.

12.3.3 Convention. In the rest of this chapter, we suppose implicitly that $(\mathcal{S}, \mathcal{T}, \mathbf{N}_t, \mathbf{T}_t, \mathbf{S}_t)_{t>0}$ is such a Poisson walk.

Basic Results

Notice that

$$\mathbb{P}(\mathbf{S}_{t+\delta} = S \,|\, \mathbf{S}_t = R)$$
$$= \sum_{k=0}^{\infty} \mathbb{P}(\mathbf{S}_{t+\delta} = S \,|\, \mathbf{N}_{t,t+\delta} = k, \mathbf{S}_t = R) \, \mathbb{P}(\mathbf{N}_{t,t+\delta} = k \,|\, \mathbf{S}_t = R)$$
$$= \sum_{k=0}^{\infty} \mathbb{P}(\mathbf{S}_{t+\delta} = S \,|\, \mathbf{N}_{t,t+\delta} = k, \mathbf{S}_t = R) \, \mathbb{P}(\mathbf{N}_{t,t+\delta} = k). \quad (12.25)$$

Indeed, by the Poisson process Axiom [T], the arrival of the tokens after time t are independent of any events in the process occurring before t, and so $\mathbb{P}(\mathbf{N}_{t,t+\delta} = k \,|\, \mathbf{S}_t = R) = \mathbb{P}(\mathbf{N}_{t,t+\delta} = k)$.

12.3.4 Lemma. *For any positive real numbers t, t', δ and δ', any $k \in \mathbb{N}$, and any two states S and $T \in \mathcal{S}$, we have*

$$\mathbb{P}(\mathbf{S}_{t+\delta} = S \,|\, \mathbf{N}_{t,t+\delta} = k, \mathbf{S}_t = R)$$
$$= \mathbb{P}(\mathbf{S}_{t'+\delta'} = S \,|\, \mathbf{N}_{t',t'+\delta'} = k, \mathbf{S}'_t = R). \quad (12.26)$$

Proof. We use induction on k. For $k = 1$, Eq. (12.26) follows from Axioms [T] and [L] via

$$\mathbb{P}(\mathbf{S}_{t+\delta} = S \,|\, \mathbf{N}_{t,t+\delta} = 1, \mathbf{S}_t = R)$$
$$= \sum_{\tau \in \mathcal{T}} \mathbb{P}(\mathbf{S}_{t+\delta} = S \,|\, \mathbf{T}_t = \tau, \mathbf{N}_{t,t+\delta} = 1, \mathbf{S}_t = R) \, \vartheta_\tau = \sum_{\tau \in \mathcal{T}} \mathfrak{d}(S, R\tau) \, \vartheta_\tau,$$

independent of t and δ. If (12.26) holds for a certain $k = j$ we use the abbreviation

$$\xi_j(R, S) = \mathbb{P}(\mathbf{S}_{t+\delta} = S \,|\, \mathbf{N}_{t,t+\delta} = j, \mathbf{S}_t = R).$$

So, suppose that (12.26) holds for $k = 1, \ldots, j$. Denote by $t + \mathbf{V}_{j,t}$ the time of the occurrence of the j-th Poisson event to be realized after time t. Thus, $\mathbf{V}_{j,t}$ is a Gamma distributed random variable with parameters j and λ. We write $g_{j,t}$ and $G_{j,t}$ for the density function and the distribution function of $\mathbf{V}_{j,t}$, respectively. Notice that for $0 < \delta' < \delta$ the three formulas

$$(\mathbf{N}_{t,t+\delta} = j+1, \mathbf{V}_{j,t} = \delta'), \qquad (\mathbf{N}_{t+\delta',t+\delta} = 1, \mathbf{V}_{j,t} = \delta'),$$
$$\text{and} \quad (\mathbf{N}_{t,t+\delta'} = j, \mathbf{N}_{t+\delta',t+\delta} = 1, \mathbf{V}_{j,t} = \delta')$$

denote the same event. With this in mind, we have successively

$$\mathbb{P}(\mathbf{S}_{t+\delta} = S \,|\, \mathbf{N}_{t,t+\delta} = j+1, \mathbf{S}_t = R)$$

$$= \sum_{Q \in \mathcal{S}} \int_0^\delta \mathbb{P}(\mathbf{S}_{t+\delta} = S \,|\, \mathbf{N}_{t,t+\delta} = j+1, \mathbf{S}_{t+\delta'} = Q, \mathbf{V}_{j,t} = \delta', \mathbf{S}_t = R)$$

$$\mathbb{P}(\mathbf{S}_{t+\delta'} = Q \,|\, \mathbf{N}_{t,t+\delta'} = j, \mathbf{S}_t = R) \frac{g_{j,t}(\delta')}{G_{j,t}(\delta)} \, d\delta'$$

$$= \sum_{Q \in \mathcal{S}} \int_0^\delta \mathbb{P}(\mathbf{S}_{t+\delta} = S \,|\, \mathbf{N}_{t+\delta',t+\delta} = 1, \mathbf{S}_{t+\delta'} = Q)$$

$$\mathbb{P}(\mathbf{S}_{t+\delta'} = Q \,|\, \mathbf{N}_{t,t+\delta'} = j, \mathbf{S}_t = R) \frac{g_{j,t}(\delta')}{G_{j,t}(\delta)} \, d\delta'$$

$$= \sum_{Q \in \mathcal{S}} \xi_1(Q, S) \xi_j(R, Q) \int_0^\delta \frac{g_{j,t}(\delta')}{G_{j,t}(\delta)} \, d\delta'$$

$$= \sum_{Q \in \mathcal{S}} \xi_1(Q, S) \xi_j(R, Q),$$

independent of t, and δ. □

Lemma 12.3.4 justifies the following definition.

12.3.5 Definition. For any $t > 0$, $\delta > 0$, $k \in \mathbb{N}$, and $R, S \in \mathcal{S}$, define

$$\xi_k(R, S) = \mathbb{P}(\mathbf{S}_{t+\delta} = S \,|\, \mathbf{N}_{t,t+\delta} = k, \mathbf{S}_t = R). \tag{12.27}$$

12.3.6 Theorem. *The stochastic process $(\mathbf{S}_t)$ is a homogeneous Markov process, with transition probability function $\mathfrak{p}$ defined by*

$$\mathfrak{p}_\delta(R, S) = \mathbb{P}(\mathbf{S}_{t+\delta} = S \,|\, \mathbf{S}_t = R)$$

$$= \sum_{k=0}^\infty \xi_k(R, S) \frac{(\lambda \delta)^k e^{-\lambda \delta}}{k!}. \tag{12.28}$$

Proof. Apply (12.25), (12.27), and Eq. (12.23) of Axiom [T]. □

Thus, $\mathfrak{p}_\delta(R, S)$ is the probability of a transition from state R to state S in δ units of time.

12.3.7 Definition. We recall that two distinct states R and S of a medium are called adjacent if there exists some token τ such that $R\tau = S$ or, equivalently, $S\tilde{\tau} = R$. Since the only possible transitions are those occurring between adjacent states, we refer to the Markov process $(\mathbf{S}_t)$ as a *random walk* process on the medium $(\mathcal{S}, \mathcal{T})$.

In the vein of Lemma 12.3.4, we have the following lemma, generalizing Axiom [I].

12.3.8 Lemma. *For any $t > 0$, $t' > 0$, $n \in \mathbb{N}_0$, and $S \in \mathcal{S}$, we have*

$$\mathbb{P}(\mathbf{S}_t = S \,|\, \mathbf{N}_t = n) = \mathbb{P}(\mathbf{S}_{t'} = S \,|\, \mathbf{N}_{t'} = n). \tag{12.29}$$

The proof is in the style of that of Lemma 12.3.4 (see Problem 12.4).

We now turn to the asymptotic probabilities of the states. To this end, we define an embedded discrete random walk keeping track of the state of the process on the arrival of each of the tokens.

12.3.9 Definition. We define a collection $(\mathbf{M}_n)_{n \in \mathbb{N}_0}$ specifying the state of the process when either $n = 0$ or the nth token occurs. Using Axiom [I] and Lemma 12.3.4, Eq. 12.29, we have for any $S \in \mathcal{S}$

$$\mathbb{P}(\mathbf{M}_n = S) = \mathbb{P}(\mathbf{S}_t = S \,|\, \mathbf{N}_t = n); \tag{12.30}$$

thus, in particular,

$$\mathbb{P}(\mathbf{M}_0 = S) = \mathbb{P}(\mathbf{S}_t = S \,|\, \mathbf{N}_t = 0) = \eta_S. \tag{12.31}$$

For convenience of reference to the Markov chain results recalled in the early part of this chapter, we write

$$\mathfrak{t}(R, S) = \mathfrak{t}_1(R, S) = \xi_1(R, S) \qquad (R, S \in \mathcal{S})$$
$$\mathfrak{t}_n(R, S) = \xi_n(R, S) \qquad (n \in \mathbb{N}, R, S \in \mathcal{S})$$
$$\mathfrak{T} = (\mathfrak{t}(R, S))_{R, S \in \mathcal{S}}$$
$$\mathfrak{T}^n = (\mathfrak{t}_n(R, S))_{R, S \in \mathcal{S}}.$$

12.3.10 Example. An example of a transition matrix in the case of the random walk on the permutohedron medium Π_2 of Example 3.5.9 and Figure 12.1 is displayed below. We recall that $\mathcal{L}_3$ denotes the set of the six strict linear orders 1,2,3, and $\mathcal{T}_3$ the corresponding set of tokens τ_{ij}, with $i, j \in \{1, 2, 3\}$, $i \neq j$, defined by the equation

$$\tau_{ij} : \mathcal{L}_3 \to \mathcal{L}_3 : L \mapsto L\tau_{ij} = \begin{cases} (L \setminus \{ji\}) \cup \{ij\} & \text{if } j \text{ covers } i \\ L & \text{otherwise.} \end{cases}$$

Writing $\vartheta_{ij} > 0$ for the probability of token τ_{ij}, we have the transition matrix

$$\mathfrak{T} = \begin{array}{c} \\ 123 \\ 132 \\ 213 \\ 231 \\ 312 \\ 321 \end{array} \begin{array}{c} \begin{array}{cccccc} 123 & 132 & 213 & 231 & 312 & 321 \end{array} \\ \left[\begin{array}{cccccc} * & \vartheta_{32} & \vartheta_{21} & 0 & 0 & 0 \\ \vartheta_{23} & * & 0 & 0 & \vartheta_{13} & 0 \\ \vartheta_{12} & 0 & * & \vartheta_{31} & 0 & 0 \\ 0 & 0 & \vartheta_{13} & * & 0 & \vartheta_{32} \\ 0 & \vartheta_{13} & 0 & 0 & * & \vartheta_{21} \\ 0 & 0 & 0 & \vartheta_{23} & \vartheta_{12} & * \end{array} \right] \end{array} \quad (12.32)$$

in which the '*' are space saving symbols denoting expressions ensuring that the entries on each line add up to 1 (thus, the '*' symbol in the cell (123,123) represents $1 - \vartheta_{32} - \vartheta_{21}$).

12.3.11 Theorem. *The stochastic process* $(\mathbf{M}_n)_{n \in \mathbb{N}_0}$ *is a regular Markov chain with initial probabilities and transition probabilities defined, for any R and S in $\mathcal{S}$, respectively by*

$$\mathbb{P}(\mathbf{M}_0 = S) = \eta_S, \tag{12.33}$$
$$\mathbb{P}(\mathbf{M}_{n+1} = S \,|\, \mathbf{M}_n = R) = \mathfrak{t}(R, S) \qquad (n \in \mathbb{N}_0). \tag{12.34}$$

Proof. We only prove that $(\mathbf{M}_n)_{n \in \mathbb{N}_0}$ is regular. (For the rest of Theorem 12.3.11, see Problem 12.5.) According to Definition 12.1.7, we have to prove that for some positive integer N, all the entries of the matrix $\mathfrak{T}^n$ are positive if $n \geq N$. We denote by τ_{VW} the token producing W from V, where V and W are distinct states. For any two states S and R, there is a at least one concise message $\mathbf{m}_{SR} = \tau_{S_1 S_2} \ldots \tau_{S_{m-1} S_m}$, with $S = S_1$ and $R = S_m$, producing R from S. This fact implies that $\mathfrak{t}_m(S, R) > 0$. Indeed, let us denote by $\vartheta_{VW} > 0$ the probability of token τ_{VW}. The entry in the cell (S, R) of the matrix $\mathfrak{T}^m$ must satisfy

$$\mathfrak{t}_m(S, R) = \mathbb{P}(\mathbf{M}_m = S_m \,|\, \mathbf{M}_0 = S_0)$$
$$\geq \mathbb{P}(\mathbf{M}_m = S_m, \mathbf{M}_{m-1} = S_{m-1}, \ldots, \mathbf{M}_1 = S_1 \,|\, \mathbf{M}_0 = S_0) \quad (12.35)$$
$$= \vartheta_{S_0 S_1} \times \ldots \times \vartheta_{S_{m-1} S_m} > 0, \tag{12.36}$$

with an equality replacing the inequality in (12.35) if there is only one concise message[7] producing R from S. In general, let $N(S, R)$ be the length of a concise message producing R from S, and let N be the smallest common multiple of all the numbers $N(S, R)$, for all pairs of states (S, R). Clearly, all the entries of $\mathfrak{T}^n$ are positive if $n \geq N$. □

---

[7] If there is more than one concise message producing R from S, then all these messages have the same content and length $N(S, R) = m$ (cf. Theorem 2.3.4), and so have the same probability expressed in the l.h.s. of (12.36), but this fact need not be used in this proof.

12.4 Asymptotic Probabilities

12.4.1 Theorem. *The asymptotic probabilities of the states of the random walk* $(\mathbf{M}_n)_{n \in \mathbb{N}_0}$ *exist and coincide with those of the random walk process* $(\mathbf{S}_t)$. *We have in fact, for any* $S \in \mathcal{S}$,

$$\lim_{n \to \infty} \mathbb{P}(\mathbf{M}_n = S) = \lim_{t \to \infty} \mathbb{P}(\mathbf{S}_t = S) = \frac{\prod_{\tau \in \widehat{S}} \vartheta_\tau}{\sum_{R \in \mathcal{S}} \prod_{\mu \in \widehat{R}} \vartheta_\mu}. \qquad (12.37)$$

Proof. Since $(\mathbf{M}_n)$ is regular, we can use Theorem 12.1.9 and assert that the powers $\mathfrak{T}^n$ of the transition matrix $\mathfrak{T}$ are converging to a stochastic matrix $\mathfrak{A}$ having each of its $q = |\mathcal{S}|$ rows equal to the same vector $\boldsymbol{\alpha} = (\alpha_1, \ldots, \alpha_q)$, the stationary distribution. It remains to show that this stationary distribution is that specified by 12.37. We use Lemma 12.1.12, and prove that for any $R, S \in \mathcal{S}$, and writing K for the denominator in (12.37),

$$\frac{1}{K} \left(\prod_{\tau \in \widehat{R}} \vartheta_\tau \right) \mathfrak{t}(R, S) = \frac{1}{K} \left(\prod_{\tau \in \widehat{S}} \vartheta_\tau \right) \mathfrak{t}(S, R). \qquad (12.38)$$

Recalling the notation $\triangle$ for the symmetric difference between sets, we consider three cases:

$$\begin{array}{lll}
\mathfrak{t}(R, S) = \mathfrak{t}(S, R) = 0, & |R \triangle S| > 2, & \text{(Case 1)}, \\
\mathfrak{t}(R, S) = \mathfrak{t}(S, R) = \mathfrak{t}(S, S) > 0, & R = S, & \text{(Case 2)}, \\
\mathfrak{t}(R, S) = \vartheta_\mu > 0, \ \mathfrak{t}(S, R) = \vartheta_{\tilde{\mu}} > 0, & R \triangle S = \{\mu, \tilde{\mu}\}, & \text{(Case 3)}
\end{array}$$

(for some $\mu \in \mathcal{T}$ in Case 3). In Case 1, both sides of (12.38) vanish, whereas in Case 2, the two expressions on both sides of the equation are identical. Equation (12.38) also holds in Case 3 since it is equivalent to

$$\left(\prod_{\tau \in (\widehat{R} \cap \widehat{S})} \vartheta_\tau \right) \vartheta_\mu \vartheta_{\tilde{\mu}} = \left(\prod_{\tau \in (\widehat{R} \cap \widehat{S})} \vartheta_\tau \right) \vartheta_{\tilde{\mu}} \vartheta_\mu.$$

Since we have $\lim_{n \to \infty} \mathbb{P}(\mathbf{M}_n = S) = \lim_{t \to \infty} \mathbb{P}(\mathbf{S}_t = S)$ (cf. Problem 12.8), this completes the proof of the theorem. □

12.4.2 Remark. In some situations, it makes sense to suppose that the conditional probability of a transition from some state R of the random walk to some state $S \neq R$ depends not only on the probability ϑ_τ of the token τ such that $R\tau = S$, but also on R. For example, we might have both $R\tau = S$ and $W\tau = V$ but the conditional probabilities of a transition to state R would be different and denoted, for example, by $\vartheta_\tau \nu_R$ and $\vartheta_\tau \nu_W$, the first factor representing the contribution of some external source, and the second one the specific effect on a particular state. A close reading of the above proof indicates that the argument used to justify Eq. (12.38) would also apply in such a case. A version of Theorem 12.4.1 would thus hold, with the products in Eq. (12.37) having many more factors. This would result in a model with $|\mathcal{T}| \times |\mathcal{S}| - 1$ independent parameters, which might be of some interest in a case where extensive sequential data are available. Note that submodels may be considered in which the number of parameters ν_R might be substantially smaller than $|\mathcal{S}|$.

In some experimental situation, organisms may be observed repeatedly at short intervals of time[8]. We limit ourselves here to two successive observations separate by a short interval of time δ. Using Theorem 12.4.1 and Theorem 12.3.6, an explicit expression can be obtained for the asymptotic joint probability of observing the states R and S at time t and time $t+\delta$, respectively. We write, as before, $\xi_{R,S}(k)$ for the k-step transition probability from the state R to the state S in the companion Markov chain on $\mathcal{S}$, and $p_{R,S}(\delta)$ for the probability of a transition between the same two states, in δ units of time, in the stochastic process $(\mathbf{S}_t)$. Defining

$$\mathfrak{q}(R) = \lim_{t \to \infty} \mathbb{P}(\mathbf{S}_t = R),$$

we get

$$\lim_{t \to \infty} \mathbb{P}(\mathbf{S}_t = R, \mathbf{S}_{t+\delta} = S) = \lim_{t \to \infty} \left(\mathbb{P}(\mathbf{S}_t = R) \mathbb{P}(\mathbf{S}_{t+\delta} = S \mid \mathbf{S}_t = R) \right)$$
$$= \mathfrak{q}(R) \, \mathfrak{p}_\delta(R, S).$$

From Theorem 12.4.1 and Theorem 12.3.6, we derive:

12.4.3 Theorem. *For all states R and S and all $\delta > 0$,*

$$\lim_{t \to \infty} \mathbb{P}(\mathbf{S}_t = R, \mathbf{S}_{t+\delta} = S) = \frac{\prod_{\tau \in \widehat{R}} \vartheta_\tau}{\sum_{V \in \mathcal{S}} \prod_{\mu \in \widehat{V}} \vartheta_\mu} \sum_{k=0}^{\infty} \xi_{R,S}(k) \frac{(\lambda \delta)^k e^{-\lambda \delta}}{k!}.$$

We leave the details of the argument as Problem 12.10.

12.5 Random Walks and Hyperplane Arrangements

The results of the previous section are obviously valid for any special case of a medium. For example, the asymptotic probabilities of a random walk on the regions of a hyperplane arrangement[9] are exactly those described by Theorems 12.4.1 and 12.4.3, provided of course that the region-to-region transition

---

[8] If the interval of time between successive observation is long, then the observations become essentially independent, and asymptotic results such as Theorem 12.4.1 can be used.

[9] Cf. Definition 9.1.7 and Theorem 9.1.8.

12.5 Random Walks and Hyperplane Arrangements

probabilities are consistent with Axioms [T], [I], and [L]. This application is of interest in view of a similar random walk on the regions of a hyperplane arrangement investigated by Brown and Diaconis (1998) (see also Bidigare et al., 1999). We briefly spell out here the differences between the two random walks. We only discuss the discrete case, which is that considered in the Brown and Diaconis's paper (referred to below as [BR]).

The following attributes are common to both random walks.

1. The states of the random walk are the regions of some hyperplane arrangement $\mathcal{A} = \{H_i\}_{i \in J}$.
2. In fact, these states are those of a token system and the set of tokens is defined from some features of the arrangement $\mathcal{A}$.
3. A probability distribution ϑ is postulated on the set of those tokens, and a region-to-region transition occur when a token is selected that is effective on the current state. The values of all the probabilities are assumed to be positive (by definition for us and in an important special case for [BR]).

The essential difference lies in the definition of the tokens. In our case, each hyperplane gives rise to two mutually reverse tokens which govern the back and forth transitions between two adjacent regions. As a result, the token system turns out to be a medium, which leads to the asymptotic formulas of Theorems 12.4.1 and 12.4.3. In the [BD] model, it is assumed that a token is associated to each of the faces of the regions defined by $\mathcal{A}$. Let us denote by $\mathcal{F}$ the collection of all these faces, and by $\mathcal{T}_\mathcal{F} = \{\tau_F \mid F \in \mathcal{F}\}$ the corresponding collection of the tokens. Suppose that the random walk is in region S and that some token τ_F has been randomly selected from the distribution ϑ. We have then $S\tau_F = T$ for some region T if T is the closest region to S having F as a face. Notice that if S and T are two adjacent regions, then

$$S\tau_T = T \quad \text{and} \quad T\tau_S = S. \tag{12.39}$$

This implies that the graph of the medium at the core of our random walk on the regions of the arrangement is a subgraph of the graph of the [BD] token system. However, (12.39) does not mean that whenever S and T are adjacent regions, then τ_S and τ_T are mutual reverses. Indeed, the pair of equations (12.39) hold even if the regions S and T are not adjacent. Thus, our random walk could not arise as a special case of the [BD] random walk simply by assuming that the probability distribution ϑ on $\mathcal{T}_\mathcal{F}$ vanishes on all the faces that are not regions. As an illustration of these remarks, Figure 12.2 displays a non-central arrangement defined by three lines. There are 19 faces:

7 regions	$F_1, \ldots, F_7$,
6 rays	$F_8, \ldots, F_{13}$,
3 segments	F_{14}, F_{15} and F_{16},
3 points	F_{17}, F_{18} and F_{19}.

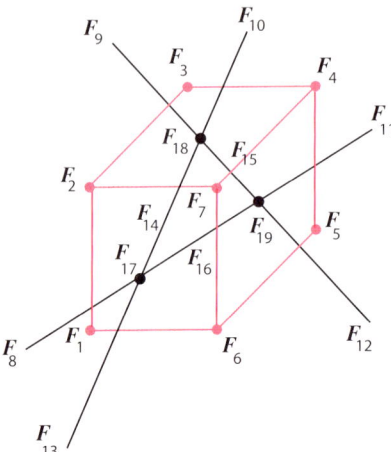

Figure 12.2. The non-central arrangement of three lines; in red the corresponding mediatic graph. Nineteen faces, labelled $F_1, \ldots, F_{19}$, are defined by the four regions.

The mediatic graph defined by the arrangement is pictured in red. The three segments emerging from region F_2 represent the three pairs of mutually reverse tokens linking region F_2 to the three adjacent regions F_3, F_7 and F_1. By contrast, in the token system defined in the [BD] model, the region F_2 can produce all the other regions of the arrangement by application of the appropriate tokens, with several different tokens producing the same state in some cases. (So, the graph of this stoken system is not a simple graph.) There is one token per face in this model. Denoting by τ_i the token corresponding to face F_i, $1 \leq i \leq 19$, we have indeed:

$$F_2\tau_3 = F_2\tau_{10} = F_3, \quad F_2\tau_4 = F_2\tau_{11} = F_4,$$
$$F_2\tau_5 = F_5, \quad F_2\tau_6 = F_2\tau_{12} = F_6, \quad F_2\tau_1 = F_2\tau_{13} = F_1,$$
$$F_2\tau_7 = F_2\tau_{15} = F_2\tau_{19} = F_2\tau_{16} = F_7.$$

So, the token system defined by the [BD] model is not a medium. Still, the asymptotic probabilities of the states exists, but their description is more complicated (cf. Brown and Diaconis, 1998; Bidigare et al., 1999).

Problems

The Markov chain concepts introduced in the first part of this chapter have been kept to the minimum useful for the derivation of the asymptotic results states in Theorems 12.4.1 and 12.4.3, which will be used in some of the applications described in Chapter 13. In Problems 12.11-12.16, we introduce a few other standard concepts and results.

Problems for Chapter 12

12.1 Prove that the two formulations (12.1) and (12.2) are equivalent.

12.2 Axiomatize directly—that is, not as an embedded process—a discrete random walk on a medium.

12.3 Would regularity still hold with the following weakened version of Axiom [Ma]: For any two states R and S, there exists a message producing R from S?

12.4 Prove Lemma 12.3.8.

12.5 Prove that the stochastic process $(\mathbf{M}_n)_{n \in \mathbb{N}_0}$ is a Markov chain satisfying 12.33 and 12.34 (cf. Theorem 12.3.11).

12.6 Show that for all $n \geq N$ all the entries of the matrix $\mathfrak{T}^n$ are positive, where $\mathfrak{T}$ is defined by (12.32). What is the minimum N for which this is true, and why? (Explain in terms of media concepts.)

12.7 Axiomatize and investigate the random walk sketched in Remark 12.4.2. Restate and prove for this new model the most important results, such as the asymptotic probabilities of the states.

12.8 Prove that we have indeed $\lim_{n \to \infty} \mathbb{P}(\mathbf{M}_n = S) = \lim_{t \to \infty} \mathbb{P}(\mathbf{S}_t = S)$ as stated at the end of the proof of Theorem 12.4.1.

12.9 By Theorem 12.1.9, the successive powers $\mathfrak{T}^n$ of the transition matrix $\mathfrak{T}$ converge (as $n \to \infty$) to a stochastic matrix $\mathfrak{A}$ having identical rows. The speed of convergence is surprisingly fast. Simulate this convergence by computation on an example (using a software such as Mathematica). Take the matrix $\mathfrak{T}$ to be consistent with the probabilities of the tokens of a medium.

12.10 Prove Theorem 12.4.3.

12.11 Let $\rightarrowtail$ be a binary relation on the set of states $\mathcal{S}$ of an homogeneous Markov chain, defined by: $i \rightarrowtail j \Leftrightarrow i = j$ or $\mathsf{t}_n(i,j) > 0$ for some $n \in \mathbb{N}$, with t_n as in Definition 12.3.9. When $i \rightarrowtail j$, we say that j is *accessible* from i, or i *communicates* with j. This relation will be used to gather the states into classes of mutually accessible states. The relation $\rightarrowtail$ is the *accessibility* relation of the Markov chain $(\mathbf{X}_n)$. A state j such that $j \rightarrowtail k \rightarrowtail j$ for some state k is called a *return* state. Prove that the relation $\rightarrowtail$ is a quasi-order, that is $\rightarrowtail$ is transitive and reflexive (cf. 1.8.3, page 14).

12.12 (Continuation.) Let $\mathfrak{S}$ be the set of equivalence classes induced by the accessibility relation $\rightarrowtail$, and let $\precsim$ be the corresponding partial order on the set $\mathfrak{S}$. For any state i, we denote by $[i]$ the class of all states j such that both $i \rightarrowtail j$ and $j \rightarrowtail i$ hold. Thus, $\mathfrak{S} = \{[i] \mid i \in \mathcal{S}\}$, and for all $i, j \in \mathcal{S}$, $[i] \precsim [j] \iff i \rightarrowtail j$. The elements of $\mathfrak{S}$ are thus classes of states. Every maximal element of $\mathfrak{S}$ (for the partial order $\precsim$) is an *ergodic class*, and the elements of an ergodic class are called *ergodic states*. An element of $\mathfrak{S}$ which is not ergodic is said to be a *transient class*. An element of a transient class is called a *transient* state. We denote by $\mathcal{E}$ and $\mathfrak{E}$, respectively, the set of all ergodic states and the family of all ergodic classes. The letters $\mathcal{U}$ and $\mathfrak{U}$ stand for the set of all transient states and the family of all transient classes, respectively. Thus, $\mathcal{S} = \mathcal{E} + \mathcal{U}$ and $\mathfrak{S} = \mathfrak{E} + \mathfrak{U}$. When an ergodic class contains a single state, this state is called *absorbing*. Prove that any Markov chain has at least one ergodic state, but there may not be any transient state.

12.13 (Continuation.) Prove that a state i of a homogeneous Markov chain is absorbing if and only if $\mathsf{t}(i, i) = 1$.

12.14 (Continuation.) A finite Markov chain tends to evolve towards the ergodic classes. Prove the following: For a finite Markov chain $(\mathbf{X}_n)$ with the family $\mathfrak{E}$ of ergodic classes, we have

$$\lim_{n \to \infty} \mathbb{P}(\mathbf{X}_n \in \cup \mathfrak{E}) = 1.$$

12.15 (Continuation.) As indicated above, a state j is a return state if $j \rightarrowtail i \rightarrowtail j$ for some state i. Thus, j is a return state if the set $R(j) = \{n \in \mathbb{N} \mid \mathsf{t}_n(j, j)\}$ is not empty. In such a case, the set $R(j)$ is necessarily infinite since we obviously have $n + m \in R(j)$ for any $n, m \in R(j)$. The *period* of any return state j is the greatest common divisor of $R(j)$. A state is said to be *aperiodic* if it has period 1. A homogeneous Markov chain is *aperiodic* if all its states are aperiodic. Prove that if two states are mutually accessible, then they necessarily have the same period.

12.16 (Continuation.) A chain without transient classes is called *ergodic*. Such a chain may have several ergodic classes, however. When a chain has a unique ergodic class, then this chain is said to be *irreducible*. The justification for this terminology is that a chain with several ergodic classes may be decomposed into components chains, each with a single ergodic class. Each of these subchains may be studied separately since there is no communication between the ergodic classes. Prove that a finite homogeneous Markov chain is regular if and only if it is ergodic, irreducible, and aperiodic.

13
Applications

In this chapter, we describe two quite different extended examples of media. The first one was encountered several times already (in Chapter 1, 4, and 11) under the title 'Learning Spaces.' It deals with a situation where the states of the medium are the possible knowledge states of individuals in some academic subject such as arithmetic or algebra. We begin by discussing a method for constructing a learning space on the basis of a teacher's expertise on the topic, and then outline a probabilistic algorithm for the assessment of knowledge in such learning spaces. The second example is also probabilistic. The states of a medium are taken to represent the latent opinion of potential voters. The evolution of these opinions over (continuous) time are modeled as a random walk whose states are exactly the states of the medium. This type of model has been used extensively to analyze the changes over time observed in the voting behavior of respondents to a political poll. The application described here, which relies on the real time random walk theory developed in Chapter 12, is based on extensive data pertaining to the 1992 presidential election context opposing Bill Clinton, George H.W. Bush, and Ross Perot.

13.1 Building a Learning Space

Consider a topic in mathematical education, such as elementary algebra[1]. From the standpoint of assessing the students' competence, this topic can be delineated by a finite set Q of problem types, or *items*, that a student must learn to solve. We call *instance* a particular case of a problem type, obtained for example by choosing the numbers involved in the problem, or the exact phrasing of a 'word problem[2].' The set Q specifies the curriculum.

---

[1] We closely follow Falmagne et al. (2006b) in our description of this first application.
[2] Our use of the term *item* is consistent with the meaning in Doignon and Falmagne (1999), but differs from the usage in psychometrics, where 'item' is referred to what we call 'instance.'

An examination of a representative sample of textbooks indicates that, for beginning algebra (sometimes called "Algebra 1"), the set Q contains approximately 250 problem types (see Remark 13.1.1(a)). Recalling Definition 1.6.1, a pair $(Q, \mathcal{K})$ is a knowledge structure if $\mathcal{K}$ is a family of subsets of Q containing all the knowledge states that are feasible, that is, that could characterize some individual in a population of reference. In other words, an individual in knowledge state K in $\mathcal{K}$ can, in principle[3] solve all the problem types in K and would fail any problem in $Q \setminus K$. It is assumed that a knowledge structure $\mathcal{K}$ contains $\varnothing$ and $\cup \mathcal{K} = Q$: it is possible for someone to know everything, and for somebody else to know nothing at all in Q. In practice, because algebra is a highly structured topic, whose concepts are practically always taught in approximately the same order, $|\mathcal{K}|$ is considerably smaller than 2^{250}, the number of subsets in set of size 250. Typically[4], $|\mathcal{K}|$ is of the order of 10^8, which is within the capabilities of modern P.C.'s. Further constraints are imposed on $\mathcal{K}$ in the form of the two axioms recalled below, making the pair $(Q, \mathcal{K})$ a learning space (cf. 1.6.1).

[K1] If $K \subset L$ are two states, with $|L \setminus K| = n$, then there is a chain of states $K_0 = K \subset K_1 \subset \cdots \subset K_n = L$ such that $K_i = K_{i-1} + \{q_i\}$ with $q_i \in Q$ for $1 \leq i \leq n$.

[K2] If $K \subset L$ are two states, with $K \cup \{q\} \in \mathcal{K}$ and $q \notin L$, then $L \cup \{q\} \in \mathcal{K}$.

Some reflection shows that both of these axioms are quite sensible from a pedagogical standpoint. Taken together, they are formally equivalent to the hypothesis that $\mathcal{K}$ is a wg-family which is closed under unions (cf. Theorem 4.2.1). In fact, $\mathcal{K}$ can be regarded as the collection of positive contents of a rooted, closed medium (Theorem 4.2.2).

13.1.1 Remarks. (a) For concreteness, six examples of items and instances in beginning algebra are given in Table 13.1. Since some of these items obviously cover a considerable number of instances, the number of specific problems implicitly involved in an assessment based on Q and $\mathcal{K}$ is very large.

(b) The construction of the collection $\mathcal{K}$ of knowledge states in a practical situation is for the time being a very demanding task, relying in part on the judgement of expert teachers responding to probing questions about the curriculum. These questions can be generated systematically by the QUERY routine developed by Koppen (1993), Dowling (1993a), Dowling (1993b) and Cosyn and Thiéry (2000) (see also Villano, 1991; Müller, 1989; Kambouri et al., 1994; Dowling, 1994; Doignon and Falmagne, 1999). The resulting structure is only a preliminary step, potentially plagued by inconsistencies in the experts' responses (see Kambouri, 1991), and in need of refinements and corrections dictated by a statistical analysis of students' data. Nevertheless, because the theoretical basis of the QUERY routine is intimately related

---

[3] Discounting careless errors, for example.
[4] For topics in elementary mathematics such as arithmetic, geometry or algebra.

to the very definition of a learning space, and so of a closed, rooted medium (cf. Theorem 4.2.2) we will briefly review its principles here.

13.1.2 The Query Routine. From the standpoint of an expert, such as an experienced teacher, the first step, or *Block 1*, of the the QUERY routine takes the form a series of questions such as

[Q1] *Suppose that a student is not capable of solving problem* **p**. *Would this student also fail problem* **p'**?

The respondents are told to discount chance factors such as careless errors and lucky guesses. More complex questions are asked in Blocks 2, 3, etc. which are discussed in Section 13.2. Assuming that the experts are both logically consistent and reliable, the positive responses to questions of the type [Q1] induce a quasi order on the set Q of items. In practice, only a small subset of the set of possible questions needs to be asked. This is due in part to the transitivity of the quasi order: in QUERY, each of the successive questions is chosen so as to maximize the number of potential inferences. An example of the quasi orders obtainable at the end of Block 1 is represented by its Hasse diagram[5] in Figure 13.1 which is reproduced from Falmagne et al. (2006b), as is Table 13.1.

Thus, an answer "Yes" to [Q1] is represented by a sequence of arcs beginning at the vertex representing **p'** and ending at the vertex representing **p** in the Hasse diagram of Figure 13.1. For concreteness, the vertices corresponding to six actual problem types are represented in the diagram by the red dots labelled (a),...,(f). The text of some of the instances of the six problems types is given in the second column of Table 13.1.

We will show in this section that such a quasi order defines a wg-family of sets closed under both union and intersection, thereby inducing a medium which is both u-closed and i-closed (see Definition 4.1.2).

13.1.3 Definition. We recall from Definition 3.1.5 that a knowledge structure $\mathcal{K}$ on a set Q is discriminative if

$$q = t \quad \Longleftrightarrow \quad (\forall K \in \mathcal{K} : q \in K \Leftrightarrow t \in K) \qquad (q, t \in Q). \qquad (13.1)$$

A knowledge structure is called a *knowledge space* if it is closed under union. It is called *quasi ordinal* if it is closed under both union and intersection. A quasi ordinal knowledge structure is *partially ordinal* if it is discriminative. We also recall the notation $\mathcal{K}_q = \{K \in \mathcal{K} \mid q \in K\}$ for all $q \in Q$.

---

[5] We assume that the items have been gathered in the equivalence classes of the quasi order, inducing a partial order (see Definition 13.1.3). Actually, this particular Hasse diagram, which only covers part of beginning algebra, was obtained from the combined results of Block 1 of QUERY with some experts, with an extensive analysis of students' data.

288   13 Applications

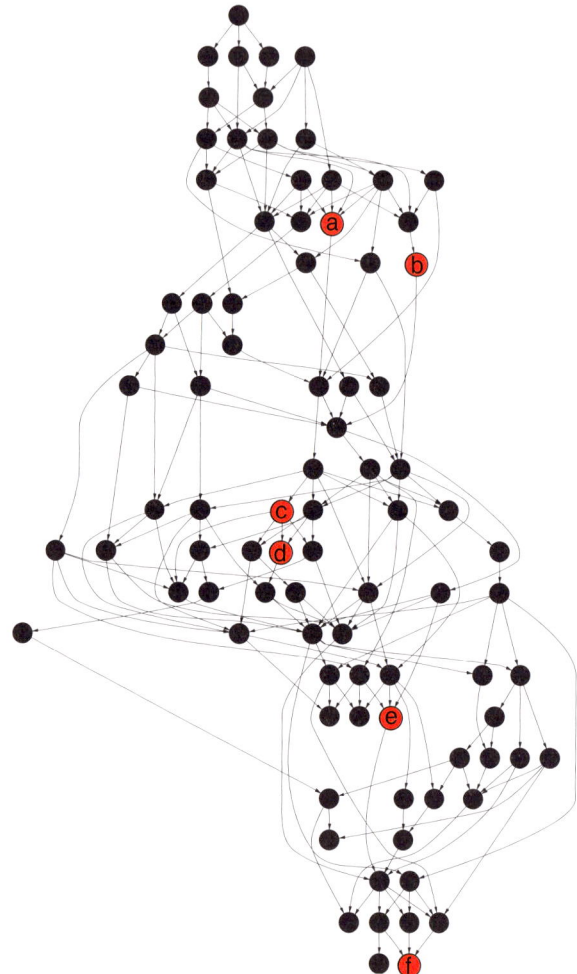

Figure 13.1. Hasse diagram of the partial order obtained from Block 1 of the QUERY routine for part of Beginning Algebra (from Falmagne et al., 2006b). The vertices marked a-f refer to Problems types (a)-(f) of Table 13.1.

Notice that the r.h.s. of (13.1) defines an equivalence relation on Q. Thus, for any knowledge structure $\mathcal{K}$, the elements of Q can be gathered into equivalence classes so as to form a discriminative knowledge structure. The main tool to establish that a quasi order defines a well-graded knowledge structure closed under union and intersection is Theorem 13.1.5, which is due to Birkhoff (1937). The proof given below is that of Doignon and Falmagne (1999) and is included for completeness. We rely on a preparatory lemma.

Table 13.1. Six types of problems in Elementary Algebra

	Name of problem type	Example of Problem
(a)	Word problem on proportions (Type 1)	A car travels on the freeway at an average speed of 52 miles per hour. How many miles does it travel in 5 hours and 30 minutes?
(b)	Plotting a point in the coordinate plane	Using the pencil, mark the point at the coordinates $(1, 3)$.
(c)	Multiplication of monomials	Perform the following multiplication: $4x^4y^4 \cdot 2x \cdot 5y^2$ and simplify your answer as much as possible.
(d)	Greatest common factor of two monomials	Find the greatest common factor of the expressions $14t^6y$ and $4tu^5y^8$. Simplify your answer as much as possible.
(e)	Graphing the line through a given point with a given slope	Graph the line with slope -7 passing through the point $(-3, -2)$.
(f)	Writing the equation of the line through a given point and perpendicular to a given line	Write an equation for the line that passes through the point $(-5, 3)$ and is perpendicular to the line $8x + 5y = 11$.

13.1.4 Lemma. *If $\mathcal{K}$ and $\mathcal{K}'$ are two quasi ordinal knowledge structures on the same set Q, then*

$$(\forall q, t \in Q : \mathcal{K}_q \subseteq \mathcal{K}_t \Leftrightarrow \mathcal{K}'_q \subseteq \mathcal{K}'_t) \quad \Longleftrightarrow \quad \mathcal{K} = \mathcal{K}'. \qquad (13.2)$$

Proof. The sufficiency is immediate. Suppose that $K \in \mathcal{K}$. This implies

$$K = \cup_{s \in K} (\cap \mathcal{K}'_s) = K'. \qquad (13.3)$$

Note that $K' \in \mathcal{K}'$ because $\mathcal{K}'$ is quasi ordinal. We will prove that in fact $K = K'$. Take any $t \in K'$. By definition of K', we must have $q \in K$ such that $t \in \cap \mathcal{K}'_q$. This gives $\mathcal{K}'_q \subseteq \mathcal{K}'_t$. By the equivalence in the l.h.s. of (13.2), we get $\mathcal{K}_q \subseteq \mathcal{K}_t$; thus $t \in \cap \mathcal{K}_q$, yielding $t \in K$. We obtain $K' \subseteq K$, and by (13.3), $K = K'$. We conclude that $\mathcal{K} \subseteq \mathcal{K}'$, and by symmetry, $\mathcal{K} = \mathcal{K}'$. □

13.1.5 Theorem. *There is a 1-1 correspondence between the collection of all quasi ordinal knowledges structures $\mathcal{K}$ on a set Q and the collection of all quasi orders $\mathcal{Q}$ on Q. This correspondence is specified by the two equivalences*

$$q \mathcal{Q} t \quad \Longleftrightarrow \quad (\forall K \in \mathcal{K} : t \in K \Rightarrow q \in K) \qquad (q, t \in Q) \quad (13.4)$$

$$K \in \mathcal{K} \quad \Longleftrightarrow \quad (\forall q, t \in \mathcal{Q} : t \in K \Rightarrow q \in K) \qquad (K \subseteq Q). \quad (13.5)$$

Moreover, under this correspondence, partial orders are mapped onto partially ordinal knowledge structures.

Note that (13.4) can be written compactly as

$$q\mathrel{\mathcal{Q}} t \iff \mathcal{K}_q \supseteq \mathcal{K}_t, \qquad (q, t \in Q). \qquad (13.6)$$

Proof. The compact form (13.6) of the equivalence (13.4) clearly defines the relation $\mathcal{Q}$ as a quasi order on Q. Conversely, for any quasi order $\mathcal{Q}$ on Q, the equivalence (13.5) defines a family $\mathcal{K}$ of subsets of Q. We show that such family is a partially ordinal knowledge structure. It obviously contains Q and $\varnothing$ (the latter because in this case the implication in the r.h.s. of (13.5) holds vacuously). Thus $\mathcal{K}$ is a knowledge structure. Suppose that $\{K_i \mid i \in \mathcal{I}\}$ is a subfamily of $\mathcal{K}$. By (13.5), we have

$$\cup_{i \in \mathcal{I}} K_i \in \mathcal{K} \iff (\forall q, t \in Q : (t \in \cup_{i \in \mathcal{I}} K_i) \Rightarrow (q \in \cup_{i \in \mathcal{I}} K_i)),$$

So, suppose that $q\mathcal{Q}t$ with $t \in \cup_{i \in \mathcal{I}} K_i$ for some $t \in Q$. We obtain $t \in K_j$ for some index $j \in \mathcal{I}$. This implies $q \in K_j$ by (13.5) yielding $q \in \cup_{i \in \mathcal{I}} K_i$; and so $(\cup_{i \in \mathcal{I}} K_i) \in \mathcal{K}$. The proof that we also have $(\cap_{i \in \mathcal{I}} K_i) \in \mathcal{K}$ is similar. Thus, to each quasi order $\mathcal{Q}$ on Q corresponds a quasi ordinal knowledge structure $\mathcal{K}$ on Q. We now show that the correspondence is 1-1.

Let $\mathfrak{K}$ be the collection of all the quasi ordinal knowledge structures $\mathcal{K}$ on Q, and let $\mathfrak{Q}$ be the collection of all quasi orders $\mathcal{Q}$ on Q. We prove that the two mappings

$$f : \mathfrak{K} \to \mathfrak{Q} : \mathcal{K} \mapsto f(\mathcal{K}) = \mathcal{Q}, \qquad g : \mathfrak{Q} \to \mathfrak{K} : \mathcal{Q} \mapsto g(\mathcal{Q}) = \mathcal{K},$$

defined by (13.4) and (13.5) are mutual inverses. The function f must be injective by Lemma 13.1.4 and the equivalence (13.6) restating (13.4). Take any quasi order $\mathcal{Q}$ on Q, with say $\mathcal{K} = g(\mathcal{Q})$ and $f(\mathcal{K}) = \mathcal{Q}'$. We will show that $\mathcal{Q} = \mathcal{Q}'$; thus any quasi order is in the range of the function f. Since f is injective, f and g must be mutual inverses.

Suppose that $p\mathcal{Q}q$. With (13.5) defining g, this implies $\mathcal{K}_q \subseteq \mathcal{K}_p$, yielding $p\mathcal{Q}'q$ by (13.6); thus, $\mathcal{Q} \subseteq \mathcal{Q}'$. Conversely, if $p\mathcal{Q}'q$, then since $f^{-1}(\mathcal{Q}') = \mathcal{K}$, we get again $\mathcal{K}_q \subseteq \mathcal{K}_p$, which gives $p\mathcal{Q}q$ by (13.6). We get $\mathcal{Q}' \subseteq \mathcal{Q}$ and so $\mathcal{Q} = \mathcal{Q}'$.

The last statement concerning ordinal spaces is immediate. □

13.1.6 Theorem. *Any partially ordinal knowledge structure is well-graded, and so can be regarded as the family of positive contents of a u-closed and i-closed medium.*

Proof. We prove that any partially ordinal knowledge structure $\mathcal{K}$ is a learning space and use Theorem 4.2.1. Suppose that L and $K \subset L$ are in $\mathcal{K}$. If $K + \{q\}$ is also in $\mathcal{K}$, then $L + \{q\} = L \cup (K + \{q\}) \in \mathcal{K}$ because $\mathcal{K}$ is closed under union. So, Axiom [K2] of a learning space holds (see Definition 1.6.1). Turning to Axiom [K1], recall that $\mathcal{K}$ is discriminative. Thus, for two distinct items in $L \setminus K$ there is some $N \in \mathcal{K}$ containing one but not both of them. (If $|L \setminus K| = 1$, there is nothing to prove.) Since $\mathcal{K}$ is closed under intersection and union, $K\cup(L\cap N)$ is in $\mathcal{K}$, with $K\subset K\cup(L\cap N)\subset L$. Axiom [K1] follows by induction. Using the implication (i)⇒(ii) of Theorem 4.2.1, we conclude that $\mathcal{K}$ is well-graded. □

To sum up, collecting all the responses of an expert to the questions of type [Q1] gives a partial order which by Theorems 13.1.5 and 13.1.6, provides a learning space closed under intersection. The interest of such a structure is that it can be conveniently summarized by a Hasse diagram. On the negative side, there are no pedagogically sound arguments supporting the closure under intersection. Moreover, stopping the QUERY routine after Block 1 has the effect of retaining in the knowledge structure a potentially large number of states that are never encountered in practice. Obviously, the presence of such useless states lengthen the assessment. The remaining blocks of the QUERY routine solve this difficulty, at least in principle[6]. Its output is a family of sets closed under union, but not necessarily well-graded nor closed under intersection. This part of the QUERY routine, although of theoretical interest, is not used in practice with human experts. We review the results in the next section.

The data resulting from questions of type [Q1] are only the first step in the construction of a satisfactory learning space. This initial learning space is generally adequate and can profitably be used by the schools, but is far from fully satisfactory. During Step 2, this preliminary learning space is gradually (and manually) altered on the basis of students' data. This second phase takes several months. While this method has been shown to yield, in practice, very reliable assessment results (see Falmagne et al., 2006a), it pertains more to art than to technique, and is prohibitively labor intensive. Much progress remains to be done in this area. One promising, realistic possibility would consist in automatizing the refinement of a learning space via an algorithm computing a statistical index of the fit of a current learning space with relevant statistics on students's responses, and implementing an change of the learning space whenever whenever a threshold is reached by such an index. Such a development is in the works.

13.2 The Entailment Relation

The remaining blocks of the QUERY routine ask more demanding questions of the following type.

[Q2] Suppose that a student has not mastered problems $\mathbf{p_1}, \mathbf{p_2}, \ldots, \mathbf{p_n}$. Would this student also fail problem $\mathbf{p}'$?

As in the case of [Q1], the expert is told to discount careless errors and lucky guesses. Obviously, [Q2] generalizes [Q1]; thus $n = 1, 2, \ldots$ in [Q2]

---

[6] If not in fact, because the questions of type [Q2] are much more difficult and the experts tend to be much less reliable.

corresponds to Block 1, 2, ... of the QUERY routine. The positive responses to all the questions covered by [Q2] define a relation $\mathcal{R} \subseteq \mathfrak{P}(Q) \times Q$, which is consistent with the (unknown) family $\mathcal{K}$ if $\mathcal{R}$ satisfies the equivalence

$$S\mathcal{R}q \iff (\forall K \in \mathcal{K} : S \cap K = \varnothing \Rightarrow q \notin K) \tag{13.7}$$
$$(S \in \mathfrak{P}(Q) \setminus \{\varnothing\}, q \in Q).$$

The meaning of the relation $\mathcal{R}$ is captured by the following two theorems, due to Doignon and Falmagne (1999). We omit both proofs. We ask the reader to provide the proof of Theorem 13.2.1 in Problem 13.3; for the proof of Theorem 13.2.3, see Doignon and Falmagne (1999).

13.2.1 Theorem. *Let $(Q, \mathcal{K})$ be a knowledge structure and suppose that the relation $\mathcal{R}$ is defined by (13.7); then,*

(i) $(\forall s \in S, T\mathcal{R}s) \wedge S\mathcal{R}t \implies T\mathcal{R}t$;
(ii) $q \in S \subseteq Q \implies S\mathcal{R}q$.

Thus, Condition (ii) states that $\mathcal{R}$ extends the inverse membership relation.

13.2.2 Definition. A relation $\mathcal{R} \subseteq (\mathfrak{P}(Q) \setminus \{\varnothing\}) \times Q$ satisfying Conditions (i) and (ii) of Theorem 13.2.1 is called an *entailment* for Q.

The next theorem generalizes the result of Birkhoff (1937) recalled as our Theorem 13.1.5.

13.2.3 Theorem. *Let Q be a nonempty set. There exists a 1-1 correspondence between the family of all knowledge spaces $\mathcal{K}$ on Q and the collection of all entailments $\mathcal{R}$ for Q. This correspondence is specified by the two equivalences*

$$S\mathcal{R}q \iff (\forall K \in \mathcal{K} : S \cap K = \varnothing \Rightarrow q \notin K) \tag{13.8}$$
$$K \in \mathcal{K} \iff (\forall (S, q) \in \mathcal{R} : S \cap K = \varnothing \Rightarrow q \notin K) \tag{13.9}$$
$$(S \in \mathfrak{P}(Q) \setminus \{\varnothing\}, q \in Q).$$

We omit the proof (see Doignon and Falmagne, 1999, Theorem 5.5). Interestingly, in the kind of situations considered here, that is, topics in elementary mathematics, the questioning of the experts by the QUERY routine terminates relatively early. Experiments have shown that the full knowledge space may be obtained, in some cases, with $n \leq 5$ in [Q2] (see Kambouri, 1991). Nevertheless, asking questions of type [Q2] to expert teachers is not a very profitable exercise because, as mentioned earlier, their responses are not reliable. In particular, the agreement between teachers is relatively poor at least for questions of type $n \geq 2$ in [Q2]. To palliate this unreliability of the experts, Cosyn and Thiéry (2000) have developed a 'PENDING STATUS' variation of the QUERY routine in which the responses of an expert are only implemented after a confirmation. In other words, a response r to a question q is temporarily stored in a stack until a response to a later question either contradicts r (by implication), in which case q and r are deleted, or that response confirms r, which is then implemented. This technique can be used for both [Q1] and [Q2] types of questions. It has obvious advantages, and can even be modified to compute a personal index of reliability for an expert, enabling the selection of competent experts whose judgement may be more trustworthy. So far, however, it has not been used in practice.

13.3 Assessing Knowledge in a Learning Space

Taking for granted that we have a suitable learning space, the goal of an assessment is to uncover, by efficient questioning the state of a student under examination[7]. The situation is similar to that of *adaptive testing*—i.e. the computerized forms of the GRE and other standardized tests (see, for example Wainer et al., 2000)—except that the outcome of the assessment is a knowledge state, rather than a numerical estimate of a student's competence in the topic. The procedure follows a scheme outlined in Figure 13.2, which is reproduced from Falmagne et al. (2006b) (as are the other figures in this section[8]).

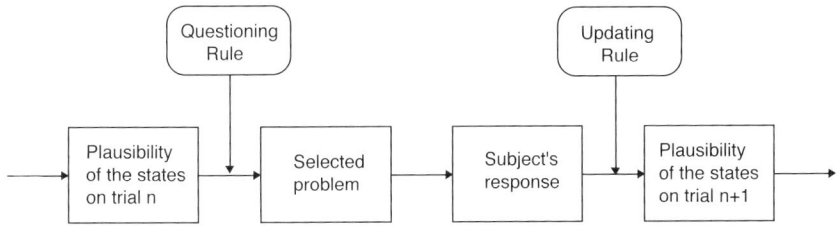

Figure 13.2. Transition diagram of the assessment procedure.

At the outset of the assessment procedure (trial 1), an *a priori* probability (referred to as 'plausibility' in Figure 13.2) is assigned to each of the knowledges states, depending on what is known about the student, such as the age or the school year. The sum of these probabilities is thus equal to 1. They play no role in the final result of the assessment but may be helpful in shortening it. If no useful information is available, then all the states are assigned the same probability. The first problem $\mathbf{p_1}$ is selected so as to be

---

[7] In this section, we follow the description of the algorithm given in Falmagne et al. (2006b,a). The algorithm was originally proposed by Falmagne and Doignon (1988a). A discrete procedure, based on a finite Markov chain, can be found in Falmagne and Doignon (1988b).
[8] With the permission of Falmagne's co-authors.

'maximally informative.' More than one interpretation can be given to this term. It means here that, on the basis of the *a priori* probability distribution on the set of states, the student has about a 50% chance of knowing how to solve $\mathbf{p_1}$. In other words, the sum of the probabilities of all the states containing $\mathbf{p_1}$ is as close to .5 as possible[9]. If several problems are maximally informative in the above sense (as may happen at the beginning of an assessment), one of these problems is randomly selected. The student is then asked to solve an instance of that problem, also picked at random. The student's response is then checked by the system, and the probability distribution is modified according to the following *updating rule*. If the student gave a correct response to $\mathbf{p_1}$, the probability of each of the states containing $\mathbf{p_1}$ is increased and, correspondingly, the probability of each of the states not containing $\mathbf{p_1}$ is decreased (so that the overall probability, summed over all the states, remains equal to 1). A false response given by the student has the opposite effect: the probability of all the states not containing $\mathbf{p_1}$ is increased, and that of the remaining states decreased. The exact formula of the operator modifying the probability distribution will not be recalled here; see Definition 10.10 in Doignon and Falmagne (1999). It is proved there that this operator is commutative, in the sense that its cumulative effect in the course of a full assessment does not depend upon the order in which the problems have been proposed to the student. This commutativity property is consistent with the fact that, as observed by Mathieu Koppen[10], this operator is Bayesian. If the student does not know how to solve a problem, he or she can choose to answer "I don't know" instead of guessing. This results in a substantial[11] increase in the probability of the states **not** containing $\mathbf{p_1}$, thereby decreasing the total number of questions required to uncover the student's state. Problem $\mathbf{p_2}$ is then chosen by a mechanism identical to that used for selecting $\mathbf{p_1}$, and the probability values are increased or decreased according to the student's response via the same updating rule. Further problems are dealt with similarly. In the course of the assessment, the probability of some states gradually increases. The assessment procedure stops when two criteria are fulfilled: (1) the entropy of the probability distribution, which measures the uncertainty of the assessment system regarding the student's state, reaches a critical low level, and (2) there is no longer any useful question to be asked (all the problems have either a very high or a very low probability of being solved correctly). At that moment, a few likely states remain and the system selects the most likely one among them. Note that, because of the stochastic nature of the assessment procedure, the final state may very well contain a problem to which

---

[9] A different interpretation of 'maximally informative' was also investigated, based on the minimization of the expected entropy of the probability distribution. This method did not result in an improvement, and was computationally more demanding (Problem 13.4).

[10] Personal communication.

[11] As compared to a false response, in which case the increase would be less: a false response could be imputed to a careless error.

the student gave a false response. Such a response is thus regarded as due to a careless error. As mentioned earlier, because all the problems have either open-ended responses or multiple choice responses with a large number of possible solutions, the probability of lucky guesses is negligible.

To illustrate the evolution of an assessment, a graphic representation is used in the form of a *probability map* of the learning space, which evolves in the course of an assessment. For practical reasons, the example given by Falmagne et al. (2006b) and reproduced here is a small scale one, which concerns a part of arithmetic consisting in 108 problems, rather that the full arithmetic domain whose large number of states would render the graphic representation computationally more difficult. In principle, each colored pixel in the oval shape of Figure 13.3 represents one of the 57,147 states of the learning space for that part of arithmetic. (Because of graphics limitations, some grouping of similar states into a single pixel was necessary.)

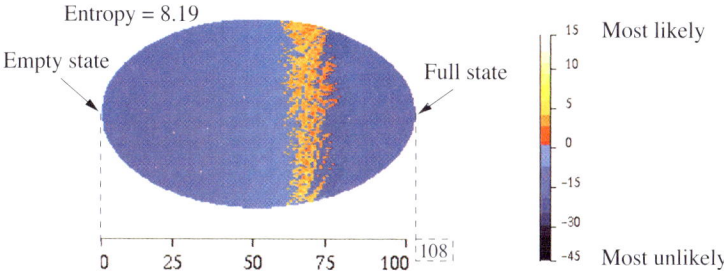

Figure 13.3. Probability map of the learning space representing the exemplary part of arithmetic under discussion.

Knowledge states are ordered along the abscissa according to the number of problems they contain, from 0 problems on the far left to 108 problems on the far right. The leftmost point stands for the empty knowledge state, which is that of a student knowing nothing at all in that part of arithmetic. The rightmost point represents the full domain of 108 problem. The points located on any vertical line within the oval represent knowledge states containing exactly the number of problems indicated on the abscissa.

The oval shape given to the map reflects, with some idealization, the fact that, by and large, there are many more states around the middle of the scale than around the edges. For instance, there are 1,668 states containing exactly 75 problems, but fewer than 100 states containing either more than 100 problems or fewer than 10 problems. The arrangement of the points on any vertical line is arbitrary.

The color of a pixel indicates the probability of the corresponding state. A color coded logarithmic scale, pictured on the right of Figure 13.3, is used to represent the values of the probabilities. Red, orange, and yellow-white

296   13 Applications

indicate states with a probability exceeding the mean of the distribution, with yellow-white marking the most likely states. Conversely, dark blue, blue, and light blue represent states that are less likely than the mean, with dark blue marking the least likely states.

The spiral display of Figure 13.4 represents a sequence of probability maps describing the evolution of an assessment from the initial state, before the first problem, to the end, after the response to the last problem is recorded by the system and acted upon to compute the last map. The complete assessment took 24 questions, which is close to the average for this part of arithmetic. The initial map results from preliminary information obtained from that student. The reddish strip of that map represents the *a priori* relatively high probabilities of the knowledge states containing between 58 and 75 problems: as a six grader, this student can be assumed to have mastered about two thirds of this curriculum.

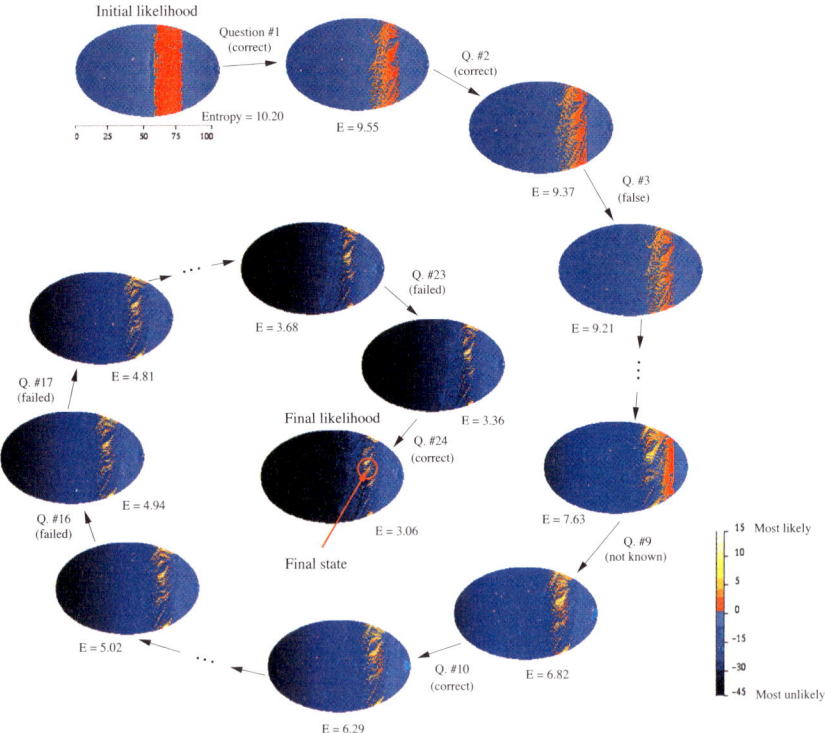

Figure 13.4. Sequence of probability maps representing an assessment converging toward the student's knowledge state. In the final map, the slected state is marked by the long arrow pointing to the circle.

Next to each of the maps in Figure 13.4 figure the entropy of the corresponding probability distribution and the student's response to the question (correct, false, or not known). The initial entropy is 10.20, which is close to the theoretical maximum of 10.96 obtained for a uniform distribution on a set of 57,147 knowledge states. The entropy decreases gradually as more information is gathered by the system via the student's responses to the questions. Eventually, after 24 questions have been answered, a single very bright point remains (indicated by the red arrow) among mostly dark blue points and a few bright points. This very bright point indicates the most likely knowledge state for that student, based on the responses to the problems. The assessment stops at that time because the entropy has reached a critical low level and the next 'best' problem to ask has only a 19% chance of being solved, and so would not be very informative. In this particular case only 24 problems have sufficed to pinpoint the student's knowledge state among 57,147 possible ones. This striking efficiency is achieved by the numerous inferences implemented by the system in the course of the assessment. With the full arithmetic curriculum from the 4th grade up, the assessment takes around 30-35 questions.

In Falmagne et al. (2006a), the authors analyze the reliability/validity of such an assessment in beginning algebra by the following method. In the course of each assessment, an extra problem is a posed to the student, the response to which is not taken into account in gauging the knowledge state of the student. The student's response can be predicted from the assessed state and correlated with the actual response. The observed correlations (in the form of odds ratios) are quite high.

13.4 The Stochastic Analysis of Opinion Polls

In our last example, we review a very different type of application of stochastic media theory in which the states of a medium coincide with those of a random walk describing the evolution of voters' opinions, as revealed by political polls. The states are typically order relations of some sort, such as semiorders or weak orders. Our discussion in this section draws on the theoretical approach and the empirical results presented in Falmagne and Doignon (1997), Regenwetter et al. (1999), Falmagne et al. (2007), Hsu et al. (2005), and Hsu and Regenwetter (in press) (see also Falmagne, 1997; Falmagne et al., 1997). The analysis of the data relies essentially on the results presented in Chapter 12, in particular Theorems 12.4.1 and 12.4.3.

The exemplary data considered here consists of the responses given by potential voters, called the *respondents*, to surveys of their opinions concerning the three major candidates in the 1992 Presidential Election in the US. The candidates in that election were George Bush, Bill Clinton and Ross Perot. Two surveys, one performed before the election, and the other just after, are analyzed jointly, the respondent being the same in both surveys. These respondent were classified in terms of their political affiliation (independent, democrat, or republican). The set of respondents is referred as a *panel* and is regarded as a random sample in the population of potential voters. For each survey, the opinion of a respondent is expressed in the form of a *thermometer score*, attributed to each of the three candidates, and ranging from 0 to 100. For one particular respondent at some time t of a survey, these scores might be, with obvious abbreviations for the names of the candidates,

$$(b, 45), \quad (c, 50), \quad \text{and} \quad (p, 37). \tag{13.10}$$

13.4.1 A Random Walk on Weak Orders. In this model, which was developed by Falmagne et al. (1997) and tested experimentally by Regenwetter et al. (1999) and Hsu and Regenwetter (in press), only the (strict) weak order relation (cf. 1.8.3, page 14) induced by the thermometer scores is retained, so that (13.10) gets translated in a straightforward manner into $p \prec b \prec c$. The relation $\prec$ is one of the 6 possible linear orders on the set $\{b, c, p\}$. Overall, there are 13 weak orders on that set, which are represented by their Hasse diagrams in each of the 13 black frames Figure 13.5. These 13 weak orders are regarded as the states of a medium, which is represented by the black digraph in Figure 13.5. (Disregard the pale red part of the figure for the moment.) There are two pairs of tokens $(\tau_i, \tilde{\tau}_i)$ and $(\tau_{-i}, \tilde{\tau}_{-i})$ per candidate, with $i \in \{b, c, p\}$. The effect of the tokens can be inferred from the digraph. It can be seen from the graph that the tokens τ_i "push" a candidate up in the weak order, inducing a different weak order in which candidate i has a different relative position, while the token τ_{-i} has the opposite effect. More precisely, let $\mathfrak{W}$ be the set of all weak orders on the set $\{b, c, p\}$; thus, $\mathfrak{W}$ is the set of states of the medium. For any tokens τ_i and τ_{-i} and any weak order $\prec \in \mathfrak{W}$, define for any $j \neq k$ in $\{b, c, p\}$ distinct from i:

$$\prec \tau_i = \prec' \iff \begin{cases} \prec = \varnothing, j \prec' i, k \prec' i, \\ \text{or } j \prec i, j \prec k, j \prec' k \prec' i, \\ \prec = \prec' \text{ otherwise;} \end{cases} \tag{13.11}$$

$$\prec \tau_{-i} = \prec' \iff \begin{cases} \prec = \varnothing, i \prec' j, i \prec' k, \\ \text{or } i \prec j, k \prec j, i \prec' k \prec' j, \\ \prec = \prec' \text{ otherwise.} \end{cases} \tag{13.12}$$

The reverses $\tilde{\tau}_i$ and $\tilde{\tau}_{-i}$ are then automatically defined. Writing $\mathfrak{T}$ for the collection of all such tokens, it is easily verified that $(\mathfrak{W}, \mathfrak{T})$ is a medium. (The digraph of Figure 13.5 represented in black defines a mediatic graph.)

As two polls have been taken, the data consists of the numbers of respondents having given thermometer scores consistent with a pair $(\prec, \prec')$ for each pair of weak orders on $\mathfrak{W}$. We are thus in the framework of the continuous random walks on media developed in Chapter 12. We suppose that the first poll was taken at time t, long after the beginning of the campain, and the second poll at time $t + \delta$, with δ relatively short. The asymptotic Theorems

13.4 The Stochastic Analysis of Opinion Polls

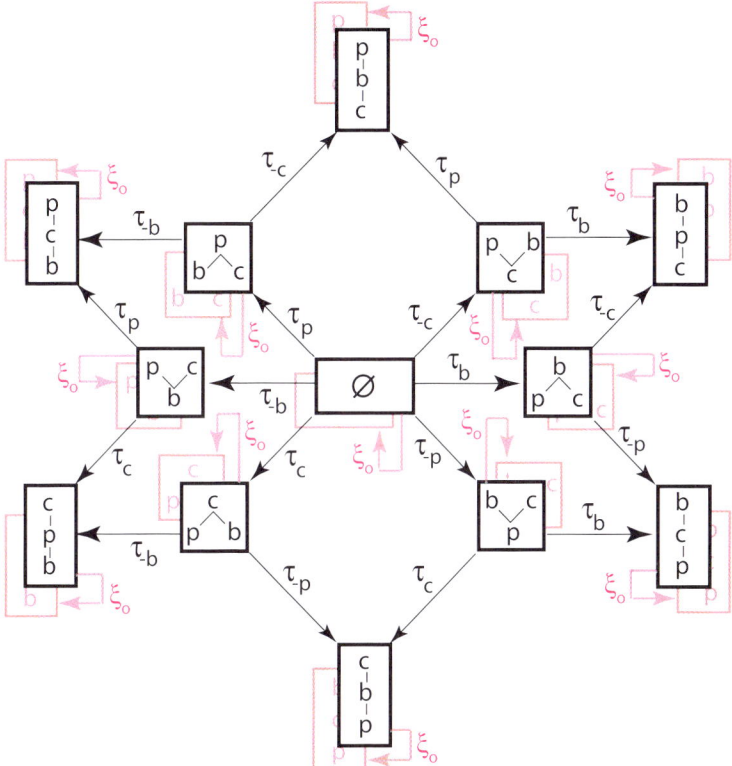

Figure 13.5. In black only: digraph of the medium of all the weak orders on the set $\{b, c, p\}$. The full digraph represents the medium of all those weak orders together with their 'frozen siblings', indicated in pale red. Each weak order is represented by its Hasse diagram. To simplify the figure, only the centrifugal tokens are indicated on the graph. The effect of a token is to 'push' a candidate one step up or down in the weak order (see the text for details).

12.4.1 and 12.4.3 are thus especially relevant. Applied to the current situation, Theorem 12.4.3 states that, for any two $\prec$, $\prec'$ in $\mathfrak{W}$,

$$\lim_{t \to \infty} \mathbb{P}(\mathbf{S}_t = \prec, \mathbf{S}_{t+\delta} = \prec')$$

$$= \underbrace{\frac{\prod_{\tau \in \widehat{\prec}} \vartheta_\tau}{\sum_{\prec'' \in \mathfrak{W}} \prod_{\mu \in \widehat{\prec''}} \vartheta_\mu}}_{A} \underbrace{\sum_{k=0}^{\infty} \xi_{\prec, \prec'}(k) \frac{(\lambda \delta)^k e^{-\lambda \delta}}{k!}}_{B}, \quad (13.13)$$

in which:

ϑ_τ is the probability of token $\tau \in \mathcal{T}$, with $\tau = \tau_i$, $\tau = \tilde{\tau}_i$, $\tau = \tau_{-i}$ or $\tau = \tilde{\tau}_{-i}$, $i \in \{b, c, p\}$ (with a similar convention for ϑ_μ);

$\xi_{\prec,\prec'}(k)$ is the probability of transition from state $\prec$ to state $\prec'$ in k steps of the random walk (that is, the entry for the cell $(\prec, \prec')$ in the kth power of the transition matrix);

λ is the parameter of the Poisson process governing the delivery of the tokens.

Note that, because the election took place between the two polls, the token probabilities involved in the factor $\xi_{\prec,\prec'}(k)$ in B, which represents the cell $(\prec, \prec')$ in the kth power of the transition matrix of the random walk, should not be assumed to be identical to those entering in the computation of the factor A in (13.13). Also, it makes sense to allow all the parameters to be different for the three political affilliations. The hypothesis that these parameters were identical was tested and soundly rejected.

Actually, in the application of the model to the polling data performed by Regenwetter et al. (1999), the political affiliation was represented by a random variable[12] with the effect that the actual equation used for the prediction is somewhat more complicated than (13.13). Also, the summation factor B in (13.13) was replaced by an approximation. The parameters were estimated by two methods: by minimizing the Chi-square statistic and by maximizing the log-likelihood of the data, using a Conjugate Gradient Algorithm in both cases. An example of the results is given in Table 13.2, which displays both the data at the top of the cell, that is the number of responses for each pair of weak orders, and the predictions of the model at the bottom[13]. There were, obviously, a similar table for each of the two other affiliations of the respondents.

While the fit of the model to the data was satisfactory from a statistical standpoint, it was noticed that most of the discrepancy between the model and the data came from the diagonals of the three tables, which contained relatively large numbers for the data entries. This can be verified in Table 13.2, in which these numbers are set in bold font (see the diagonal cells). In fact, these numbers are maximal for their row and column in all cases except three. In other words, the respondents—or at least some of them—had a striking aversion against changing their mind about the candidates, a feature which was not represented in the weak order model. To account for such a phenomenon, Hsu et al. (2005) imagined a mechanism, operating in conjunction with the weak order model, by which some respondents might at a certain point "tune

---

[12] The random selection of the panel of respondents was performed without any *a priori* control of the political affiliation.

[13] A dash in a cell indicates an empirical frequency of 0. Such cells were pooled with others for the analysis. We do not enter into such details of the analysis here (see Regenwetter et al., 1999; Hsu and Regenwetter, in press).

13.4 The Stochastic Analysis of Opinion Polls

Table 13.2. Results of the two polls for the respondents classifying themselves as independents. The rows refer to the first poll. Each cell contains the observed number of respondents (top) and the number predicted by the weak order model (bottom).

	b/p/c	bp / c	b / p/c	pb / c	p / b/c	p/c / b	cp / b	c / p/b	c / b/p	cb / p	bc / p	b / c/p	∅
b/p/c	26 / 21	3 / 7	6 / 4	–	–	2 / 1	1 / 0	–	2 / 1	4 / 2	10 / 9	14 / 10	3 / 2
bp / c	4 / 4	5 / 5	10 / 7	1 / 3	–	2 / 1	–	–	1 / 1	4 / 1	1 / 1	3 / 2	1 / 1
p/b/c	8 / 4	10 / 10	17 / 20	10 / 9	10 / 8	1 / 3	2 / 1	–	–	2 / 2	–	–	1 / 3
p / b/c	3 / 1	3 / 3	7 / 7	10 / 6	7 / 11	5 / 5	–	3 / 1	–	2 / 1	1 / 1	1 / 1	1 / 2
p/c / b	1 / 0	1 / 2	3 / 5	9 / 8	38 / 30	12 / 14	12 / 9	–	–	2 / 1	2 / 0	2 / 0	–
c/p / b	1 / 0	1 / 1	–	1 / 2	8 / 10	6 / 7	13 / 10	3 / 2	2 / 2	2 / 1	–	3 / 0	–
c / p/b	–	–	–	1 / 1	6 / 5	11 / 8	58 / 71	9 / 10	9 / 9	–	2 / 0	–	2 / 1
c / b/p	–	–	–	–	2 / 1	3 / 3	13 / 14	10 / 7	10 / 11	–	1 / 1	–	3 / 1
c/b / p	1 / 0	–	–	1 / 1	3 / 2	3 / 4	33 / 19	15 / 18	37 / 35	5 / 8	3 / 3	1 / 1	–
b/c / p	–	2 / 2	3 / 1	3 / 2	2 / 2	5 / 3	2 / 5	1 / 5	11 / 10	17 / 7	3 / 7	2 / 3	3 / 2
b/c/p	13 / 11	2 / 3	4 / 1	3 / 2	1 / 1	–	–	3 / 2	10 / 6	11 / 10	32 / 32	11 / 18	2 / 4
b / c/p	9 / 8	–	2 / 1	1 / 1	1 / 1	–	1 / 1	1 / 1	2 / 2	3 / 3	6 / 11	15 / 8	3 / 2
∅	1 / 1	2 / 1	1 / 1	2 / 1	–	3 / 2	2 / 3	3 / 1	–	3 / 1	1 / 1	1 / 1	4 / 1

out" and stop paying attention to the stimuli relevant to the campaign. This could happen under the influence of some unobservable or unrecorded event, such as excessively negative campaigning, or simply "election fatigue." Such respondents might of course at some point "tune in" again for some reason or other. This mechanism was implemented in a framework very similar to that of the weak order model by the introduction, for each weak order $\prec \in \mathfrak{W}$, of a *frozen sibling* $\prec^*$ representing the same weak order on $\{b, c, p\}$ but a different state of the medium, in which the respondent was momentarily impervious to events of the campaign. Two new types of mutually reverse tokens were also introduced, the effect of which was to induce a respondent to change his or her state from $\prec$ to $\prec^*$ or vice versa. To avoid a substantial increase in the number of parameters in the model, they assumed that the same pair of parameters ξ_o and $\tilde{\xi}_o$ were governing the probabilities of transition between any weak order of a standard kind and its frozen sibling. The resulting medium, whose graph is indicated in Figure 13.5, was referred to as the *tune-in-and-out extension* of the weak order model, or *TIO extension*. Note that each frozen sibling is pictured in pale red and partly hidden behind its corresponding 'responsive' sibling. Following our previous convention, only the centrifugal tokens are represented in the graph.

The idea behind the TIO extension can obviously be applied to other order relations, such as the semiorders or the linear orders. The TIO extension of the weak order model was systematically tested, together with a variety of other models, by Hsu and Regenwetter (in press). It was shown to provide the best explanation, among those considered, for the panel data collected for the three US presidential election of 1992, 1996 and 2000.

13.5 Concluding Remarks

The two applications described this chapter have led to extensive empirical implementation. The learning spaces discussed in the first part of the chapter enable the assessment engine of a mathematical education software used today by several hundreds thousands students in the US and elsewhere. The random walks on media have been successfully and repeatedly used for the study of opinion polls, based on large sets of data.

On the face of it, these applications differ widely. The most important difference lies in the conceptual meanings of the states of the medium in the two cases. In the learning spaces, a state is a collection of problem types that a student has mastered in a mathematical topic, while in the case of the opinion polls, a state is an order relation, such as a weak order, a semiorder or a linear order. Another important distinction concerns the way probability and stochastic processes are used in the two cases.

Yet, the concept of a medium is at the core of the structure in both cases, and the role played by the axioms is essential. It is our view that these two examples are only the beginning of a potentially long list of applications. To identify those empirical cases susceptible of an interpretation in terms of media, some criteria have to be fulfilled, which we review here as our parting pointer. There are four such criteria.

1. The set of features. There must be a fundamental set of defining features or units: the problem types in the learning spaces are such features. When the states are order relations, the features may be the ordered pairs of some basic set. But not always. Sometimes, the features may be quite subtle. What could be the units in the case of the weak orders? Why shouldn't the ordered pairs form these units in such a case (Problem 13.5)?

2. The states or the organisms made of those features. Each of the states must be definable by a unique subset of those features. Media theory provides the algorithm for finding this subset: it consists of the content of the state.

3. The links between the states or the organisms. Even though the states can be very large, even uncountable, there is always at least one direct, finite path—a sequence of features—from on state S to another state V. This path tells you exactly in what way S and V differ. Actually, the path can be viewed as an algorithm for constructing a state from another state.

The above description begins to suggest possible applications in biology (genetics), chemistry, or other fields in which the basic objects are made of parts assembled in a certain way. However, there is a critical restriction:

4. In general, there is no golden path between two states. When there are several paths—we have called those 'messages'—all of them must have the same length, and must go through the same steps—we have called them 'tokens'. Only the order of those steps may differ.

Problems

13.1 On the basis of the Hasse diagram of Figure 13.1, specify completely the entailment relation restricted to the six problem types (a)-(f).

13.2 Verify that the r.h.s. of (13.1) defines an equivalence relation on Q, for any knowledge structure $\mathcal{K}$. Clarify the relationship between that equivalence relation and the quasi orders in the statement of Theorem 13.1.5.

13.3 Prove Theorem 13.2.1.

13.4 Design an algorithm for selecting p_1 (and $p_2, p_3, \ldots$) which is maximally informative in a sense different from that described on page 294, and based on the idea of minimizing the expected entropy of the probability distribution on the set of knowledge states on the next trial of the assessment.

13.5 What are the 'units' making the states in the example of the opinion polls analyzed in terms of weak orders (cf. our question on page 302). How is the definition of a unit changed in the TIO extension of the weak order model?

13.6 How many parameters must be estimated in the weak order model with a panel selected with a fixed number of respondents, but random subsamples of independents, democrat and republicans?

13.7 Write an expression based on Eq. (13.13) for the joint probability of observing (after recoding the thermometer scores) the weak orders $\prec$ and $\prec'$ at times t and $t + \delta$ (respectively) for large t and for a respondent selected randomly in a panel composed of N respondents, with n_i independents, n_d democrats, and n_r republicans.

13.8 Can you think of plausible application of stochastic media theory in sport, based on one or more rankings of a fixed set of top athletes, based on their performance and updated weekly? How about developing the equations of a model for such an aplication?

Appendix: A Catalog of Small Mediatic Graphs

p	connected bipartite graphs	mediatic graphs	trees
2	1	1	1
3	1	1	1
4	3	3	2
5	5	4	3
6	17	12	6
7	44	25	11
8	182	78	23

Table A. The number of connected bipartite graphs, mediatic graphs, and trees with 2–8 vertices.

Any tree T with q edges and m leaves is a mediatic graph of isometric dimension $\dim_I(T) = q$ and of lattice dimesion $\dim_{\mathbb{Z}}(T) = \lceil m/2 \rceil$.

All mediatic graphs with less than nine vertices that are not trees are shown in Figures A.1–A.3. The numbers $p, q; r, s$ at the bottom of cells stand for:

p - the number of vertices of the graph,
q - the number of edges of the graph,
r - the isometric dimension of the graph,
s - the lattice dimension of the graph.

The coloring of vertices corresponds to the bipartition of the vertex set.

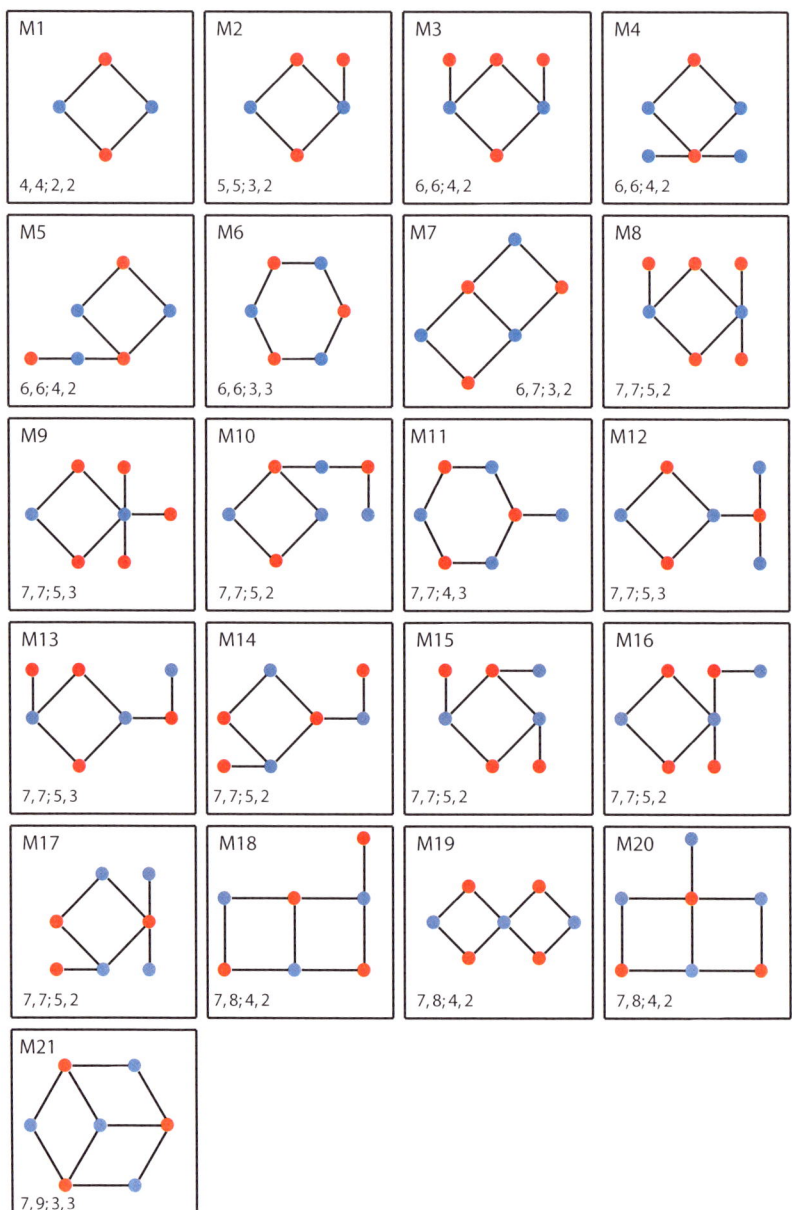

Figure A.1. Mediatic graphs with 4–7 vertices; not trees.

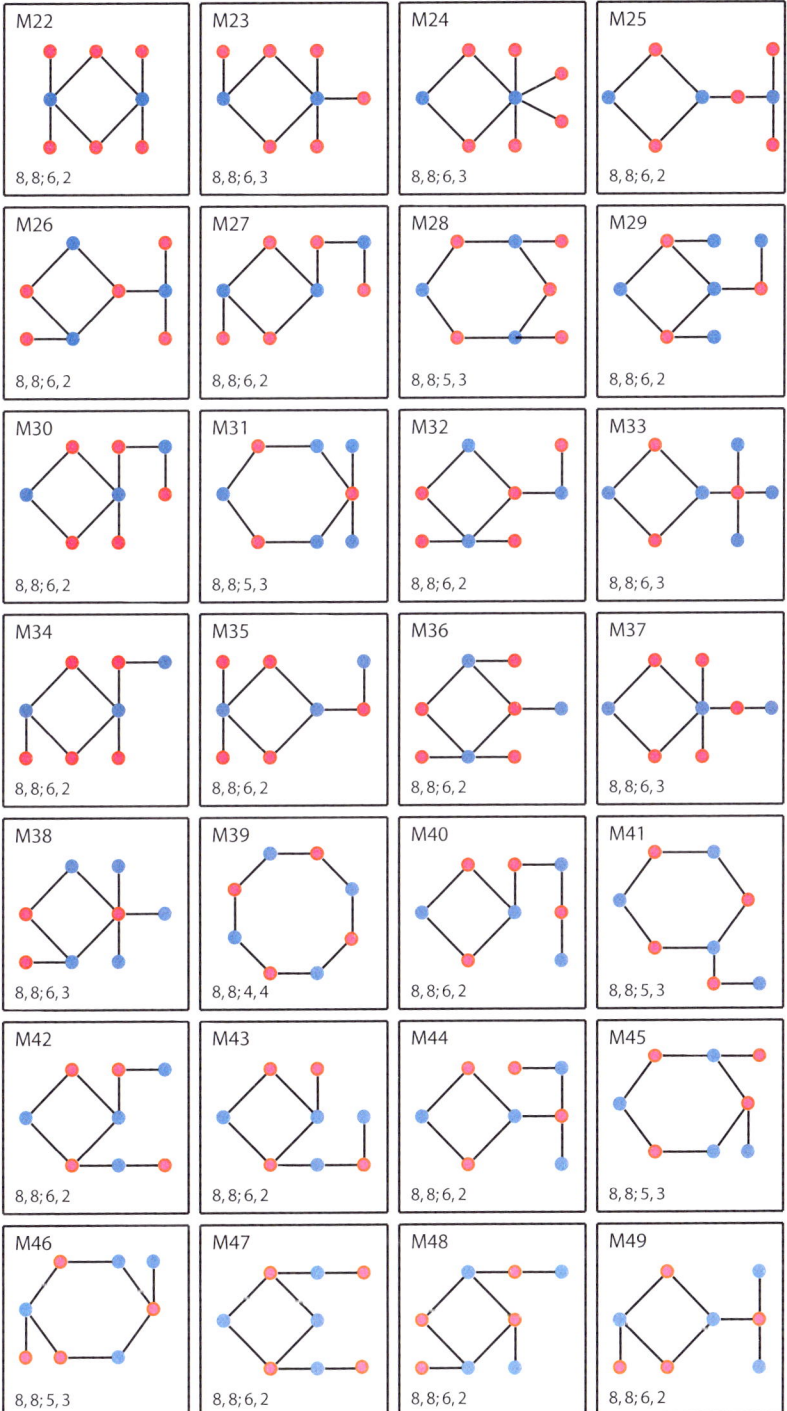

Figure A.2. Mediatic graphs with 8 vertices, part I; not trees.

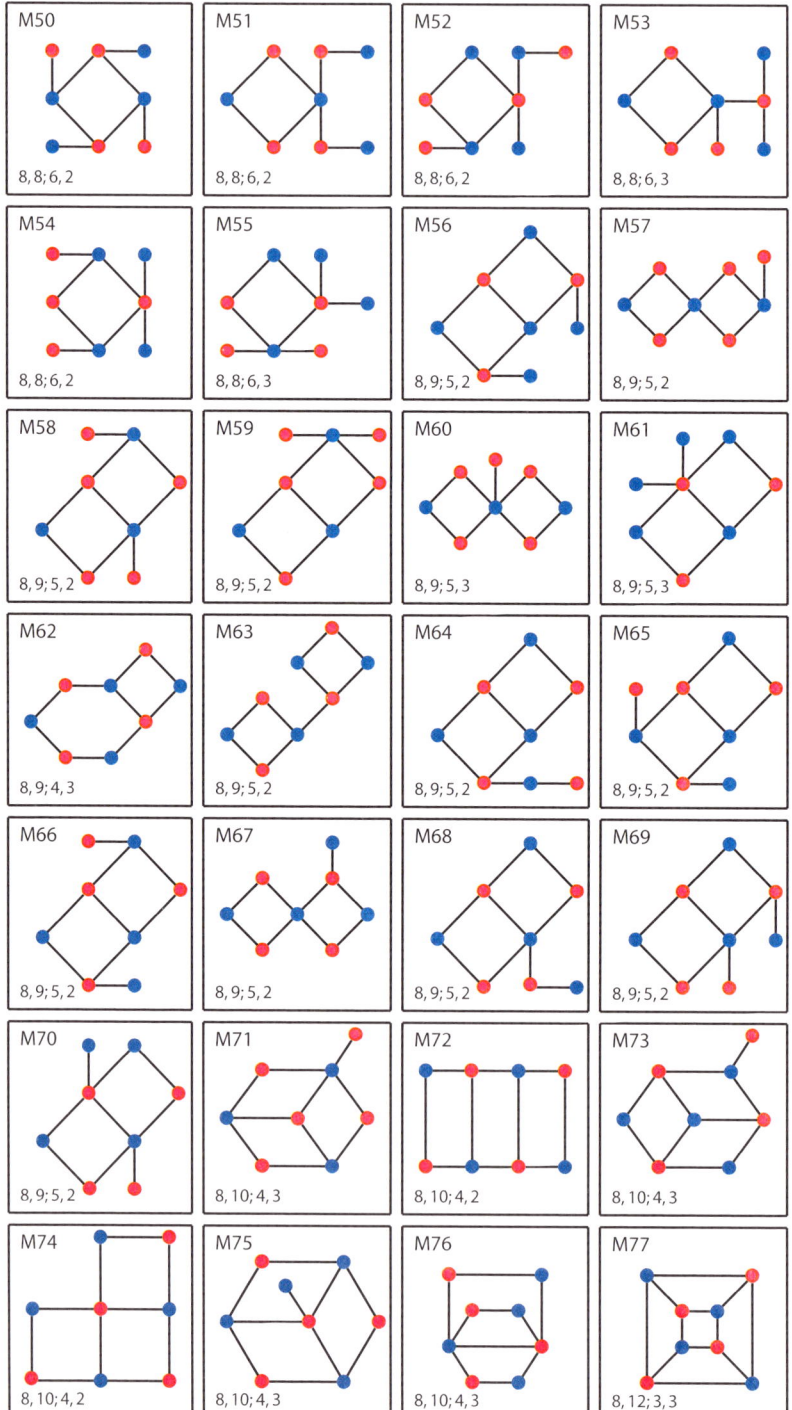

Figure A.3. Mediatic graphs with 8 vertices, part II; not trees.

Glossary

$\Leftrightarrow$	the logical equivalence standing for 'if and only if'
$\&, \neg$	the logical conjunction 'and' and negation 'not'
$\exists, \forall$	existential and universal quantifiers
$\varnothing$	empty set
$\in$	set membership
$\subseteq, \subset$	set inclusion, and proper (or strict) set inclusion
$\cup, \cap, \setminus$	union, intersection, and difference of sets
$+, \sum$	ordinary addition or the union of disjoint sets
$f(B)$	if B is a set and f a function, the image of B by f
$\|X\|$	number of elements (or cardinality) of a set X
$\mathfrak{P}(X)$	power set of the set X (i.e., the set of all subsets of X)
$\mathfrak{P}_F(X)$	the set of all finite subsets of X
$X_1 \times \ldots \times X_n$	Cartesian product of the sets $X, \ldots, X_n$
$\mathbb{N}$	the set of all natural numbers $1, \ldots, n, \ldots$
$\mathbb{N}_0$	the set of nonnegative integers $0, 1, \ldots, n, \ldots$
$\mathbb{Q}$	the set of all rational numbers
$\mathbb{R}$	the set of all real numbers
$\mathbb{R}_+$	the set of all nonnegative real numbers $[0, \infty[$
$\mathbb{Z}$	the set of all integers
$\mathbb{P}$	a probability measure
$\mathcal{S}$	typically, the set of states of a medium
$\widehat{S}$	content of a state S
$\widehat{\mathcal{S}}$	family of all the contents of the states in $\mathcal{S}$
token	a mapping $\tau : \mathcal{S} \to \mathcal{S} : S \mapsto S\tau$
message	a composition of tokens
$\mathcal{T}$	typically, the set of tokens of a medium
$\{\mathcal{T}^+, \mathcal{T}^-\}$	a bipartition of $\mathcal{T}$ defining an orientation
$S\tau$	the image of the state S by the token τ
$S\boldsymbol{m}$	the image of the state S by the message $\boldsymbol{m}$

$]x,y[$	open interval of real numbers $\{z \in \mathbb{R} \mid x < z < y\}$
$[x,y]$	closed interval of real numbers $\{z \in \mathbb{R} \mid x \leq z \leq y\}$
$]x,y], [x,y[$	real, half open intervals
$\check{R}$	Hasse diagram of a partial order R
R^*	transitive closure of a relation R
$\mathbb{S}(\mathcal{F}), \mathbb{S}_\mathrm{F}(\mathcal{F})$	span and finite span of a family $\mathcal{F}$ of sets
$\square$	marks the end of a proof
l.h.s., r.h.s.	'left-hand side' and 'right-hand side' (of a formula)
$\langle \ldots \rangle$	betweenness relation for triples of points; or the class containing "..." in a partition; or a weak order in terms of its indifference classes (cf. Theorem 9.4.6)

Bibliography

B. Aspvall, M.F. Plass, and R.E. Tarjan. A linear time algorithm for testing the truth of certain quantified Boolean formulas. *Information Processing Letters*, 8:121–123, 1979.

A.S. Asratian, T.M.J. Denley, and R. Häggkvist. *Bipartite Graphs and their Applications*. Cambridge University Press, Cambridge, London, and New Haven, 1998.

D. Avis and K. Fukuda. Reverse search for enumeration. *Discrete Applied Mathematics*, 65:21–46, 1996.

L. Babai and E. Szemerédi. On the complexity of matrix group problems I. In *Proceedings of the Twenty Fifth Annual IEEE Symposium on Foundations of Computer Science*, pages 229–240, Los Alamitos, Calif., 1984. IEEE Computer Society Press.

H.J. Bandelt. Retracts of hypercubes. *Journal of Graph Theory*, 8(4):501–510, 1984.

H.J. Bandelt and V. Chepoi. Metric graph theory and geometry: a survey. Manuscript, 2005.

M. Barbut and B. Monjardet. *Ordre et Classification*. Collection Hachette Université, Paris, 1970.

A.T. Barucha-Reid. *Elementary Probability with Stochastic Processes*. Springer-Verlag, Berlin, Heidelberg, and New York, 1974.

M.A. Bender and M. Farach-Colton. The LCA problem revisited. In *LATIN 2000: Theoretical Informatics, 4th Latin American Symposium, Punta del Este, Uruguay, April 10-14, 2000, Proceedings*, volume 1776 of *Lecture Notes in Computer Science*, pages 88–94, Berlin, Heidelberg, and New York, 2000. Springer-Verlag.

L.W. Berman. Symmetric simplicial pseudoline arrangements. Manuscript, submitted for journal publication., 2007.

S.N. Bhatt and S.S. Cosmodakis. The complexity of minimizing wire lengths in VLSI layouts. *Information Processing Letters*, 25:263–267, 1987.

P. Bidigare, P. Hanlon, and D. Rockmore. A combinatorial description of the spectrum for the Tsetlin library and its generalization to hyperplane arrangements. *Duke Mathematical Journal*, 99(1):135–174, 1999.

G. Birkhoff. Rings of sets. *Duke Mathematical Journal*, 3:443–454, 1937.

G. Birkhoff. *Lattice Theory*. American Mathematical Society, Providence, R.I., 1967.

A. Björner, M. Las Vergnas, B. Sturmfels, N. White, and G.M. Ziegler. *Oriented Matroids*. Cambridge University Press, Cambridge, London, and New Haven, second edition, 1999.

K.P. Bogart. Maximal dimensional partially ordered sets. I. Hiraguchi's theorem. *Discrete Mathematics*, 5:21–31, 1973.

K.P. Bogart and W.T. Trotter. Maximal dimensional partially ordered sets. II. Characterization of $2n$-element posets with dimension n. *Discrete Mathematics*, 5:33–43, 1973.

S.A. Bogatyi. Metrically homogeneous spaces. *Russian Mathematical Surveys*, 57:221–240, 2002.

J.A. Bondy. Basic graph theory: paths and circuits. In R.L. Graham, M. Grötschel, and L. Lovász, editors, *Handbook of Combinatorics*, volume 1. The M.I.T. Press, Cambridge, MA, 1995.

J.A. Bondy and U.S.R. Murphy. *Graph Theory with Applications*. North-Holland Publishing Co., Amsterdam and New York, 1976.

N. Bourbaki. *Lie Groups and Lie Algebras*. Springer-Verlag, Berlin, Heidelberg, and New York, 2002.

V.J. Bowman. Permutation polyhedra. *SIAM Journal on Applied Mathematics*, 22:580–589, 1972.

G.S. Brodal and L. Gąsieniec. Approximate dictionary queries. In *Proceedings of the 7th Annual Symposium on Combinatorial Pattern Matching*, pages 65–74, 1996.

K.S. Brown and P. Diaconis. Random walks and hyperplane arrangements. *Annals of Probability*, 26(4):1813–1854, 1998.

D. Burago, Y. Burago, and S. Ivanov. *A Course in Metric Geometry*. American Mathematical Society, Providence, R.I., 2001.

R.G. Busacker and T.L. Saaty. *Finite Graphs and Networks—An Introduction with Applications*. McGraw-Hill, Boston and New York, 1965.

D.R. Cavagnaro. Projection of a medium. Submitted to *Journal of Mathematical Psychology*, 2006.

T.M. Chan. On levels in arrangements of curves. *Discrete and Computational Geometry*, 29(3):375–393, April 2003.

V. Chepoi, F. Dragan, and Y. Vaxès. Center and diameter problems in plane triangulations and quadrangulations. In *Proceedings of the Thirteenth Annual ACM-SIAM Symposium on Discrete Algorithms*, pages 346–355, January 2002.

J.S. Chipman. Consumption theory without transitive indifference. In J.S. Chipman, L. Hurwicz, M.K. Richter, and H.F. Sonnenschein, editors, *Pref-*

erences, Utility, and Demand, pages 224–253. Harcourt Brace Jovanovich, 1971.

K.L. Chung. *Markov Chains with Stationary Transition Probabilities*. Springer-Verlag, Berlin, Heidelberg, and New York, 2nd edition, 1967.

O. Cogis. Ferrers digraphs and threshold graphs. *Discrete Mathematics*, 38: 33–46, 1982.

T.H. Cormen, C.E. Leiserson, R.L. Rivest, and C. Stein. *Introduction to Algorithms*. The M.I.T. Press, Cambridge, MA, 2nd edition, 2003.

E. Cosyn and N. Thiéry. A Practical Procedure to Build a Knowledge Structure. *Journal of Mathematical Psychology*, 44:383–407, 2000.

E. Cosyn and H.B. Uzun. Axioms for learning spaces. To be submitted to *Journal of Mathematical Psychology*, 2005.

H.S.M. Coxeter. *Regular Polytopes*. Dover, 1973.

B.A. Davey and H.A. Priestley. *Introduction to Lattices and Order*. Cambridge University Press, Cambridge, London, and New Haven, 1990.

N.G. de Bruijn. Algebraic theory of Penrose's non-periodic tilings of the plane. *Indagationes Mathematicae*, 43:38–66, 1981.

H. de Fraysseix and P. Ossona de Mendez. Stretching of Jordan arc contact systems. In *Graph Drawing: 11th International Symposium, GD 2003, Perugia, Italy, September 21–24, 2003*, volume 2912 of *Lecture Notes in Computer Science*, pages 71–85, Berlin, Heidelberg, and New York, 2004. Springer-Verlag.

M. Deza and M. Laurent. *Geometry of Cuts and Metrics*. Springer-Verlag, Berlin, Heidelberg, and New York, 1997.

M. Deza and M.I. Shtogrin. Mosaics and their isometric embeddings. *Izvestia: Mathematics*, 66:443–462, 2002.

G. Di Battista and R. Tamassia. Incremental planarity testing. In *Proceedings of the Thirtieth Annual IEEE Symposium on Foundations of Computer Science*, pages 436–441, Los Alamitos, Calif., 1989. IEEE Computer Society Press.

G. Di Battista, P. Eades, R. Tamassia, and I.G. Tollis. *Graph Drawing: Algorithms for the Visualization of Graphs*. Prentice Hall, Englewood Cliffs, New Jersey, 1999.

R. Diestel. *Graph Theory*. Springer-Verlag, Berlin, Heidelberg, and New York, 2000.

R.P. Dilworth. Lattices with unique irreducible decompositions. *Annals of Mathematics*, 41:771–777, 1940.

D.Z. Djoković. Distance preserving subgraphs of hypercubes. *Journal of Combinatorial Theory, Ser. B*, 14:263–267, 1973.

C.W. Doble, J.-P. Doignon, J.-Cl. Falmagne, and P.C. Fishburn. Almost connected orders. *Order*, 18(4):295–311, 2001.

J.-P. Doignon and J.-Cl. Falmagne. Well-graded families of relations. *Discrete Mathematics*, 173:35–44, 1997.

J.-P. Doignon and J.-Cl. Falmagne. *Knowledge Spaces*. Springer-Verlag, Berlin, Heidelberg, and New York, 1999.

J.-P. Doignon and J.-Cl. Falmagne. Spaces for the Assessment of Knowledge. *International Journal of Man-Machine Studies*, 23:175–196, 1985.

J.-P. Doignon, A. Ducamp, and J.-Cl. Falmagne. On realizable biorders and the biorder dimension of a relation. *Journal of Mathematical Psychology*, 28:73–109, 1984.

C.E. Dowling. Applying the basis of a knowledge space for controlling the questioning of an expert. *Journal of Mathematical Psychology*, 37:21–48, 1993a.

C.E. Dowling. On the irredundant construction of knowledge spaces. *Journal of Mathematical Psychology*, 37:49–62, 1993b.

C.E. Dowling. Integrating different knowledge spaces. In G.H. Fischer and D. Laming, editors, *Contributions to Mathematical Psychology, Psychometrics, and Methodology*, pages 149–158. Springer-Verlag, Berlin, Heidelberg, and New York, 1994.

A. Ducamp and J.-Cl. Falmagne. Composite measurement. *Journal of Mathematical Psychology*, 6:359–390, 1969.

B. Dushnik and E.W. Miller. Partially ordered sets. *American Journal of Mathematics*, 63:600–610, 1941.

J. Ebert. st-ordering of the vertices of biconnected graphs. *Computing*, 30: 19–33, 1983.

P.H. Edelman and R. Jamison. The theory of convex geometries. *Geometrica Dedicata*, 19:247–271, 1985.

H. Edelsbrunner. *Algorithms in Combinatorial Geometry*. Springer-Verlag, 1987.

D. Eppstein. Happy endings for flip graphs. In *Proc. 23rd Annual ACM Symp. Computational Geometry*, 2007a. Electronic preprint cs.CG/0610092, arXiv.org.

D. Eppstein. Algorithms for drawing media. In *Graph Drawing: 12th International Symposium, GD 2004, New York, NY, USA, September 29–October 2, 2004*, volume 3383 of *Lecture Notes in Computer Science*, pages 173–183, Berlin, Heidelberg, and New York, 2005a. Springer-Verlag.

D. Eppstein. The lattice dimension of a graph. *European Journal of Combinatorics*, 26(6):585–592, 2005b.

D. Eppstein. Upright-quad drawing of st-planar learning spaces. In *Graph Drawing: 14th International Symposium, GD 2006, Karlsruhe, Germany, September 18–20, 2006*, Lecture Notes in Computer Science, Berlin, Heidelberg, and New York, 2006. Springer-Verlag.

D. Eppstein. Recognizing partial cubes in quadratic time. Electronic preprint 0705.1025, arXiv.org, 2007b.

D. Eppstein and J.-Cl. Falmagne. Algorithms for media. Electronic preprint cs.DS/0206033, arXiv.org, 2002.

D. Eppstein, J.-Cl. Falmagne, and H.B. Uzun. On verifying and engineering the well-gradedness of a union-closed family. Electronic preprint 0704.2919, arXiv.org, 2007.

S. Even and R.E. Tarjan. Computing an *st*-numbering. *Theoretical Computer Science*, 2:339–344, 1976.

J.-Cl. Falmagne. *Lectures in Elementary Probability and Stochastic Processes*. McGraw-Hill, Boston and New York, 2003.

J.-Cl. Falmagne. A stochastic theory for the emergence and the evolution of preference structures. *Mathematical Social Sciences*, 31:63–84, 1996.

J.-Cl. Falmagne. Stochastic token theory. *Journal of Mathematical Psychology*, 41(2):129–143, 1997.

J.-Cl. Falmagne and J.-P. Doignon. Stochastic evolution of rationality. *Theory and Decision*, 43:107–138, 1997.

J.-Cl. Falmagne and J-P. Doignon. A class of stochastic procedures for the assessment of knowledge. *British Journal of Mathematical and Statistical Psychology*, 41:1–23, 1988a.

J.-Cl. Falmagne and J-P. Doignon. A Markovian procedure for assessing the state of a system. *Journal of Mathematical Psychology*, 32:232–258, 1988b.

J.-Cl. Falmagne and S. Ovchinnikov. Media theory. *Discrete Applied Mathematics*, 121:83–101, 2002.

J.-Cl. Falmagne and S. Ovchinnikov. Mediatic graphs. Electronic preprint 0704.0994, arXiv.org, 2007.

J.-Cl. Falmagne, M. Regenwetter, and B. Grofman. A stochastic model for the evolution of preferences. In A.A.J. Marley, editor, *Choice, Decision and Measurement: Essays in Honor of R. Duncan Luce*. Erlbaum, New Jersey and London, 1997.

J.-Cl. Falmagne, E. Cosyn, C. Doble, N. Thiéry, and H.B. Uzun. Assessing Mathematical Knowledge in a Learning Space: Validity and/or Reliability. Submitted to *Psychological Review*, 2006a.

J.-Cl. Falmagne, E. Cosyn, J.-P. Doignon, and N. Thiéry. The assessment of knowledge, in theory and in practice. In B. Ganter and L. Kwuida, editors, *Formal Concept Analysis, 4th International Conference, ICFCA 2006, Dresden, Germany, February 13–17, 2006*, Lecture Notes in Artificial Intelligence, pages 61–79. Springer-Verlag, Berlin, Heidelberg, and New York, 2006b.

J.-Cl. Falmagne, Y.-F. Hsu, F. Leite, and M. Regenwetter. Stochastic applications of media theory: Random walks on weak orders or partial orders. *Discrete Applied Mathematics*, 2007. doi: 10.1016/j.dam.2007.04.032.

W. Feller. *An Introduction to Probability Theory and its Applications*, volume 1. John Wiley & Sons, London and New York, 3rd edition, 1968.

S. Felsner and H. Weil. Sweeps, arrangements and signotopes. *Discrete Applied Mathematics*, 109(1):67–94, April 2001.

S. Fiorini and P.C. Fishburn. Weak order polytopes. *Discrete Mathematics*, 275:111–127, 2004.

P.C. Fishburn. *Utility Theory for Decision Making*. John Wiley & Sons, London and New York, 1970.

P.C. Fishburn. *Interval orders and interval graphs*. John Wiley & Sons, London and New York, 1985.

P.C. Fishburn. Generalizations of semiorders: A review note. *Journal of Mathematical Psychology*, 41:357–366, 1997.

P.C. Fishburn and W.T. Trotter. Split semiorders. *Discrete Mathematics*, 195:111–126, 1999.

P. Frankl. Extremal Set Systems. In R.L. Graham, M. Grötschel, and L. Lovász, editors, *Handbook of Combinatorics*, volume 2. The M.I.T. Press, Cambridge, MA, 1995.

P. Gaiha and S.K Gupta. Adjacent vertices on a permutohedron. *SIAM Journal on Applied Mathematics*, 32(2):323–327, 1977.

B. Gärtner and E. Welzl. Vapnik-Chervonenkis dimension and (pseudo-)hyperplane arrangements. *Discrete and Computational Geometry*, 12:399–432, 1994.

J.G. Gimbel and A.N. Trenk. On the weakness of an ordered set. *SIAM Journal of Discrete Mathematics*, 11(4):655–663, 1998.

S. Ginsburg. On the length of the smallest uniform experiment which distinguishes the terminal states of a machine. *Journal of the ACM*, 5:266–280, 1958.

S.W. Golomb. *Polyominoes: Puzzles, Patterns, Problems, and Packings*. Princeton Science Library. Princeton University Press, NJ, 2nd edition, 1994.

R.L. Graham and H. Pollak. On addressing problem for loop switching. *Bell Systems Technical Journal*, 50:2495–2519, 1971.

B. Grünbaum. *Arrangements and Spreads*. Number 10 in Regional Conference Series in Mathematics. American Mathematical Society, Providence, R.I., 1972.

B. Grünbaum and G.C. Shephard. *Tilings and Patterns*. W.H. Freeman, New York, 1987.

I. Gutman and S.J. Cyvin. *Introduction to the Theory of Benzenoid Hydrocarbons*. Springer-Verlag, Berlin, Heidelberg, and New York, 1989.

L. Guttman. A basis for scaling qualitative data. *American Sociological Review*, 9:139–150, 1944.

F. Hadlock and F. Hoffman. Manhatten trees. *Utilitas Mathematica*, 13:55–67, 1978.

K. Handa. A characterization of oriented matroids in terms of topes. *European Journal of Combinatorics*, 11:41–45, 1990.

F. Harary. *Graph Theory*. Addison-Wesley, Reading, Mass., 1969.

Y.-F. Hsu and M. Regenwetter. Application of stochastic media theory to 1992, 1996 and 2000 national election study panel data. *Chinese Journal of Psychology*, in press.

Y.-F. Hsu, J.-Cl. Falmagne, and M. Regenwetter. The tuning in-and-out model: a random walk and its application to presidential election surveys. *Journal of Mathematical Psychology*, 49:276–289, 2005.

W. Imrich and S. Klavžar. *Product Graphs*. John Wiley & Sons, London and New York, 2000.

W. Imrich, S. Klavžar, and H.M. Mulder. Median graphs and triangle-free graphs. *SIAM Journal on Discrete Mathematics*, 12(1):111–118, 1999.

M.F. Janowitz. On the semilattice of weak orders of a set. *Mathematical Social Sciences*, 8:229–239, 1984.

D. Joyner. *Adventures in Group Theory—Rubik's Cube, Merlin's Machine, and other Mathematical Toys*. The Johns Hopkins University Press, Baltimore and London, 2002. ISBN 0-8018-6947-1.

M. Kambouri. *Knowledge assessment: A comparison between human experts and computerized procedure*. PhD thesis, New York University, New York, 1991.

M. Kambouri, M. Koppen, M. Villano, and J.-Cl. Falmagne. Knowledge assessment: Tapping human expertise by the QUERY routine. *International Journal of Human-Computer Studies*, 40:119–151, 1994.

R.M. Karp, E. Upfal, and A. Wigderson. Are search and decision problems computationally equivalent? In *Proceedings of the Seventeenth Annual ACM Symposium on Theory of Computing*, pages 464–475, May 1985.

J.G. Kemeny and J.L. Snell. *Finite Markov Chains*. Van Nostrand, Princeton, N.J., 1960.

J.G. Kemeny and J.L. Snell. *Mathematical Models in Social Sciences*. The M.I.T. Press, Cambridge, MA, 1972.

D. König. Über Graphen und ihren Anwendung auf Determinanten-theorie und Mengenlehre. *Mathematische Annalen*, 77:453–465, 1916.

M. Koppen. Extracting human expertise for constructing knowledge spaces: An algorithm. *Journal of Mathematical Psychology*, 37:1–20, 1993.

M. Koppen. *Ordinal Data Analysis: Biorder Representation and Knowledge Spaces*. Doctoral dissertation, Katholieke Universiteit te Nijmegen, Nijmegen, The Netherlands, 1989.

B. Korte, L. Lovász, and R. Schrader. *Greedoids*. Number 4 in Algorithms and Combinatorics. Springer-Verlag, 1991.

V.B. Kuzmin and S. Ovchinnikov. Geometry of preference spaces I. *Automation and Remote Control*, 36:2059–2063, 1975.

V.B. Kuzmin and S. Ovchinnikov. Geometry of preference spaces II. *Automation and Remote Control*, 37:110–113, 1976.

Cl. Le Conte de Poly-Barbut. Le diagramme du treillis permutoèdre est Intersection des diagrammes de deux produits directs d'ordres totaux. *Mathématiques, Informatique et Sciences Humaines*, 112:49–53, 1990.

G. Lo Faro. A note on the union-closed sets conjecture. *Journal of the Australian Mathematical Society, Ser. A*, 57:230–236, 1994a.

G. Lo Faro. Union-closed sets conjecture: improved bounds. *Journal of Combinatorial Mathematics*, 16:97–102, 1994b.

R.D. Luce. Semiorders and a theory of utility discrimination. *Econometrica*, 24:178–191, 1956.

R.D. Luce and E. Galanter. Psychophysical scaling. In R.D. Luce, R.R. Bush, and E. Galanter, editors, *Handbook of Mathematical Psychology*, volume 1. John Wiley & Sons, London and New York, 1963.

J. Matoušek. The number of unique-sink orientations of the hypercube. *Combinatorica*, 26(1):91–99, 2006.

D.W. Matula. Graph-theoretical cluster analysis. In S. Kotz and N.L. Johnson, editors, *Encyclopedia of Statistical Sciences*, volume 3. John Wiley & Sons, London and New York, 1983. C.B. Read, associate editor.

S. Micali and V.V. Vazirani. An $O(\sqrt{V}E)$ algorithm for finding maximum matching in general graphs. In *Proceedings of the Twenty First Annual IEEE Symposium on Foundations of Computer Science*, pages 17–27, Los Alamitos, Calif., 1980. IEEE Computer Society Press.

B.G. Mirkin. *Group Choice*. Winston, Washington, D.C., 1979.

B. Monjardet. Axiomatiques et propriétés des quasi-ordres. *Mathématique et Sciences Humaines*, 63:51–82, 1978.

E.F. Moore. Gedanken-experiments on sequential machines. *Automata Studies*, Princeton University Press, Annals of Mathematics Studies 34:129–153, 1956.

H.M. Mulder. The interval function of a graph. Mathematical Centre Tracts 142, Mathematical Centre, Amsterdam, 1980.

C.E. Müller. A procedure for facilitating an expert's judgments on a set of rules. In E.E. Roskam, editor, *Mathematical Psychology in Progress*, Recent Research in Psychology, pages 157–170. Springer-Verlag, Berlin, Heidelberg, and New York, 1989.

P. Mutzel. The SPQR-tree data structure in graph drawing. In J.C.M. Baeten, J.K. Lenstra, J. Parrow, and G.J. Woeginger, editors, *Automata, Languages and Programming, 30th International Colloquium, ICALP 2003, Eindhoven, The Netherlands, June 30–July 4, 2003*, volume 2719 of *Lecture Notes in Computer Science*, pages 34–46, Berlin, Heidelberg, and New York, June 2003. Springer-Verlag.

P. Orlik and H. Terano. *Arrangements of Hyperplanes*. Springer-Verlag, Berlin, Heidelberg, and New York, 1992.

S. Ovchinnikov. The lattice dimension of a tree. Electronic preprint math.CO/0402246, arXiv.org, 2004.

S. Ovchinnikov. Media theory: representations and examples. Electronic preprint arXiv:math/0512282v1, arXiv.org, 2006.

S. Ovchinnikov. Hyperplane arrangements in preference modeling. *Journal of Mathematical Psychology*, 49:481–488, 2005.

S. Ovchinnikov. Partial cubes: structures, characterizations and constructions. Electronic preprint arXiv:0704.0010v1v1, arXiv.org, 2007.

S. Ovchinnikov. Convexity in subsets of lattices. *Stochastica*, IV:129–140, 1980.

S. Ovchinnikov and A. Dukhovny. Advances in media theory. *International Journal of Uncertainty, Fuzziness and Knowledge-Based Systems*, 8(1):45–71, 2000.

E. Parzen. *Stochastic Processes*. Holden-Day, San Francisco, 1994.

M. Pirlot. Synthetic description of a semiorder. *Discrete Applied Mathematics*, 31:299–308, 1991.

W. Prenowitz and M. Jordan. *Basic Concepts of Geometry*. Blaisdell Publishing Company, Waltham, MA, 1965.

M. Regenwetter, J.-Cl. Falmagne, and B. Grofman. A stochastic model of preference change and its application to 1992 presidential election panel data. *Psychological Review*, 106(2):362–384, 1999.

F. Restle. A metric and an ordering on sets. *Psychometrika*, 24(3):207–220, 1959.

J. Riguet. Les relations de Ferrers. *Compte Rendus des Scéances de l'Académie des Sciences (Paris)*, 232:1729–1730, 1951.

I. Rival, editor. *Graphs and Order: The Role of Graphs in the Theory of Ordered Sets and Its Applications*. Reidel, Dordrecht, 1985.

F.S. Roberts. *Measurement Theory, with Applications to Decisionmaking, Utility, and the Social Sciences*. Addison-Wesley, Reading, Mass., 1979.

F.S. Roberts. *Applied Combinatorics*. Prentice Hall, Englewood Cliffs, New Jersey, 1984.

D.G. Sarvate and J.-C. Renaud. On the union-closed sets conjecture. *Ars Combinatoris*, 27:149–153, 1989.

D.G. Sarvate and J.-C. Renaud. Improved bounds for the union-closed sets conjecture. *Ars Combinatoris*, 29:181–185, 1990.

D. Scott and P. Suppes. Foundational aspects of theories of measurement. *Journal of Symbolic Logic*, 23:113–128, 1958.

M. Senechal. *Quasicrystals and Geometry*. Cambridge University Press, Cambridge, London, and New Haven, 1995.

P.W. Shor. Stretchability of pseudolines is NP-hard. In P. Gritzmann and B. Sturmfels, editors, *Applied Geometry and Discrete Mathematics*, volume 4 of *DIMACS Series in Discrete Mathematics and Theoretical Computer Science*, pages 531–554. American Mathematical Society, Providence, R.I., 1991.

M. Skandera. A characterization of $(3+1)$-free posets. *Journal of Combinatorial Theory, Ser. A*, 93:655–663, 2001.

R.P. Stanley. *Enumerative Combinatorics*, volume 2. Wadsworth and Brooks/Cole, Monterey, California, 1986.

R.P. Stanley. Hyperplane arrangements, interval orders, and trees. *Proceedings of the National Academy of Sciences of the United States of America*, 93:2620–2625, 1996.

A.N. Trenk. On k-weak orders: Recognition and a tolerance result. *Discrete Mathematics*, 181:223–237, 1998.

W.T. Trotter. *Combinatorics and Partially Ordered Sets: Dimension Theory*. The Johns Hopkins University Press, Baltimore and London, 1992.

M. Van de Vel. *The theory of convex structures*. North-Holland Publishing Co., Amsterdam and New York, 1993.

T.P. Vaughan. Families implying the Frankl conjecture. *European Journal of Combinatorics*, 23:851–860, 2002.

M. Villano. Computerized knowledge assessment: Building the knowledge structure and calibrating the assessment routine. PhD thesis, New York

University, New York, 1991. In *Dissertation Abstracts International*, vol. 552, p. 12B.

C. Villee. *Biology*. W.B. Saunders Company, Philadelphia, 5th edition, 1967.

H. Wainer, N.J. Dorans, D. Eignor, R. Flaugher, B.F. Green, R.J. Mislevy, L. Steinberg, and D. Thissen. *Computerized Adaptive Testing: A Primer*. Lawrence Erlbaum Associates, New Jersey and London, 2000.

D.J.A. Welsh. Matroids: Fundamental concepts. In R.L. Graham, M. Grötschel, and L. Lovász, editors, *Handbook of Combinatorics*, volume 1. The M.I.T. Press, Cambridge, MA, 1995.

P.M. Winkler. Isometric embedding in products of complete graphs. *Discrete Applied Mathematics*, 7:221–225, 1984.

T.-T. Wong, W.-S. Luk, and P.-A. Heng. Sampling with Hammersley and Halton points. *Journal of Graphics Tools*, 2(2):9–24, 1997.

T. Zaslavsky. *Facing up to arrangements: face count formulas for partitions of space by hyperplanes*, volume 154 of *Memoirs of the AMS*. American Mathematical Society, Providence, R.I., 1975.

G.M. Ziegler. *Lectures on Polytopes*. Springer-Verlag, Berlin, Heidelberg, and New York, 1995.

Index

$\mathcal{K}$-homogeneous space, **154**
$\mathfrak{P}(X)_{\bowtie}$, cluster partition, **58**
$\mathfrak{P}(\mathcal{Z})$, power set of $\mathcal{Z}$, **12**
$\mathfrak{P}_F(\mathcal{Z})$, **12**, 81, 98
$\mathcal{WG}$-homogeneous space, **154**
$\bowtie$, **58**
$\cap$-closed family, **73**, 87
$\cup$-closed family, **73**, 78, 79, 86–94
$\cup_F$-closed family, **73**
$\langle S \rangle$, cluster, **58**
∂, see Kronecker ∂
n-star, see star
2-SAT, 221

abstract simplicial complex, see independence system
accessible, see state
acyclic orientation, 181, 222
adjacency graph, **28**
adjacency list, **204**
  edge-labeled, **205**
  vertex-labeled, **205**
adjacent
  states, 4, **23**, 277
  vertices, **16**
algorithm, 199
antisymmetric, see relation
apex, **76**, 222
arc, **15**
arrangement, see hyperplane, **177**
  Boolean, **180**
  braid, **180**

central, **177**, 180, 185
deformation of, **180**
graphical, 183
non-stretchable, 238
of lines, or line arrangement, 5
of right angles, **255**
simple, 237
weak pseudoline, **238**, 246, 255
Aspvall, B., 221
Asratian, A.S., 143
asymmetric, see relation
asymptotic probabilities, 279
atom, **88**
average length, **150**
Avis, D., 224
axioms
  for a medium, **24**
  learning space, **10**

Babai, L., 223
ball, 171
Bandelt, H.J., 77, 245
Barbut, M., 195
Barucha-Reid, A.T., 263
base, 222
  of a $\cup$-closed family, **86**
Bender, M.A., 208
benzenoid graph, 197, **244**
Berman, L.W., 238
betweenness, see relation
Bhatt, S., 231
big theta, 166

binary, see relation
biorder, see relation
bipartite, see graph
bipartition, 4, **15**
Birkhoff's Theorem, 289
  generalisation, 292
Birkhoff, G., 88
Björner, A., 10, 78, 177
black box medium, **223**
Bogart, K.P., 7, 107
Bogatyi, S.A., 154
Bondy, J.A., 17
bounded
  face of a drawing, **229**
Bourbaki, N., 177
Bowman, V.J., 6
breadth-first search, 199
Brodal, G.S., 205
Brown, K.S., 281
Busacker, R.G., 17
Bush, G.H.W., 19, 285, 297

cardinal number, 12
cardinality, 161
Cartesian, see product
Cavagnaro, D.R., 40, 45
cell, **178**
center, **47**
centrally symmetric, **234**
Chan, T., 238
Chapman-Kolmogorov Equation, **267**
Chepoi, V., 47, 245
Chung, K.L, 263
circuit
  even, **124**
  in a graph, **16**, 124
  minimal, **124**
class
  absorbing, 283
  equivalence, 283
  ergodic, 283
  of a partition, **15**
  transient, 283
Clinton, W., 19, 285, 297
closed medium, see medium
closedness
  condition, see medium
closure
  of an E-star, 83

transitive, **14**
cluster, **58**
  analysis, 58
  partition, **58**
clustered
  linear orders, 62–68, 81
Cogis, O., 107
coincident
  right angles, **255**
commutativity (weak form), 28
complement
  of a set, **12**
complete, see medium
computational group theory, 223
concise, see message
connected, see graph
consistent, see message
content
  of a message, **23**
  of a state, **29**
content family, **29**
convex
  for media, 135
  hull, **171**
Cormen, T.H., 199, 203, 204
corner
  of a right angle, **255**
Cosmodakis, S., 231
Cosyn, E., 10, 78, 286–293
covering relation, see Hasse diagram
Coxeter, H.S.M., 180, 265
critical pair, 106
cube, **139**, 253
  partial, 139, 200, 201, 204, 210, 211, 218, 227
cubical complex, **194**
cycle, **16**
  length, **16**
Cyvin, S.J., 244

Davey, B.A., 88
de Bruijn, N.G., 237
deformation
  of an arrangement, 181
degree
  of a vertex, 165
Delaunay triangulation, 96
deletion, see mutation
depth-first search, 224

Deza, M., 157, 164, 165, 235, 237
Di Battista, G., 229, 247, 258
Diaconis, P., 281
Diestel, R., 172
digraph, **15**
   positive, **82**
Dilworth, R.P., 10
dimension, 158
   isometric, **158**
   lattice (integer), **161**
   lattice (partial cube), **164**
   oriented lattice, **172**
directed, see graph, edge
discrepancy, 233
discriminative, **51**, 288
distance (in a graph), **16**
distribution
   stationary, 271
Djoković, D.Ž., 139, 142, 143, 145, 147, 158
Djoković–Winkler relation, 209
Doble, C.W., 101, 114–119
Doignon, J.-P., 7, 10, 17, 30, 51, 86, 88, 92, 101–119, 285–297
Dowling, C.E., 89, 287
downgradable, 103, **118**
Dragan, F., 47, 245
drawing, **229**
   planar, **229**
   straight-line, **229**
   upright-quad, **254**
Drosophila melanogaster, 11
Ducamp, A., 107
Dukhovny, A., 18, 56
duplication, see mutation
Dushnik, B., 115

E-star, **81**
Eades, P., 229, 258
Ebert, J., 252
Edelman, P.H., 10, 78, 92
Edelsbrunner, H., 238
edge, **15**
   directed, **15**
   opposite, **124**
edge-colouring, **148**
effective, see message
embedding, see graph, **34**, 37
empty message, see message

entailment, **292**
Eppstein, D., 18, 86, 142, 161, 199, 209, 211, 218, 229, 234, 238, 252
equipollent, **12**
equivalence, see relation
ergodic, see chain, class, state
Even, S., 252
expansion
   of a token, 86
extended star (medium)
   see also E-star, **81**
exterior
   edge, **229**
   of a right angle, **255**
   vertex, **229**

face poset, **191**
Falmagne, J.-Cl., 7, 10, 17–19, 30, 51, 62, 66, 68, 86, 88, 92, 101–119, 123–137, 199, 211, 263, 273, 285–302
Farach-Colton, M., 208
Felsner, S., 234
Fiorini, S., 194
Fishburn, P.C., 107, 114, 115, 194
flip graph, 96
Frankl, P., 94
Fraysseix, H. de, 238
fringe
   of a set in a family, **103**
   inner, 30, **103**
   outer, 30, **103**
Fukuda, K., 224
fully homogeneous space, 154
function
   symmetric, **13**
Fáry's theorem, 229

Gaiha, P., 6
Galanter, E., 51
Gąsieniec, L., 205
Gauss puzzle, 1–4
Gauss, C.F., 1
genetic mutation, 11
Gimbel, J.G., 115
Ginsburg, S., 18
Golomb, S.W., 246
grading collection, **53**, 57
Graham, P.L., 139

graph, **16**, 28
  st-oriented, **252**
  adjacency, *see* adjacency
  biconnected, **246**
  bipartite, **17**
  connected, **17**
  directed, **15**
  directed acyclic, 252
  distance, **16**
  embedding, **17**
  isomorphic, **17**
  of a medium, **123**
  oriented semicube, **173**
  semicube, **167**, 231
  triconnected, **246**
graph drawing, 229
grid
  pentagrid, **237**
Grofman, B., 297
group
  isometry, 158
  of permutations, 7
Grünbaum, B., 177, 234, 238
Gupta, S.K, 6
Gutman, I., 244
Guttman scales, **107**
Guttman, L., 107

Hadlock, F., 165
Hammersley points, 233
Harary, F., 58
hash table, 203
Hasse diagram, **14**, 20, 183, 189
Heng, P.-A., 233
Hoffman, F., 165
hollows, 114
Hsu, Y.-F., 19, 297
Huzun, H.B., 10
hypercube, V, 98, 221, 233
hyperplanes
  arrangement, 4, 80, 223

i-closed, *see* medium
Imrich W., 159
Imrich, W., 77, 142, 147, 150, 244
inconsistent, *see* message
independence (of axioms), 25
independence system, **73**, 78, 192
indicator function, **162**, 163

indifference class, **190**
ineffective, *see* message
initial probabilities, **266**
input size, 199
instance, 285
integer lattice, 98, **161**
integer partition, **165**
interior
  edge, **229**, 244
  of a right angle, **255**
  vertex, **229**, 235, 244, 245
inverse of a transformation, 1
inversion, *see* mutation
isometric subgraph, 8, **16**, 26
isometry group, *see* group
isomorphic, *see* graph
isomorphism, 37
  media, **34**
  sign-isomorphism, **37**
  token systems, **34**
item, 285
item (of knowledge), 10

Jamison, R., 10, 78, 92
Janowitz, M.F., 191
jigsaw puzzle, 1
Johnson, N.L., 58
jointly consistent, *see* message
Jordan, M., 70
Joyner, D., 47

k-graded, *see* well-graded
Kambouri, M., 287, 293
Karp, R.M., 224
Kemeny, J.G., 189, 265
Klavžar, S., 77, 142, 147, 150, 159, 244
knowledge state, **10**
knowledge structure, **10**
König's Infinity Lemma, 172
König, D., 17, 172
Koppen, M., 92, 287–294
Korte, B., 10, 73
Kotz, S., 58
Kronecker ∂, **274**
Kuzmin, V.B., 51, 171

Las Vergnas, M., 177
lattice dimension, 230
Laurent, M., 157, 164, 165

Le Conte de Poly-Barbut, Cl., 6
leaf, **17**
learning space, **10**, 19, 40, 78, 79, 252, 285–293
least common ancestors, 208
Leiserson, C.E., 199, 203, 204
Leite, F., 19
length, *see* message, *see* message
   of a cycle, **16**
   of a path, **16**
length function, **186**
limit, *see* asymptotic
line arrangement, *see* arrangement
linear extension, 222
linear medium, *see* medium
linear order, *see* relation
linear programming, 223
linear trace, 64
Lo Faro, G., 94
Lovász, L., 10, 73
Luce, R.D., 51, 111
Luk, W.-S., 233

Markov chain
   aperiodic, **284**
   finite, **266**
   homogeneous, **266**
   irreducible, **284**
Markov, A.A., 265
matching, **168**
Matoušek, J., 200
Matula, D.W., 58
maximal, *see* message
median graph, **76**, 245
medium, **23**
   closed, 7, 10, 19, **74**, 218
   complete, **80**
   i-closed, **74**, 78, 192
   linear medium, **68**
   of a graph, **132**
   open, **74**
   rooted, 19, **38**
   taut, **98**
   u-closed, **74**
message, **23**
   canonical, **74**
   circuit (orderly), 128
   circuit (regular), 128
   concise, **24**, 201, 225

consistent, **24**
effective, **24**
empty, **24**, 74
inconsistent, **24**
ineffective, **24**
jointly consistent, **24**
length, **24**
maximal, **47**
mixed, **74**
negative, **37**, 74
positive, **37**, 74
producing, **24**
return, **24**
return (orderly), **31**
return (regular), 34, **34**
reverse, **24**
stepwise effective, **24**
vacuous, **24**
metric space, **13**
Micali, S., 231
Miller, E.W., 115
minimal element, **14**
Mirkin, B.G., 191
Monjardet, B., 195
Moore, E.F., 18
mosaic, **234**
   of dodecagons, hexagons
      and squares, 237
   of hexagons, 235, 237
   of octagons and squares, 237
   of squares, 235, 237
   of triangles, 235
   of triangles and hexagons, 237
Mulder, H.M., 77
Müller, C.E., 287
multigrid, **236**
Murphy, U.S.R, 17
mutation, *see* genetic mutation
   deletion, 11
   duplication, 11
   inversion, 11
   translocation, 11
Mutzel, P., 247

nonforced pair, 106
noses, 114
NP-complete, 231
number theory, 165

O-notation, 200
opposite, see token
oracle, 223
   independence, 224
order
   weak, 78, **188**
orientation, 4, 6, 7, 9, **37**
   apex, **76**
   bipolar, **252**
   induced, **37**
oriented lattice dimension, see dimension
oriented semicube graph, see graph
Orlik, P., 177
Ossona de Mendez, P., 238
Ovchinnikov, S., 5, 18, 34, 51, 56, 62, 69, 107, 123–137, 165, 171, 186, 191

partial cube, 77, see cube
partial order, see relation
partition, **15**
   integer, **165**
Parzen, E., 263, 274
path
   between two vertices, **16**
   length, 16
   tight, **50**
pending status, 293
pentagon, 96
permutation, 7
permutohedron, 6, 7
Perot, H.R., 19, 285, 297
Pirlot, M., 114
Plass, M.F., 221
Poisson walk, **275**
Pollak, H.O., 139
polygon
   centrally symmetric, 234
polyomino, **246**
polytope
   weak order, **194**
poset, 189
positive definiteness, **13**
power set, **12**, 253
prefix, see segment
Prenowitz, W., 70
Priestley, H.A., 88
problem type, 285

producing, see message
product
   Cartesian, **13**
   of relations, **13**
projection
   isometric, 230
pseudoline, **238**
puzzle, 1
Python, 227

QUERY, 287–293

Rado, R., 139
RAM model of computation, **202**
random walk, 11, 18, 263
   process, 277
Read, C.B., 58
reduction, **35**
reflexive, see relation
Regenwetter, M., 19, 297–302
region, **178**
region graph, **178**, 255
regular, see message
relation
   n-connected, **103**
   (strict) weak order, 15
   2-connected, 103, 110
   ac-order, 114–119
   accessibility, **283**
   antisymmetric, **14**, **103**
   asymmetric, **103**
   betweenness, **70**
   binary, **13**
   biorder, 17, **103**, 107–110
   communicates, 283
   equivalence, **15**, 283
   Ferrers, 107
   identity, **14**
   indifference, **186**
   interval order, **103**, 107, 110
   irreflexive, **103**
   like, **127**, 143
   linear order, 6, **14**, 62–68, 81
   partial order, **14**, 20, **103**, 283
   quasi order, **14**
   reflexive, **14**, **103**
   semiorder, 17, **103**, 110–114
   strict linear order, **14**
   strict partial order, **14**, 20

strongly connected, **14**, **103**
symmetric, **103**
transitive, **14**, **103**
weak order, **15**, 188
weak k-order, **190**
relative product, *see* product
Renaud, J.C., 94
representation theorem, **56**
representing function, 186, 189
representing medium, **55**
Restle, F., 51
restriction, 35
retraction, **141**
return, *see* message
reverse, 1, *see* token
reverse search, **224**
right angle, **255**
Riguet, J., 107
Rival, I., 94
Rivest, R.L., 199, 203, 204
Roberts, F.S., 15, 17
root, **38**
Rubik's Cube, 47

Saaty, T.L., 17
Sarvate, D.G., 94
satisfiability problem, 221
Schrader, R., 10, 73
Scott, D., 111
segment, **74**
  initial, **74**
  negative, **74**
  positive, **74**
  prefix, **74**
  suffix, **74**
  terminal, **74**
semicube, **142**
  in a medium, 149
  opposite, 142, 149
semicube graph, *see* graph
semigroup, V
  transitivity in, 17
semiorder, *see* relation, 181
Senechal, M., 234
Shephard, G.C., 234
Shor, P.W., 238, 251
Shtogrin, M.I., 235, 237
sign vector, **178**
sign-isomorphism, *see* isomorphism

simple generator protocol, 227
simplicial complex, *see* independence system
sink vertex, **252**
Skandera, M., 115
Snell, J.L., 189, 265
source vertex, **252**
span of a family of sets, **86**
SPQR tree, **247**
squaregraph, **245**
Stanley, R.P., 177, 183, 186
star (medium), **39**, 81, 212, 221, 227
state, 1
  absorbing, **284**
  accessible, **283**
  aperiodic, **284**
  ergodic, **283**
  period (of a state), **284**
  return, **283**
  transient, **283**
state (of a medium), **23**
stationary distribution, 271
Stein, C., 199, 203, 204
stepwise effective, *see* message
stochastic matrix, *see* transition
stochastic process, 263
strict linear order, *see* relation
strict partial order, *see* relation
Sturmfels, B., 10, 78, 177
subgraph, 8, **16**
  convex, **145**
  isometric, 9
submedium, 34, **36**, 61
subsystem, *see* token
suffix, *see* segment
Suppes, P., 111
support, **161**
surmise system, 92
symmetric
  difference, 7, **12**
  difference distance, **12**
system, *see* token
Szemerédi, E., 223

Tamassia, R., 229, 247, 258
Tarjan, R.E., 221, 252
taut, *see* medium
Terano, H., 177
terminal, *see* segment

thermometer score, 297
Thiéry, N., 287–293
tight, *see* path
tiling, **234**
    Penrose, 237
    zonotopal, 234
token, **23**
    opposite, **34**
    reverse, **23**
    subsystem, **36**
    system, **23**
Tollis, I.G., 229, 258
tope graph, **180**
trace
    (well-ordering), **64**
    linear, **64**
transition
    matrix, **267**
    one-step, **266**
    probabilities, **266**
    stationary probabilities, **266**
transition table, **202**
    hashed, **203**
transitive, *see* relation
transitive closure, *see* closure
transitive reduction, 183
translocation, *see* mutation
tree, **17**, 164
Trenk, A.N., 115
triangle inequality, **13**
triangulation, **96**
trie, 208
Trotter, W.T., 106, 114
truth assignment, 221

u-closed, *see* medium
Upfal, E., 224
upgradable, 103, **118**
upright quadrilateral, **254**
Uzun, H.B., 78, 86

vacuous, *see* message
Van de Vel, M., 92
Vaughan, T.P., 94
Vaxès, Y., 47, 245
Vazirani, V. V., 231
vertex connectivity, **246**
Villee, C., 11

walk, **16**
    closed, open, **16**
    length, **16**
weak order, *see* relation
Weil, H., 234
well-graded, **51**
well-ordering, **14**, 63, 64
well-orders, **14**
wellgradedness, 15, 17, 51
Welsh, D.J.A., 10, 78
wg-family, **51**, 79
White, N., 177
Wigderson, A., 224
Winkler, P., 142, 143, 147
Wong, T.-T., 233
worst-case analysis, **199**

Zaslavsky, T., 177
Ziegler, G., 67, 177, 195
zone, **257**
zonotopal tiling, **234**

Printing: Krips bv, Meppel
Binding: Stürtz, Würzburg